IET CONTROL ENGINEERING SERIES 80

Control-oriented Modelling and Identification

Other volumes in this series:

Control-oriented Modelling and Identification

Theory and practice

Edited by Marco Lovera

The Institution of Engineering and Technology

Published by The Institution of Engineering and Technology, London, United Kingdom

The Institution of Engineering and Technology is registered as a Charity in England & Wales (no. 211014) and Scotland (no. SC038698).

First published 2015

The Institution of Engineering and Technology
Michael Faraday House
Six Hills Way, Stevenage
Herts, SG1 2AY, United Kingdom

www.theiet.org

British Library Cataloguing in Publication Data
A catalogue record for this product is available from the British Library

ISBN 978-1-84919-614-7 (hardback)
ISBN 978-1-84919-615-4 (PDF)

Typeset in India by MPS Limited
Printed in the UK by CPI Group (UK) Ltd, Croydon

To Matteo, Giorgio, Paola

Contents

Chapter 1
Introduction to control-oriented modelling

*Marco Lovera**

Abstract

The process of developing control-oriented mathematical models of physical systems is a complex task, which in general implies a careful combination of prior knowledge about the physics of the system under study and information coming from experimental data. The aim of this book is to present state-of-the-art methods and tools available within the systems and control literature to support control-oriented modelling activities and to illustrate their usefulness by means of a number of case studies and applications.

1.1 Introduction

The design of automatic control systems requires the availability of some knowledge of the dynamics of the process to be controlled. In this respect, current methods for control system synthesis can be classified in two broad categories: model-free and model-based ones. The former aims at designing (or tuning) controllers solely on the basis of experimental data collected directly on the plant, without resorting to mathematical models. The latter, on the contrary, assumes that suitable models of the plant to be controlled are available, and relies on this information to work out control laws capable of meeting the design requirements.

While the research on model-free design methods is a very active field, the vast majority of control synthesis methods and tools fall in the model-based category and therefore assume that knowledge about the plant to be controlled is encoded in the form of dynamic models of the plant itself. Furthermore, more and more application areas are such that control system performance is becoming a key competitive factor for the success of innovative, high-tech systems. As examples, consider high-performance mechatronic systems (such as robots), vehicles enhanced by active integrated stability, suspension and braking control, aerospace systems, advanced energy conversion systems. All the above-mentioned applications call for accurate

*Politecnico di Milano, Milano, Italy, e-mail: marco.lovera@polimi.it

mathematical modelling for the design of the control system for one or more of the following reasons: closed-loop performance critically depends on the dynamic behaviour of the plant; the system is made of many closely interacting subsystems; advanced control systems are required to obtain competitive performance, and these in turn depend on explicit mathematical models for their design; the system is safety critical and requires extensive validation of closed-loop stability and performance by simulation.

Therefore, building control-oriented mathematical models of physical systems is a crucial prerequisite to the design process itself. Mathematical models play a number of different roles in the design of control systems. In particular, different classes of mathematical models are usually employed: *detailed* models for system simulation and *compact* models for control design. In the following sections the two model classes are presented and their respective roles in the design of control systems are described. Note, in passing, that although hybrid system control is an interesting and emerging field, this book focuses on purely continuous-time physical models, with application to the design of continuous-time or sampled-time control systems.

1.1.1 Detailed models for system simulation

Object-oriented modelling (OOM) methodologies and equation-based, object-oriented languages (EOOLs) provide very good support for the development of such models, thanks to equation-based modelling, a-causal physical ports, hierarchical system composition and inheritance. Any continuous-time EOOL model is equivalent to the set of differential-algebraic equations (DAEs):

$$F(x(t), \dot{x}(t), u(t), y(t), p, t) = 0 \tag{1.1}$$

where x is the vector of dynamic variables, u is the vector of input variables, y is the vector of algebraic variables, p is the vector of parameters and t is the time. It is interesting to highlight that the object-oriented approach (in particular, the use of replaceable components) allows defining and managing families of models of the same plant with different levels of complexity, by providing more or less detailed implementations of the same abstract interfaces. This feature of OOM allows the development of simulation models for different purposes and with different degrees of detail throughout the entire life of an engineering project, from preliminary design down to commissioning and personnel training, all within a coherent framework.

In particular, when focusing on control systems, detailed models play a double role in the control design process. On one hand they allow checking how good (or crude) the compact model is, compared to a more detailed description, thus helping to develop good compact models. On the other hand, they allow closed-loop performance verification of the controlled system, once a controller design is available.[1] To this

[1] Indeed, verifying closed-loop performance using the same simplified model that was used for control system design is not a sound practice; conversely, verification performed with a more detailed model is usually deemed a good indicator of the control system performance, whenever experimental validation is not possible for some reason.

purpose (and regardless of the actual control design methodology) once the controller has been set up, an OOM tool can be used to run closed-loop simulations, including both the plant and the controller models. Many OOM tools provide model export facilities, which allow to connect a plant model with only causal external connectors (actuator inputs and sensor outputs) to a causal controller model in a causal simulation environment. From a mathematical point of view, this corresponds to reformulating (1.1) in state space form, by means of analytical and/or numerical transformations.

1.1.2 Compact models for control design

The requirements for a control-oriented model can vary significantly from application to application. Design models can be tentatively classified in terms of two key features: complexity and accuracy. For a dynamic model, complexity can be measured in terms of its order; accuracy, on the other hand, can be measured using many different metrics (e.g. time-domain simulation or prediction error and frequency-domain matching with the real plant) related to the capability of the model to reproduce the behaviour of the true system in the operating conditions of interest.

Broadly speaking, it can be safely stated that the required level of closed-loop performance drives the requirements on the accuracy and complexity of the design model. Similarly, it is intuitive that more complex models have the potential for being more accurate. So, one might be tempted to resort to very detailed mathematical representations of the plant to be controlled in order to maximise closed-loop performance. This consideration, however, is moderated by a number of additional requirements, which actually end up driving the control-oriented modelling process. First of all, present-day controller synthesis methods and tools have computational limitations in terms of the complexity of the mathematical models they can handle, so compact models representative of the dominant dynamics of the system under study are what is really needed. Furthermore, for many synthesis methods (such as LQG or H_∞ synthesis) the complexity of the design model has an impact on the complexity of the controller, which in turn is constrained by implementation issues. Last but not least, in engineering projects the budget of the control-oriented modelling activity is usually quite limited, so the achievable level of accuracy is affected by this limitation.

It is clear from the above discussion that developing mathematical models suitable for control system synthesis is a non-trivial task, but rather corresponds to the pursuit of a careful tradeoff between complexity and accuracy. Furthermore, throughout the model development one should keep in mind the eventual control application of the model, so its mathematical structure has to be compatible with currently available methods and tools for control system analysis and design.

1.1.3 Building models for control system synthesis

The process of developing control-oriented mathematical models of physical systems is a complex task, which in general implies a careful combination of prior knowledge about the physics of the system under study and information coming from experimental data. In particular, this process can follow very different paths depending on the type of information available on the plant to be controlled. Such paths are typically classified in the literature as follows.

White-box modelling refers to the development of control-oriented models on the basis of first principles only. In this framework, one uses the available information on the plant to develop a detailed model using OOM or EOOL tools and subsequently works out a compact control-oriented model from it. If the adopted tool only supports simulation, then one can run simulations of the plant model, subject to suitably chosen excitation inputs (ranging from steps to persistently exciting input sequences such as pseudo-random binary sequences and sine sweeps), and then reconstruct the dynamics by means of system identification methods. Note that in this way the structure/order selection stage of the system identification process provides effective means to manage the complexity versus accuracy tradeoff in the derivation of the compact model. A more direct approach, presently supported by many tools, is to directly compute the A, B, C, D matrices of the linearized system around specified equilibrium points, using symbolic and/or numerical techniques. The result is usually a high-order linear system, which then can (sometimes must) be reduced to a low-order system by using model order reduction techniques (such as balanced truncation). Model reduction techniques allow to automatically obtain approximated compact models starting from much more detailed simulation models, by formulating specific approximation bounds in control-relevant terms (e.g. percentage errors of steady-state output values, norm-bounded additive or multiplicative errors of weighted transfer functions and ℓ_∞-norm errors of output transients in response to specified input signals).

Black-box modelling corresponds to situations in which the modelling activity is entirely based on input–output data collected on the plant (which therefore must be already available), possibly in dedicated, suitably designed, experiments. Regardless of the type of model to be built (i.e. linear or nonlinear, time-invariant or time-varying, discrete-time or continuous-time), the black-box approach consists of a number of well-defined steps. First of all the structure of the model to be identified must be defined: in the linear time-invariant case this corresponds to the choice of the number of poles and zeros for an input–output model or to the choice of model order for a state space representation; in the non-linear case structure selection is a much more involved process in view of the much larger number of degrees of freedom that are potentially involved. Once a model structure has been defined, a suitable cost function to measure the model performance must be selected (e.g. time-domain simulation or prediction error and frequency-domain model fitting) and the experiments to collect identification and validation data must be designed. Finally, the uncertain model parameters must be estimated from the available identification dataset and the model must be validated on the validation dataset.

Grey-box modelling (in various shades) corresponds to the many possible intermediate cases that can occur in practice, ranging from the white-box approach to the black-box one. The most critical issue in the development of an effective approach to control-oriented grey-box modelling lies in the integration of existing methods and tools for physical systems modelling and simulation with methods and tools for parameter estimation. Such integration can take place in a number of different ways depending on the relative role of data and priors on the physics of the system in the specific application. A typical situation that occurs frequently in applications is when a white-box model (developed by means of OOM or EOOL tools) contains parameters having unknown or uncertain numerical values (such as damping factors in structural models, aerodynamic coefficients in aircraft models and so on). Then, one may rely on input–output data collected in dedicated experiments on the real system to refine the white-box model by estimating the parameters using the information provided by the data. This process is typically dependent on the specific application domain as the type of experiment, the number of measurements and the estimation technique must meet application-specific constraints.

1.2 Overview of the book

Following this Introduction, the book is organised in two parts. The first part aims at presenting methods and tools covering the main steps that are usually followed in the derivation of detailed and compact control-oriented models; the second part, on the other hand, illustrates the methods by means of applications and case studies covering a wide range of engineering problems, ranging from automotive to aerospace, robotics, energy systems and many more.

1.2.1 Part 1: theory

Chapter 2, by F. Casella, introduces the basic concepts of equation-based, OOM and simulation of physical system, with specific reference to the Modelica language and outlines the key ideas and algorithms for the mathematical processing of object-oriented models. The application of OOM to various phases of the control system design process is also illustrated. Subsequently, Chapter 3, by P. Vuillemin, C. Poussot-Vassal and D. Alazard, deals with the very important problem of linear time-invariant (LTI) model approximation. In particular, two sets of recent methods are presented: truncation methods and moment matching methods. Their main characteristics are first introduced in a tutorial fashion and their relative merits are discussed using a number of numerical examples. In Chapter 4, by M. Lovera and F. Casella, OOM methods are again considered, in view of developing automated tools for control-oriented linear fractional representation (LFR) modelling and identification starting from equation-based, object-oriented descriptions of the plant dynamics (formulated using the Modelica language) and input–output data. The approach discussed in this chapter allows to reduce the gap between user-friendly model representations, based on object diagrams with physical connections, block diagrams with signal connections and generic differential-algebraic models, and the use of advanced LFR-based identification and control techniques. The interplay between model identification and robust control has been studied extensively in the systems and control literature: this problem is discussed in Chapter 5, by T. Oomen and M. Steinbuch, where a new system identification algorithm that delivers models that are suitable for subsequent robust control design and can be reliably applied to complex systems is presented. More precisely, an identification algorithm is developed that delivers system model in terms of co-prime factorizations and thereby extends classical iterative procedures to the closed-loop case. Subsequently, Chapter 6, by M. Cescon and R. Johansson, focuses on the derivation of multi-step ahead predictors oriented to model-predictive control. In particular, in this chapter subspace-based identification methods are used for the estimation of short-term predictors and one illustrative example is provided, dealing with blood glucose prediction in type 1 diabetes mellitus. Chapter 7, by G. van der Veen, J.-W. van Wingerden and M. Verhaegen, also considers subspace methods in view of predictive control with reference to systems whose dynamics can be described locally by LTI models and proposes a control approach based on the predictor-based subspace identification framework. The basic idea is to estimate the observer Markov parameters of the system using a recursive algorithm and to use those parameters to construct an output predictor, which in turn is applied to the solution of a predictive control problem subject to constraints. The identification of structured models is the subject of Chapter 8, by T. Vincent, K. Poolla and C. Novara. Indeed, the first step in the identification of a model from data is to specify the set of candidate models, or model structure. This choice is very critical: if the model structure is specified too restrictively, the identified model will be biased, and, on the other hand, if the model structure is specified too generally, the identified model can have a high variance. The structured non-linear system identification approach described in this chapter is designed to give the practitioner a very flexible model structure that can easily be configured to be as restrictive or permissive as the *a priori* information about the system warrants. Chapter 9, by D. Vizer, G. Mercère, E. Laroche and O. Prot, focuses on the identification of linear parameter varying (LPV) models rather than LTI ones, and proposes a 'glocal' approach to the problem. This allows to determine, using only experiments carried out locally in the neighborhood of a finite number of operating points, an LPV-LFR model of the system that can be structured by using the available non-linear equations describing the behavior of the system under study and standard linearization techniques.

1.2.2 Part 2: applications

The first chapter in Part 2 of this book, Chapter 10 by M. Lovera and F. Casella, discusses the main issues related to the modelling and simulation of satellite dynamics for control purposes and presents an OOM framework (along the lines of the material in Chapter 2), implemented as a Modelica library to this purpose. The proposed approach allows a unified approach to a range of problems spanning from initial mission design and actuator sizing phases, down to detailed closed-loop simulation of the control system, including realistic models of sensors and actuators. Chapter 11, by C. Poussot-Vassal, C. Roos, P. Vuillemin, O. Cantinaud and J.-P. Lacoste, comes back to the problem of model reduction introduced in Chapter 3 of the first part of the book. It presents the results obtained by applying the methods discussed therein to the problem of deriving a suitable low-order parameter-dependent model, appropriate to the control design purpose, for the dynamics of a business jet, in view of an aeroelastic control-design problem. In Chapter 12, by G. van der Veen, J.-W. van Wingerden and M. Verhaegen, the subspace predictive control framework discussed in Chapter 7 is applied to the problem of controlling vibrations in a flexible beam equipped with piezoelectric transducers. The application demonstrates that computations underlying the subspace predictive scheme can be performed in real time for a realistic system, shows how the scheme rapidly adapts when a sudden significant change in structural dynamics is introduced and finally illustrates the performance level achievable by this method. Chapter 13, by M. Bergamasco and M. Lovera, deals with model identification issues in the flight dynamics of helicopters. In particular, a novel, two-step technique is proposed, which combines the advantages of time-domain and frequency-domain methods. In the first step, the identification of a black-box model using a subspace model identification method is carried out, using a technique which can deal with data generated under feedback; subsequently, in the second step, *a priori* information on the model structure is enforced in the identified model using an $\mathscr{H}_\infty$ model matching method. Chapter 14, by A. Della Bona, G. Ferretti, E. Ficara and F. Malpei, considers the problem of parameter identification of a reduced order linear fractional transformation (LFT) model of anaerobic digestion processes. Indeed, because of their complexity, detailed models of anaerobic digestion cannot be used for on-line monitoring and control. To this aim reduced order models have to be considered, the derivation of which can be challenging. In this chapter, a modification of the well-known AMOCO model for anaerobic digestion is first proposed. Then, to perform parameter identification, an LFT formulation is derived, thanks to a suitable implementation of the general approach presented in Chapter 4. The approach has been applied first to data generated by a simulation of a fully detailed anaerobic digestion model and subsequently to data collected on a real plant, used for anaerobic digestion of agricultural wastes. In Chapter 15, by M. Corno, F. Todeschini, G. Panzani and S.M. Savaresi, modelling and identification issues arising in the design of an electro-hydraulic brake-by-wire actuator for racing motorcycles are considered. In particular, starting from a detailed description of the device, a first-principle model is derived by decomposing the plant in domain-specific subsystems. The model parameters are then identified following a grey-box approach by designing dedicated experiments for specific parameter sets. Experimental results are used to validate the model and carry out a parameter sensitivity analysis. Finally, control-oriented modelling issues in robotics are considered in Chapter 16, by D. Vizer, G. Mercère, E. Laroche and O. Prot, by applying the 'glocal' LPV parameter estimation approach presented in Chapter 9. Specifically, an analytical model of a 2-DOF robotic manipulator having structural flexibilities is first established and an LPV-LFR representation suitable for control design is identified based on the results of local experiments only.

Chapter 2
Object-oriented modelling and simulation of physical systems

Francesco Casella[1]

Abstract

Modular dynamic models based on physical first principles are often an essential part of the control system design and performance verification process. Traditional block diagrams, which are popular among control engineers, are not an appropriate modelling paradigm in this context, because they assume causal connections, while physical connections are a-causal in nature. This chapter introduces the basic concepts of equation-based, object-oriented (OO) modelling and simulation of physical system, with specific reference to the Modelica language. The key ideas and algorithms for the mathematical processing of OO models are then outlined. Finally, the application of OO modelling to various phases of the control system design process is illustrated.

2.1 Introduction

The construction of dynamic models of the plant is an essential part in the process of designing a control system. One way to build such models is to start from physical first principles, such as mass, energy or momentum conservation laws, under suitable modelling assumptions that aim at correctly capturing the control-relevant dynamic behaviour, while avoiding excessively cumbersome models.

Engineering systems are usually designed as the physical connection of basic components, the description of which may belong to different physical domains, e.g. mechanical, electrical, thermal, hydraulic and so forth. Furthermore, complex systems are hierarchically designed as the connection of interacting sub-systems, each performing some specific task or function. For the modelling of non-trivial systems it is thus very convenient to employ a modular modelling approach, in which the overall system model is obtained bottom-up by first defining basic or elementary physical components, then connecting them to form sub-systems, and eventually connecting sub-systems to form the overall system model.

[1] Politecnico di Milano, Milano, Italy, e-mail: francesco.casella@polimi.it

The block-diagram paradigm, which is very popular among control engineers, is ill-suited for this task, because in general physical components and sub-system do not interact in a causal way, through input and output variables. For example, the voltage at the two pins of an electrical resistor may be an input or an output variable, depending on the circuit this component is connected to. It is therefore desirable to follow a declarative, object-oriented (OO) modelling approach, whereby individual components are described by equations and a-causal connectors at their boundary, systems are assembled by means of a-causal connection equations and causality is only determined at the overall system level.

This approach guarantees the maximum flexibility, as it is always possible to replace a part of the system model, provided that the substitute shares the same physical connectors. It also promotes the re-use of models, since the way a component or sub-system is defined only needs to take into account its boundary connectors and is completely independent of the actual systems that will be connected to them.

Physical models suitable for a system-level description are usually lumped-parameters models, or possibly low-order, one-dimensional distributed-parameters models, which are then discretized using the finite volume, finite elements or finite differences methods. The underlying mathematical description is based on differential-algebraic equations (DAEs). A special emphasis is put by most OO modelling tools into automatic causalization techniques, which basically transform these models into state-space form in an efficient way.

OO models can have multiple uses in the control system design process, which is introduced in this chapter, and also demonstrated in other chapters of this book. They can be used as detailed references to validate simplified models, which are used for model-based control design. They can be linearized around operating points, possibly resorting to model reduction techniques to obtain models of a complexity suitable for model-based design. They can be used as a basis for grey-box parameter identification and, last but not least, for closed-loop verification and validation of the designed controller in realistic scenarios.

This chapter is organized as follows: in Section 2.2 basic, basic concepts and principles of equation-based, OO modelling are formulated; Section 2.3 describes the Modelica language, which is an emerging standard in this field, while Section 2.4 describes the fundamental steps required to bring an OO model in a form which is suitable to be used for control system design and for efficient simulation. The last three sections deal with possible uses of OO models in the control system design process: process modelling, analysis and identification (Section 2.5); control system performance verification (Section 2.6); and finally the direct use of OO models for optimal control (Section 2.7).

2.2 Basic concepts and principles

When writing software that deals with dynamic system models, there are two fundamentally different approaches for defining the systems themselves. In the *procedural* approach, the software modules corresponding to the models contain code that describes how to solve the model equations in that particular context, thus computing the trajectories of their variables over time. For instance, one could devise a simple software to simulate block diagrams using an explicit, fixed time step solver. In this case, each 'model' is actually a procedure or function that computes the outputs and updates the internal states, based on the current values of the inputs and of the state. In the *declarative* approach, instead, software modules corresponding to the models contain *equations* that relate the internal variables and the connector variables in an a-causal fashion, without any explicit mention about how those equations should be solved.

Declarative models have several advantages over procedural ones. First of all, they can be truly modular, because they do not need more or less artificial classifications of input/output causality at their physical interfaces, so it will be possible to connect them in an arbitrary fashion, as long as this makes physical sense. Second, they can be written in a way that is much closer to the original equations, as may be found in papers or textbooks. In this way, the code is somewhat self-documenting; that is, it is immediately clear by inspection what is actually being represented in the model, and how. This is usually not the case with procedural models, because the code is actually meant to *solve* the equations, which are thus not immediately apparent. It becomes also much easier to modify existing models, or write new ones from scratch. Last, but not least, when a declarative model has been built, its equations are available in symbolic form, which allows to perform many different actions on them: for example, symbolic differentiation to automatically compute Jacobian and Hessian matrices, causality analysis, symbolic solution, transformation into specific formalisms such as linear fractional transformations, using them as constraints for dynamic optimization problems, model reduction [15], and so forth. Procedural models do not usually possess this kind of flexibility, because they are already geared to a specific use, for example, time-domain simulation, possibly using a specific integration algorithm.

The first principle in OO modelling, which is made possible by the declarative modelling style, is *encapsulation*. On one hand, this means that when writing a component model by means of equations, the interaction with the outside world can only be specified in terms of the connector variables, which represent the model interfaces, without any knowledge about the actual systems that will be connected to them from the outside. As a consequence, re-usability is maximized. On the other hand, this also means that the only way models can interact with each other is through their connectors (i.e. their interface), without any direct reference to their internal variables. Consequently, it is easy to build model variants by replacing components and sub-systems with others that share the same physical interface.

The second principle is *modularity*: system models can be built by connecting components through their connectors, which can represent a-causal physical connections or causal signal connections. A-causal physical connectors contain coupled variables of two kinds, namely *effort* and *flow variables*. When two or more connectors are bound together in a connection set, new equations are added to the system, stating that the effort variables with the same name are equal, and that the sum of flow variables with the same name is zero. Unconnected connectors are considered as trivial cases of connection sets with cardinality one.

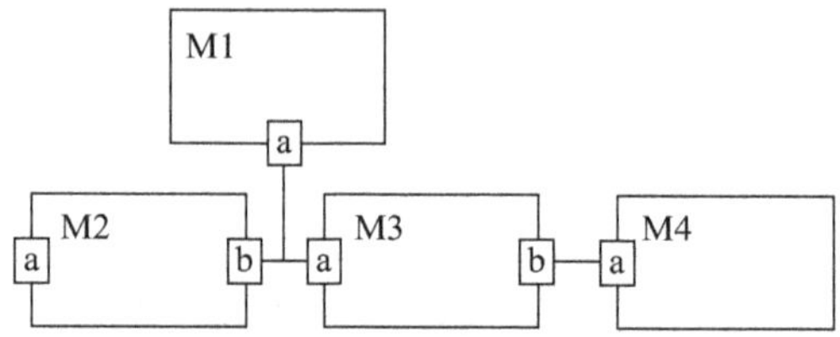

Figure 2.1 A simple modular model

Referring to the example shown in Figure 2.1, the following connection sets are formed: {M2.a}, {M1.a, M2.b, M3.a}, {M3.b, M4.a}. Assuming the connectors contain a flow variable f and an effort variable e, the following connection equations are generated:

$$\mathrm{M2.a.f} = 0 \tag{2.1}$$

$$\mathrm{M1.a.e} = \mathrm{M2.b.e} \tag{2.2}$$

$$\mathrm{M1.a.e} = \mathrm{M3.a.e} \tag{2.3}$$

$$\mathrm{M1.a.f} + \mathrm{M2.b.f} + \mathrm{M3.a.f} = 0 \tag{2.4}$$

$$\mathrm{M3.b.e} = \mathrm{M4.a.e} \tag{2.5}$$

$$\mathrm{M3.b.f} + \mathrm{M4.a.f} = 0 \tag{2.6}$$

Connectors can also contain input and output variables that can be used for causal signal connections. For each connection set, new equations are added, stating that all input and output variables having the same name are equal; furthermore, each connection set must contain one and only one output variable having a certain name; that is, there must be only one single signal source in a connection set.

Depending on the specific physical domain, flow and effort variables must be selected so that the connection equations correspond to physical laws. For electrical systems, the pin voltage and current entering the pin can be used as effort and flow variables respectively, so that connection equations correspond to Kirchhoff's laws. For thermal systems, temperature is the effort variable and heat flow is the flow variable; connection equations state the thermal equilibrium and the conservation of energy over the connection set. For one-dimensional translational (or rotational) systems, displacement (angle) in the direction of the common reference axis is the effort variable and cut force (torque) applied on the connector in the direction of the common reference axis is the flow variable; connection equations state that the displacement (angle) is the same and that the sum of cut forces (torques) is zero, that is Newton's third law.

In order to exploit the modularity principle to its fullest extent, and promote re-use of models, it is essential to define standard connectors for each physical domain, so that components and models developed by different people and organizations are interoperable, as long as they all employ the standard connectors.

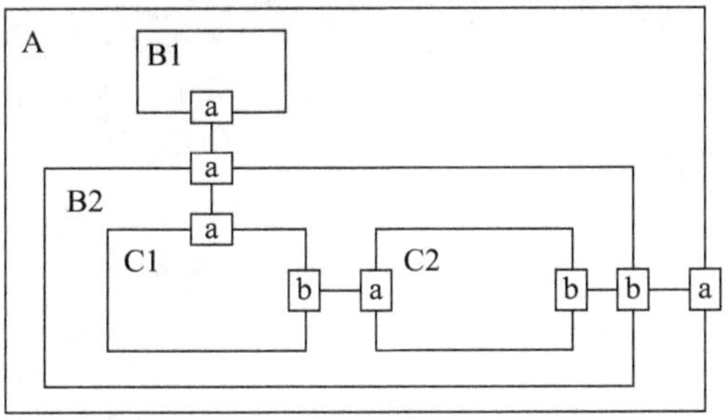

Figure 2.2 A hierarchically structured model

A related principle is *hierarchical decomposition* that can be applied to build hierarchically structured systems. A model can be defined by declaring its outer connectors (the interface), and then declaring its behaviour by connecting internal sub-models to the interface (the implementation). A recursive application of this principle results in hierarchical models, such as the one shown in Figure 2.2. When tackling a complex modelling problem, it is then possible to first decompose it in a top-down fashion, defining the interfaces of all the sub-models, and then providing the implementations in a bottom-up fashion, possibly by re-using components taken from existing model libraries, and also possibly changing the degree of detail of the implementations depending on the specific modelling needs.

Connection sets are defined at each hierarchical level. In order to handle flow variables correctly, a distinction is made between *inside connectors* (connectors belonging to the sub-models that are contained within a certain model M) and *outside connectors* (the connectors of model M itself). In the connection equations, flow variables belonging to outside connectors are taken with the minus sign.

With reference to the example shown in Figure 2.2, the following connection sets are formed: {A.a.in} at the top level; {B1.a.in, B2.a.in}, {B2.b.in, A.a.out} inside model A; {C1.a.in, B2.a.out}, {C1.b.in, C2.a.in}, {C2.b.in, B2.b.out} inside model B2. These connection sets correspond to the following connection equations:

$$\mathrm{A.a.f} = 0 \tag{2.7}$$

$$\mathrm{B1.a.e} = \mathrm{B2.a.e} \tag{2.8}$$

$$\mathrm{B1.a.f} + \mathrm{B2.a.f} = 0 \tag{2.9}$$

$$\mathrm{B2.b.e} = \mathrm{A.a.e} \tag{2.10}$$

$$\mathrm{B2.b.f} - \mathrm{A.a.f} = 0 \tag{2.11}$$

$$\mathrm{C1.a.e} = \mathrm{B2.a.e} \tag{2.12}$$

$$\mathrm{C1.a.f} - \mathrm{B2.a.f} = 0 \tag{2.13}$$

$$\mathrm{C1.b.e = C2.a.e} \tag{2.14}$$

$$\mathrm{C1.b.f + C2.a.f = 0} \tag{2.15}$$

$$\mathrm{C2.b.e = B2.b.e} \tag{2.16}$$

$$\mathrm{C2.b.f - B2.b.f = 0} \tag{2.17}$$

The third principle, which is shared with OO *programming* languages like C++ or Java, is *inheritance*. One can define a base model M, which might also be incomplete, and then define one or more models that extend M by inheriting all of its features (connectors, parameters, variables, equations, sub-models) and by adding specific ones. For instance, one might first define an electrical one-port component, which has two pins and defines the voltage v as the difference between the positive pin voltage and the negative pin voltage and the current i as the current entering the positive pin. Then, the resistor model can be defined by extending the one-port model and by adding Ohm's law $v = Ri$ to it, while the capacitor model can be obtained by adding the corresponding voltage-current law $i = Cdv/dt$.

Inheritance allows to avoid code duplications and define families of related models in a compact way. A possible drawback of this approach, in particular if multiple levels of inheritance are used, is that the code of any usable model is scattered among a complex hierarchy of ancestors in some library, so it might be difficult to understand what the definition of a model is, if one is not familiar with the architecture of the library. A reasonable balance should then be struck between these two aspects.

Another important use of inheritance is to define reconfigurable models. In this case, one defines a base model A for a certain sub-system, which only contains the interface (i.e. the connectors). Then, different implementations B and C can be defined that extend from the base model; these can be either models of the same physical object with different levels of detail or models of different physical objects sharing the same interface. An abstract system model can then be defined that uses A, and later be specialized by re-declaring A to be either B or C.

For example, one might define an abstract model of a four-wheeled vehicle using abstract models for the front suspension rear suspension and powertrain. On one hand, each of these can then be re-declared to be one specific implementation, in order to obtain the model of a specific type of car. On the other hand, they can also be re-declared to change the degree of detail of some sub-system, for example, by swapping a simple suspension model based on kinematic curves with a full 3D mechanical model with elastic, non-linear joints. This mechanism allows to manage families of models that are consistent with each other, for instance during different phases of development of an engineering project, or depending on the use scenario of the model itself. An interesting example of application of these concepts for the modelling of spacecraft attitude and orbit dynamics is discussed in Chapter 10.

2.3 Modelica

Modelica [21, 27, 18] is an OO modelling language based on the principles described in the previous section. It is defined by the non-profit Modelica Association [26], supported by a growing number of modelling and simulation tools, both commercial and open source, and widely used in industry and academia. The remainder of this chapter focuses on this language, although there are others, such as VHDL-AMS [12], gPROMS [4] or SimScape, that are based on the same principles and are similar in many respects.

Some of the features of the language will be described by means of a simple example, namely an RC electrical circuit with a sinusoidal voltage source. The first step to define such a model is to define specialized data types for the physical values, which can be used for automatic unit checking and for display and documentation purposes:

```
type Voltage = Real (unit = "V");
type Current = Real (unit = "A");
type Resistance = Real (unit = "Ohm");
type Capacitance = Real (unit = "F");
type Frequency = Real (unit = "Hz");
```

The electrical connector, containing one effort and one flow variable, is then defined as

```
connector Pin "Electrical pin"
  Voltage v "Voltage of the pin";
  flow Current i "Current entering the pin";
end Pin;
```

It is now possible to define all the components involved in the circuit; according to the encapsulation principle, their equations refer only to local variables and connector variables:

```
model Ground
  Pin p;
equation
  p.v = 0;
end Ground;

partial model OnePort
  Pin p,n;
  Voltage v "Voltage across the component";
  Current i "Current entering the positive pin";
equation
  v = p.v - n.v;
  i = p.i;
  0 = p.i + n.i;
end OnePort;

model SinusoidalVoltageSource
  extends OnePort;
  constant Real pi = 3.1415927;
```

```
  parameter Voltage Vmax "Amplitude";
  parameter Frequency f "Frequency";
equation
  v = Vmax*sin(2*pi*f);
end SinusoidalVoltageSource;

model Resistor
  extends OnePort;
  parameter Resistance R;
equation
  v = R*i;
end Resistor;

model Capacitor
  extends OnePort;
  parameter Capacitance C;
equation
  i = C*der(v);
end Capacitor;
```

In this case, inheritance has been used to factor out the common structure of all the one-port components. The `OnePort` model is marked as partial, to declare that it cannot be used directly in a model, as it is incomplete. Also, note the a-causal formulation of the models, which is particularly apparent in the last equation of the `OnePort` component. In general, there is no causality in the equation, so it is by no means implied that the right-hand-side terms must be used to compute the left-hand-side terms; the computational flow will depend on the structure of the overall system, as will be discussed in the next chapter.

We can now build a simple RC circuit model, by declaring its components, giving them suitable parameter values, and connecting them through their ports:

```
model Circuit
  SinusoidalVoltageSource Vs(Vmax = 10, f = 50);
  Resistor R1(R = 1e4);
  Resistor R2(R = 1e3);
  Capacitor C1(C = 1e-6);
  Ground G;
equation
  connect(G.p,  Vs.n);
  connect(Vs.p, R1.n);
  connect(R1.p, R2.n);
  connect(R2.p, C1.p);
  connect(C1.n, G.p);
end Circuit;
```

It is worth noting at this point that there are some physical domains where the concept of coupled flow and effort variables (which was used here for electrical circuits) is too restricted in order to handle the modelling problems in a fully modular, yet numerically robust, way. A first notable case is that of 3D multibody systems, where mechanical connectors need to carry the orientation of their frame of reference; this has three degrees of freedom, but it is not possible to uniquely identify a set of three variables to describe it in a non-singular way. A second case is that of thermo-hydraulic systems, where the specific energy content and the composition of the fluid is carried along with the fluid flow through connectors, leading to singularities when the flow rate goes to zero. Over-constrained connectors (see [23] and [26], section 9.4) and stream connectors (see [17] and [26], section 15) have been defined in Modelica to address these cases.

While textual code is very convenient for writing equations inside basic models, composite models are more easily built and understood if a drag-and-drop graphical user interface (GUI) is used. The Modelica language allows to define the graphical appearance of components (icons), as well as their placement in object diagrams, through graphical annotations that are automatically generated by the GUI. In such a graphical environment, the circuit model will look like in Figure 2.3.

The Modelica language also allows to write hybrid models that mix continuous-time variables, governed by DAEs, and discrete variables, which only change their values at event instants. The values of discrete variables are governed by DAE, contained within *when-clauses* that are activated only when the event is triggered; at all other times, these equations are turned off and the corresponding discrete variables are held constant. Time-events, for which triggering times are known a priori, and state-events, which depend on a boolean condition involving the values of other system variables, can be defined. In the latter case, event detection algorithms are used to locate the exact instant when the boolean conditions become true.

Two self-explanatory simple examples of discrete models are shown here: the `sample(T0,Ts)` operator generates equally spaced time-events every `Ts` seconds,

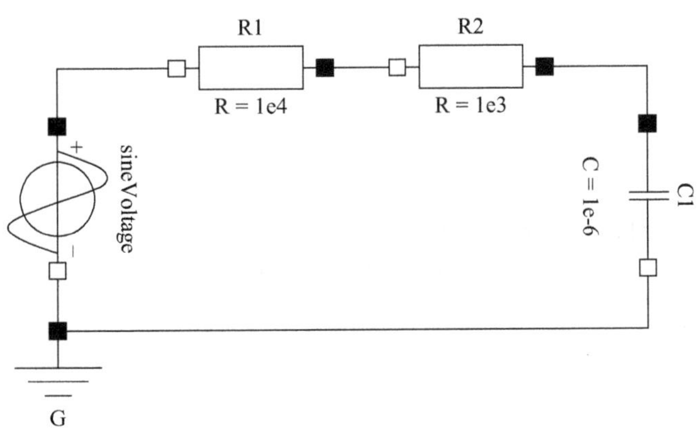

Figure 2.3 Object diagram of the circuit model

starting at time T0, while the pre() operator denotes the value of the discrete variable immediately before the event has been triggered.

```
connector InputReal = input Real;
connector OutputReal = output Real;
connector OutputBoolean = output Boolean;

model DigitalPI
  parameter Time Ts "Sampling time";
  parameter Real Kp "Proportional gain";
  parameter Time Ti "Integral time";
  InputReal yr "Set point";
  InputReal y "Measured variable";
  OutputReal u "Control variable";
  discrete Real e "Error";
  discrete Real e_int "Integral error";
equation
  when sample(0, Ts) then
    e = yr - y;
    e_int = pre(e_int) + Ts*e;
    u = Kp*(e + e_int/Ti);
  end when;
end DigitalPI;

model OnOffSwitch
  parameter Real thr "Triggering threshold";
  InputReal u "Input analog signal";
  OutputBoolean y "Output boolean signal";
equation
  when u > thr then
    y = not pre(y);
  end when;
end OnOffSwitch;
```

This kind of models can be used to describe digital controllers, which can then be connected in closed-loop to the continuous-time physical model of the plant.

For a rigorous handling of digital multi-rate controller models, the concept of *clocked* variables has been recently introduced in Modelica, borrowing concepts from the theory of synchronous languages such as Lustre [10] and Esterel [5]. Clocked variables are only defined at evenly spaced events, whose period is defined as the ratio between two integers. Clocked variables can be generated from continuous-time variables by using the overloaded sample(u,c) operator, which returns a clocked variable that has the values of u at the events specified by the clock c. The hold(u) operator is used to return a continuous-time variable that is defined at all times and changes with the clocked variable u at each clock tick.

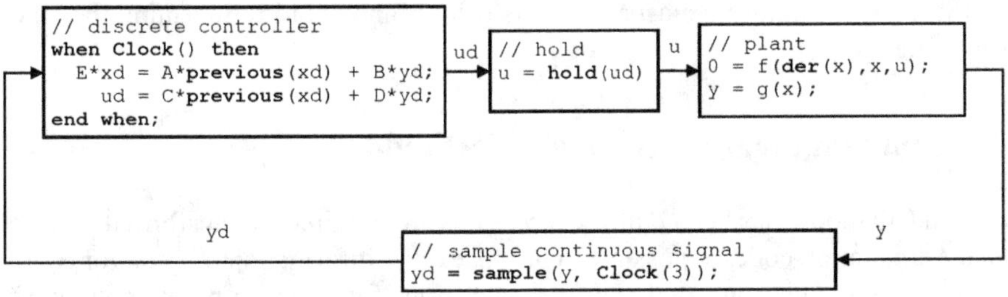

Figure 2.4 Block diagram of a control system model using clocked variables, taken from the Modelica Language Specification v. 3.3 (with permission, © The Modelica Association)

The fundamental mechanism with clocked variables is *clock inference*; that is, every equation involving clocked variables implicitly states that all their clocks are equal. It is then possible to specify the clock of the digital system model at one place (i.e. the `sample()` operator), and then automatically propagate the clock information to all other clocked variables by means of structural analysis of the equations. Multiple, inconsistent clock settings can be easily detected and reported. Figure 2.4 shows an example of a linear feedback control system model, using clocked variables for the controller description, and making use of this feature.

It is possible to rigorously define slower and faster clocks by means of the `subSample(u,factor)` and `superSample(u,factor)` operators, where `factor` is an integer value. For example, one can define a base clock of 100 ms for an inner loop in a cascaded controller, and a sub-sampled clock that ticks every ten base clock ticks (i.e. every second) for the outer loop controller. The interesting feature here is that it is possible to rigorously infer that every tick of the sub-sampled clock is synchronous with one tick out of ten of the base clock, without the issues and ambiguities that arise in other modelling and simulation environments that rely on numerical comparisons with small tolerances to detect synchronous events. It is also worth noting that it is easily possible to translate the controller code corresponding to the clocked partition(s) of a system model into real-time, code for embedded, multi-rate controllers.

As a concluding remark, it is important to mention that the Modelica Association also maintains an open-source library of models, the Modelica Standard Library. This library contains basic components and standardized connectors for all the physical domains found in engineering systems: electrical, covering analog and digital electronics as well as three-phase electrical machinery; thermal; thermo-hydraulic; 1D mechanical and 3D multibody mechanical. The standardized connectors are essential to ensure interoperability between models developed by different people and different organizations, as already noted in the previous section. The basic components provide a head start when tackling new modelling problems, as it is often the case that most of the required components are already available in the library, or can be derived from them by small modifications or customizations, so the modeller can focus on the innovative aspects of his project and avoid re-inventing the wheel every time.

2.4 Mathematical processing of OO models

Once an OO model has been built, several steps are required to transform it into executable simulation code, or possibly to produce other information relevant for control system design. These are outlined in this section, focusing on purely continuous-time system models for simplicity. The reader can refer to [11] for a comprehensive in-depth discussion.

The first step in the processing of a model, such as the `Circuit` model shown in Figure 2.3, is the so-called *flattening* procedure. First, all the required model definitions involving inheritance are expanded. Then, the models are instantiated, and all their variables, parameters and equations are added to the sets of variables, parameters and equations defined at the top level of the model, using dot notation (e.g. Ohm's law for `R1` becomes `R1.v=R1.R*R1.i`). The binding equations for parameters are also added (e.g. `R1.R=1e4`). Finally, the connect statements are processed to determine the connection sets; for each connection set, the connection equations for input, output, flow and effort variables are added to the set of equations. The end result is a set of variables, with a corresponding set of DAEs, and a set of parameters, with a corresponding set of binding equations.

Structural analysis is then performed on these sets, with the goal of ordering the equations so that they can be solved sequentially as much as possible, instead of being solved simultaneously. This is done by means of graph-theoretical algorithms, starting from bipartite equations-variables (E-V) graphs, where the left (E) nodes correspond to equations, the right (V) nodes correspond to variables, and an edge is added between an E node and a V node if variable V shows up in equation E.

First, the parameter-binding equations are ordered, so that they can be solved sequentially, one at a time (cyclic dependencies among these equations are forbidden by the Modelica language specification). It is also possible to find out if some variables depend only on parameters, in which case they can be computed only once during the initialization phase, together with the parameters.

Since OO models are usually quite redundant due to connection equations, trivial equations such as $a = \pm b$ and $a \pm b = 0$ can be symbolically solved for one of the two unknowns, substituting the result in the remaining set of equations. This process, known as *alias elimination*, typically removes more than 50% of equations and variables. The remaining set of differential equations can be written as

$$F(x, \dot{x}, v, u, p, t) = 0 \tag{2.18}$$

where $F(\cdot)$ is a vector-valued residual function, x is the vector of variables appearing under derivative sign, v is the vector of all other algebraic variables, u is the vector of top-level input variables, p is the vector of parameters and t is time. The next step is *causalization*, which means solving (2.18) for $\dot{x}$ and v, given x, u, p and t, thus bringing the system into state-space form:

$$\dot{x} = f(x, u, p, t) \tag{2.19}$$

$$v = g(x, u, p, t) \tag{2.20}$$

This step is performed by applying graph-theoretical algorithms to the E-V graph. The matching algorithm is first run, returning a subset of matching edges such that each variable and each equation is touched by one and only one edge, spanning the entire graph. Then, E and V nodes are collapsed, and each remaining edge is transformed into an oriented arc going from the E node to the V node, obtaining a directed graph that expresses the mutual dependencies between variables and equations: an arc going from node A to node B means that equation A depends on a variable that is matched to equation B. This means that equation B should be solved first, if possible, in order to use its value when solving equation A for its matched variable. Tarjan's algorithm is then run on this graph, to identify its strong components, which correspond to systems of implicit equations that cannot be solved sequentially by just re-ordering the equations. Finally, the obtained equation dependency graph is used to re-order the equations and the variables, bringing the system into the so-called block-lower-triangular (BLT) form, that is having a BLT incidence matrix.

If the BLT is strictly lower triangular, and each equation can be solved symbolically with respect to its matched unknown, then system (2.18) can be brought in the state-space form (2.19) and (2.20) by explicitly computing the unknowns one at a time.

In most cases, however, there will be blocks on the diagonal having size greater than 1, which correspond to systems of algebraic equations that need to be solved: these correspond to *algebraic loops* in block-diagram models. For small, sparse systems, it might be possible to solve such algebraic equations symbolically in an efficient way, otherwise linear or non-linear numerical solvers need to be used for that purpose. Provided that the physical model is well posed (i.e. it is not singular), linear systems have a unique solution that can be computed reliably using state-of-the-art linear algebra solvers; conversely, non-linear systems might have multiple solutions and require iterative solvers, which might need good initial guesses to converge to the right solution. The result of the above-described process is an algorithmic procedure to compute all the derivatives and algebraic variables, given the state variables, the parameters and the time, that is the state-space form (2.19).

It might be possible to apply further optimization steps to make the procedure more efficient. Common subexpression elimination techniques can be used to identify recurring expressions; these can be computed only once, storing the result in an auxiliary variable that is then used in place of all other expression instances. Variables that are set to zero in some equation can be removed from summations in other equations, or allow to remove products in other equations, in which they are involved as a factor. Tearing can be used to transform non-linear systems with many unknowns into equivalent systems with less unknowns; the idea is to suitably select a subset of tearing variables so that, by assuming they are known, it is possible to sequentially solve the majority of equations explicitly (the torn equations), leaving a smaller set of implicit equations in the tearing variables to be solved by iterative numerical techniques. For example, consider this system of three equations in three unknowns:

$$x_1 + \log(x_3) = 0 \tag{2.21}$$

$$x_1 - 5x_2^3 = 0 \tag{2.22}$$

$$x_1^2 + x_2^2 - x_3^2 = 0 \tag{2.23}$$

If x_3 is selected as a tearing variable, then the first two equations can be sequentially and explicitly solved, allowing to compute the residual of the third equation as a function of the tearing variable:

$$x_1 = -\log(x_3) \tag{2.24}$$

$$x_2 = \sqrt[3]{\frac{x_1}{5}} \tag{2.25}$$

$$f_3 = x_1^2 + x_2^2 - x_3^2 \tag{2.26}$$

so it is possible to solve the system by solving the *scalar* equation $f_3(x_3) = 0$ by some iterative method. Note that, in general, finding the set of tearing variables of minimum cardinality is a combinatorial, NP-complete problem, so appropriate heuristic algorithms need to be devised for this purpose, which also need to ensure that singularities, such as division by zero, never take place in any case.

Once the state-space form (2.19) has been obtained, it can be coupled with state-of-the-art ordinary differential equation solvers to obtain efficient simulation code. The Jacobian $\partial f/\partial x$, which is required by implicit solvers for stiff systems, can be obtained by applying automatic differentiation techniques to the above-described procedure, or by numerical differentiation. Other Jacobians, for example, for parameter sensitivity analysis, can also be obtained in the same way.

A few small examples are presented to further illustrate the symbolic processing stages. Consider the RC circuit described in the previous section. After flattening, alias elimination and BLT transformation, the following DAEs are obtained:

$$v_{\mathrm{G}} = 0 \tag{2.27}$$

$$v_{\mathrm{C}} = v_{\mathrm{C.p}} - v_{\mathrm{G}} \tag{2.28}$$

$$v_{\mathrm{S}} = v_{\mathrm{max}} \sin(2\pi f) \tag{2.29}$$

$$v_{\mathrm{R1}} = v_{\mathrm{R1.p}} - v_{\mathrm{S}} \tag{2.30}$$

$$v_{\mathrm{R1}} = R_1 i \tag{2.31}$$

$$v_{\mathrm{R2}} = v_{\mathrm{C.p}} - v_{\mathrm{R1.p}} \tag{2.32}$$

$$v_{\mathrm{R2}} = R_2 i \tag{2.33}$$

$$i = C\frac{dv_{\mathrm{C}}}{dt} \tag{2.34}$$

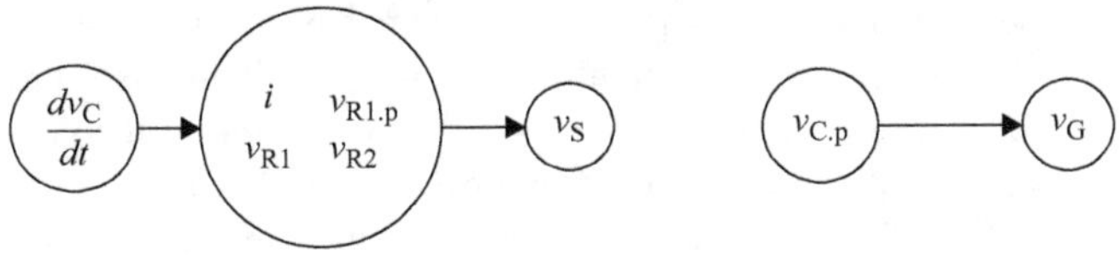

Figure 2.5 Equation dependency graph of the RC circuit model

note that the state v_C is a known variable in this context, so it need not be solved for. The equation dependency graph is shown in Figure 2.5.

The first three equations can be solved sequentially one at a time, allowing to compute v_G, $v_{C.p}$ and v_S. The following four equations form a strong component in the original E-V bipartite graph, indicating a coupled system of implicit equations, which is linear in this case. This can be solved numerically or via symbolic manipulation, yielding

$$i = \frac{v_{C.p} - v_S}{R_1 + R_2} \tag{2.35}$$

$$v_{R1} = R_1 i \tag{2.36}$$

$$v_{R2} = R_2 i \tag{2.37}$$

$$v_{R1.p} = v_S + v_{R1} \tag{2.38}$$

Finally, the last equation can be symbolically solved for its unknown, giving

$$\frac{dv_C}{dt} = \frac{i}{C} \tag{2.39}$$

There is one special case, which often takes place in a-causal OO models, when algebraic equations induce constraints on the differentiated variables. For example, consider a simple electrical circuit, in which a time-varying current source S, with current $i_0(t)$, feeds two capacitors C1 and C2 connected in parallel. After removing trivial equations that are inessential for the current analysis, the following equations are obtained:

$$i_S = i_0(t) \tag{2.40}$$

$$i_{C1} = C_1 \frac{dv_{C1}}{dt} \tag{2.41}$$

$$i_{C2} = C_2 \frac{dv_{C2}}{dt} \tag{2.42}$$

$$i_S - i_{C1} - i_{C2} = 0 \tag{2.43}$$

$$v_{C1} = v_{C2} \tag{2.44}$$

The matching algorithm fails on this system: the first equation can only be matched with i_S, the second and the third with dv_{C1}/dt and dv_{C2}/dt respectively.

The fourth can be matched to i_{C1} or i_{C2}, but then it is not possible to match the other current to the last equation, which only contains voltages, so it is apparently not possible to determine how the current is partitioned among the two capacitors. These systems, where the Jacobian of the function $F(\cdot)$ in (2.18) is singular, are called high-index DAEs. Note that the model is perfectly legitimate and makes perfect sense from a physical point of view.

One way to solve this problem is to use symbolic index reduction techniques. First, the constraints among states are identified and symbolically differentiated until the missing variables appear, allowing to successfully match them to the equations. Pantelides' algorithm [24] can be used for this purpose. In the case of the example shown above, the last equation will be differentiated and added to the system of equations:

$$\frac{dv_{C1}}{dt} = \frac{dv_{C2}}{dt} \tag{2.45}$$

and it is apparent that, by coupling this last equation to the second and the third of the original system, it is now possible to determine the partitioning of the two currents. Unfortunately, the system obtained so far is overdetermined, since it has five unknowns and six equations. Furthermore, it is clear that v_{C1} and v_{C2} cannot be really considered as 'state variables', because their initial values cannot be assigned independently, due to the constraint equation.

These issues can be solved by applying the so-called dummy derivatives (DD) algorithm [22]. The algorithm inspects the differentiated equations and tries to get rid of the constrained state variables by replacing one of the derivatives that appear in the differentiated equations, e.g. dv_{C2}/dt, by an extra algebraic variable $v_{C2,der}$ throughout the entire system. This leads to a well-posed square system of six equations in six unknowns, with only one differentiated state variable left – v_{C1}. Similar problems arise in mechanical systems, whenever a rigid connection is established between two bodies with inertia. In that case, two consecutive differentiation stages will be needed in order to solve the problem. The DD algorithm also handles cases where the choice of which state variable should be eliminated is not trivial as in the simple example discussed above. *Dynamic* state selection might be required in some cases, whereby the condition number of the Jacobian corresponding to the current state selection is monitored during the simulation, and a new, non-singular choice of states is made during simulation, when the Jacobian becomes close to singular.

This kind of symbolic manipulation can also handle another similarly special case in electrical circuits, namely a voltage source connected to a capacitor:

$$v_S = v_0(t) \tag{2.46}$$

$$i = C\frac{dv_C}{dt} \tag{2.47}$$

$$v_S = v_C \tag{2.48}$$

The matching algorithm fails here, because if v_S is matched to the first equation, there is no other unknown to match with the last equation, and vice versa. In this case, the first and third equations are differentiated, and two dummy derivatives are introduced, leading to the following system:

$$v_S = v_0(t) \tag{2.49}$$

$$v_{S,der} = \frac{dv_0}{dt} \tag{2.50}$$

$$i = Cv_{C,der} \tag{2.51}$$

$$v_{S,der} = v_{C,der} \tag{2.52}$$

Again, symbolic index reduction reduces the order of the system, which in this case becomes a state-less, algebraic system of equations. This example shows another notable feature of time-varying, high-index problems; that is, they might require the differentiation of known functions of time, which usually represent exogenous inputs to the system. Of course, index reduction is only possible in this case if the signal $v_0(t)$ is differentiable, otherwise the system cannot be handled, as the current flowing through the voltage source would contain Dirac impulses, which cannot be represented in a continuous-time numerical simulation.

2.5 Plant modelling, analysis and identification

Model-based control design techniques often require compact low-order models of the plant, which can be obtained from first principle laws by severely simplifying modelling assumptions. One first possible use for the OO physical model of the plant is to provide reference trajectories in some simulation scenarios, which can be compared against the ones obtained from the compact models to check if the simplifying assumptions are legitimate.

OO models can also be used to generate linearized, state-space models around some operating points, which might then be subject to linear model order reduction techniques. In order to do so, it is necessary to compute the Jacobians of equations (2.19) and (2.20), which can be done symbolically, by means of automatic differentiation techniques, or numerically, by computing finite differences. Note that these Jacobians are already needed for time-domain simulations when implicit integration algorithms are used, so this task can be implemented in OO simulation tools with no extra cost.

A key feature of OO models is that the modelling knowledge is symbolically encoded in the form of equations. This allows to carry out a number of activities, besides the straightforward time-domain integration of the DAEs, that are not possible with more traditional procedurally-oriented modelling and simulation tools [9].

One option is to symbolically transform the equations of the model into some specific formalism, which is the starting point for advanced control system analysis and design. For example, under suitable conditions it is possible to automatically transform the model in linear fractional representation form (see Figure 2.6), whereby the system is described in terms of a linear time-invariant model with feedback connections

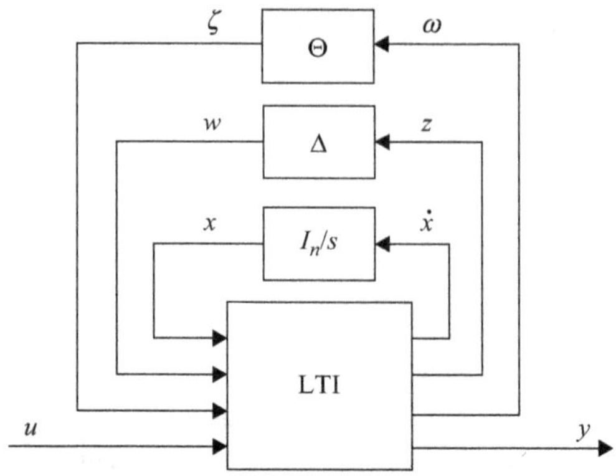

Figure 2.6 Linear fractional representation of a dynamical system

involving uncertain and/or time-varying known parameters (collected in the Δ block), as well as known non-linear functions (collected in the Θ block).

Another option is to use the model for grey-box identification of unknown parameters starting from experimental measurements, possibly by first transforming the model in some form that is more suitable for this purpose. Both options are discussed in detail in Chapter 4.

Other activities might be thought of, starting from the system DAEs in symbolic form. For this purpose, Modelica tools allow to export the system equations after the flattening phase, and possibly also after some manipulation has already taken place, such as index reduction or alias variable elimination. This can be done using XML files, which can then be easily parsed by any post-processing programme, or directly transformed into the input files for other symbolic and numerical analysis tools by means of simple XSLT transformations.

2.6 Control system performance verification

Another important application of OO physical models in control system design is the validation of the control system performance by simulation of the closed-loop system. High-fidelity system models are often used, to make sure that the controller performance is robust with respect to the simplifying assumptions that might have been taken during the control system design phase. It is also possible to exploit the flexibility of the OO approach to use different variants of the physical system model for specific simulation scenarios, e.g. on-design operation, off-design operation, system start-up, faulty operating conditions.

The closed-loop controlled system can be modelled in its entirety in Modelica, and simulated by using a Modelica tool. Continuous-time controller models can be used, as well as digital controller models that can be built using discrete and clocked variables. In many cases, continuous-time controller models are used in the early phases of system design, while accurate digital controller models (requiring much longer simulation runs due to the many time-events they generate) are used later on to make sure that the performance is not significantly degraded compared to the continuous-time design.

It is also worth mentioning that *expandable connectors* are available in Modelica to model idealized bus connections, possibly carrying dozens or hundreds of signals across complex system architectures in a convenient way. Every sub-system model contains connections from inputs and outputs of local components (e.g. sensors and actuators, or control blocks) to a bus connector, where the name of the variable on the bus is implicitly defined by the connect statement. Bus connectors are then connected together when assembling the physical system model (and possibly also the distributed control system model) in a structured hierarchical way, mimicking the way modern digital communication buses work. Finally, the compiler automatically resolves all the connections, making sure the bus signals are carried from the local component outputs to the local component inputs as expected.

Some basic functional blocks for this purpose are already available in the Modelica Standard Library for controller modelling, but it is also possible to build specialized control libraries for specific application domains. An interesting option in this case is to eventually generate the real-time code of the digital controller automatically from the Modelica model. Alternatively, control libraries can be built with function blocks that emulate the behaviour of existing control libraries on the target real-time digital control architecture.

In many cases the control engineer is more comfortable with his or her own computer-aided control system design (CACSD) tools, including block-oriented simulation functionality. Assuming that the only connections of the physical system model to the outside world are causal input and output connectors, corresponding to actuator and sensor signals, the Modelica tool can transform the physical system model in causal state-space form (2.19) and (2.20), as explained in Section 2.4. It is then possible to export this model to other CACSD tools, using suitable interfaces.

At each time step, the CACSD simulation tool can set the current time and state variables, then query the values of the state derivatives and of the outputs, using its own internal integration algorithm to solve the differential equations with respect to time, together with the other sub-systems that constitute the overall closed-loop system. The Jacobian $\partial f / \partial x$ might also be queried, if implicit integration algorithms are used. Alternatively, the Modelica tool can export a causal model to which some suitable fixed-time-step integration algorithm has been applied, so that the CACSD simulation tool can directly query the values of the states at the next time step, as well as the values of the outputs. The open functional mock-up interface (FMI) standard [6] has been developed for this purpose, and is widely supported by Modelica tools and CACSD tools.

Last, but not least, OO models of physical systems can be automatically translated into real-time, fixed-time-step simulation code. One way to do this is to link the code that computes the system's state equations (2.19) and (2.20) to an explicit fixed-time-step integration algorithm. As explained in Section 2.4, computing the derivatives and the outputs might involve solving non-linear implicit equations (algebraic loops), which introduces uncertainty in the time required to compute one step, because the non-linear solvers might require more or less iterations depending on the operating conditions of the system. A possible remedy to this problem is to limit the number of iterations of such solvers, which works fine if the system never operates too far from equilibrium conditions. More elaborate, possibly application-specific strategies can be devised to deal with this problem, see for example [25].

An interesting approach for the generation of real-time simulation code from OO models is the so-called *inline integration* method [16]. The idea is to apply simple integration rules (e.g. forward Euler) symbolically to the DAE system (2.18) after symbolic index reduction, then use the same symbolic and numerical techniques that were used for deriving (2.19) and (2.20) to compute the next values of the state variables, as well as the system outputs. In this way, symbolic processing is used to optimize the computational flow, and might lead to significant advantages in terms of CPU time savings.

In most cases, the generation of real-time code requires some iteration on the structure of the model, in order to bring down the worst-case computational time for one simulation step to an acceptable level. If a certain model eventually leads to a large system of implicit algebraic equations, alternative formulations for some of its components might be sought, which mitigate the problem. If a system is stiff, requiring short time steps to avoid numerical instabilities, some simplifications might be applied to get rid of the fast dynamics. This activity requires a good insight into the inner structure of the models and into the symbolic processing performed by the Modelica tool, but can nevertheless be advantageous compared to writing the real-time simulation C-code manually from scratch.

Given the recent trend of computer technology towards massively parallel multi-core CPUs, the evolution of OO tools will increasingly exploit parallelism for fast simulations. Some simple strategies, such as the one described in [8], can be very effective on some classes of system models. It is reasonable to expect that more and more sophisticated symbolic and numerical techniques will emerge in the future, thus significantly increasing the complexity of the models that can be simulated in real time without extensive ad-hoc tuning.

2.7 Direct use of OO models for optimal control

The use of direct methods to numerically solve constrained optimal control problems is extremely attractive, because it allows to formulate the actual control objectives and the operational constraints of the plant in a transparent way in the problem specification. The optimization problem is first transcribed into a finite-dimensional, large non-linear programming (NLP) problem, following three possible alternative strategies, namely single-shooting, multiple-shooting or direct collocation; see [14] for a comparative survey. Then, the NLP is solved using iterative numerical methods. The ever-increasing availability of computing power is also making the solution of these problems in real time increasingly feasible, e.g. for receding-horizon model-predictive control applications (see, e.g. [13]).

A key issue when tackling these problems is to provide the most appropriate formulation of the plant model, which needs to be as simple as possible to avoid excessive computing times and convergence problems, and yet capable of capturing the relevant dynamic behaviour with sufficient accuracy. The correct formulation of the control objectives is also a crucial issue; sometimes, an apparently reasonable control objective turns out to cause unanticipated, unwanted behaviour, for example, wide control oscillations, requiring further modifications to obtain a satisfactory system performance.

The optimal control design activity is therefore an iterative procedure, whereby the model, the control objective and constraints, and the underlying numerical optimization strategies need to be refined, until a satisfactory result is obtained. Each iteration requires the formulation of the dynamical model, its translation in a suitable low-level programming language (e.g. C or C++) and the interfacing with some state-of-the-art code for NLP. The generation of code to compute first derivatives (Jacobian matrices) and second derivatives (Hessian matrices) of the system equations is usually involved. In an abstract sense, this is similar to the kind of activities that need to be carried out for simulation. It seems therefore interesting to adopt a high-level, declarative approach also for this class of applications, freeing the control engineer from the need of writing low-level code and allowing him or her to concentrate on the model and control objectives formulation.

For this purpose, the Optimica language has been designed as an extension to the Modelica language that allows to define dynamic optimization problems [1, 2]. The dynamic constraints are given by a declarative Modelica model, and it is also possible to formulate the control objective, as well as inequality constraints on the system variables. The idea is then to off-load to the Optimica tool the tasks of generating the low-level C code and interfacing it to the NLP solver. This also makes it easy to try different optimization strategies (e.g. direct collocation, single-shooting, multiple-shooting) starting from the same high-level code, in order to quickly find out the most effective one on the specific problem at hand.

A small example is provided here to illustrate the features of the Optimica language, namely, the well-known problem of moving a point mass from point A to point B, from rest to rest, in minimum time. The plant dynamics is given by a Modelica model with unspecified control inputs, then an optimization problem is formulated by extending the Modelica model and adding the description of the optimization problem, with the goal of finding out the optimal trajectory of the unspecified controls that satisfies all the equality and inequality constraints.

```
model Inertia
  parameter Mass M = 10;
  Position x;
  Velocity v;
  input Force F;
equation
  der(x) = v;
  M*der(v) = F;
end Inertia;
```

```
optimization RestToRest(
  objective=finalTime,
  startTime=0,
  finalTime(free=true))
  extends Inertia(F(free = true));
  parameter Real Fmax=10.0;
initial equation
  x=0;
  v=0;
constraint
  x(finalTime)=1;
  v(finalTime)=0;
  F >= -Fmax;
  F <=  Fmax;
end RestToRest;

optimization RestToRest2(
  objectiveIntegrand=(x-1)^2,
  startTime=0,
  finalTime=2.5)
  extends Inertia(F(free = true, min = -Fmax, max = Fmax));
  parameter Real Fmax=10.0;
initial equation
  x=0;
  v=0;
constraint
  x(finalTime)=1;
  v(finalTime)=0;
end RestToRest2;
```

The first formulation (`Rest2Rest`) is straightforward: the final time is explicitly stated as the optimization objective, and all inequality constraints are stated explicitly in the constraint section. However, solving minimum time problems with direct methods can be numerically more problematic than solving problems over a fixed time horizon. The alternative formulation `Rest2Rest2` minimizes the Lagrange-type control objective:

$$J = \int_0^{t_f} (x-1)^2 dt \tag{2.53}$$

that is the average square distance from the target end point, which actually leads to the same result. Also, upper and lower bounds on the variable F are explicitly declared as such, allowing for more efficient numerical handling than generic inequality constraints.

One way to solve the optimal control problem is to directly transcribe it into a finite-dimensional optimization problem, by means of collocation methods [20]. The solution can be described in terms of Lagrange interpolation polynomials, defined by their values at some interpolation points in time. The DAEs of the system and the constraints are then transcribed into a large set of very sparse equality and inequality constraints, involving the values at the interpolation points, which become the unknowns of the numerical optimization problem. This is then solved by a state-of-the-art NLP code.

In general, if the model or the constraints are non-linear, or if the control objective is not quadratic, a non-linear, possibly non-convex optimization problem will ensue, which might have many local minima besides the global optimum solution and might also be numerically ill conditioned. Ensuring the convergence of the solver to the desired solution is therefore a non-trivial problem. Some numerical issues can be addressed automatically by the tool in a fairly straightforward way, for example the correct scaling of variables and equations. In other cases, more involved transformations would be beneficial. For example, in a declarative context one may want to formulate an upper limit on the pressure ratio of a compressor as $p_{out}/p_{in} \leq 10$, which would then be translated into a non-linear constraint. Assuming both $p_{out} > 0$ and $p_{in} > 0$, this constraint can be easily reformulated as a linear equivalent constraint $p_{out} \leq 10p_{in}$, which will be more easily handled by the solver. As another example, the problem:

$$\begin{aligned} \min_{u(t)} \quad & \int_{t_0}^{t_f} p^2(y - y_0)^2 + q^2(u - u_0)^2 dt \\ \text{s.t.} \quad & \dot{x} = ax + bu \\ & y = cx \end{aligned} \tag{2.54}$$

is a convex, linear-quadratic problem that can be solved easily. However, the very same problem could be written as

$$\begin{aligned} \min_{u(t)} \quad & \int_{t_0}^{t_f} J dt \\ \text{s.t.} \quad & \dot{x} = ax + bu \\ & y = cx \\ & J = p^2(y - y_0)^2 + q^2(u - u_0)^2 \end{aligned} \tag{2.55}$$

which has a non-linear constraint and is much harder to solve numerically.

Ideally, in a declarative context, all the transformations that lead to equivalent (or approximately equivalent) *optimization* problems that are easier to solve should be applied automatically by the tool, as it is the case with *simulation* problems, see Section 2.4. Unfortunately, to which extent this might be done automatically, or at least in an assisted way, is still an open research question, as OO tools dealing with optimization problems are still in their infancy. One possibility could be to try to combine ideas that have been successful for optimized simulation code generation, such as the ones described in [11], with ideas and methods used in convex optimization [7]. The current state of the art is that the modeller needs to be careful as to how the model equations are written, bearing in mind that they will be subject to optimization, following certain transcription rules. Nevertheless, the high-level approach can be still beneficial compared to direct coding in low-level programming languages, and there are significant results already reported in the field of power plant control [19] and chemical engineering [3]. It is reasonable to expect some development on this front in the future, which will allow a wider use of high-level mathematical models for the solution of optimal control problems.

A very interesting application of this methodology is during the early design phases of a system, as a tool for fast prototyping of optimal controllers, helping the designer to find the best possible model and the best possible control objectives and constraints, as well as to get an idea of the limit performance in ideal conditions. Whether or not robust and fast NMPC real-time control code could also be generated in a fully automated way, starting from such a high-level description, is still an open research question.

2.8 Conclusions

Equation-based, OO modelling languages are ideally suited for modular, system-level modelling of multi-domain physical systems, supporting control system design in many ways. The declarative nature of the OO models ultimately means that the DAE describing the model are symbolically available, which allows to automatically carry out a host of activities that are relevant to model-based control system design: model analysis, grey-box parameter identification, closed-loop performance validation, model reduction, optimal and model predictive control. The Modelica language, which has been introduced in this chapter, is particularly attractive in this field, since it is a non-proprietary language that is supported by multiple software tools, both commercial and open-source, and now has a wide base re-usable model libraries available. Applications of OO modelling in control design that are essentially based on simulation are now mature and well established. The use of such models for other purposes such as system identification and optimal control is still rather new, and thus a very interesting subject for future research.

References

[1] J. Åkesson. *Languages and Tools for Optimization of Large-Scale Systems*. PhD thesis, Lund University, Department of Automatic Control, 2007.

[2] J. Åkesson, C. E. Årzén, M. Gäfvert, T. Bergdahl, and H. Tummescheit. Modeling and optimization with Optimica and JModelica.org – Languages and tools for solving large-scale dynamic optimization problems. *Computers and Chemical Engineering*, 34(11):1737–1749, 2010.

[3] J. Åkesson, C. D. Laird, G. Lavedan, K. Prölss, H. Tummescheit, S. Velut, and Y. Zhu. Nonlinear model predictive control of a CO_2 post-combustion absorption unit. *Chemical Engineering & Technology*, 35(3):445–454, January 2012.

[4] P. I. Barton and C. C. Pantelides. The modeling of combined continuous and discrete processes. *AIChE Journal*, 40:966–979, 1994.

[5] G. Berry and G. Gonthier. The esterel synchronous programming language: Design, semantics, implementation. *Science of Computer Programming*, 19(2):87–152, November 1992.

[6] T. Blockwitz et al. The functional mockup interface for tool independent exchange of simulation models. In *Proceedings 8th International Modelica Conference*, pages 105–114, Dresden, Germany, Mar. 20–22, 2011. Modelica Association.

[7] S. Boyd and L. Vandenberghe. *Convex Optimization*. Cambridge University Press, New York, NY, USA, 2004.

[8] F. Casella. A strategy for parallel simulation of declarative object-oriented models of generalized physical networks. In Henrik Nilsson, editor, *Proceedings of the 5th International Workshop on Equation-Based Object-Oriented Modeling Languages and Tools (EOOLT)*, pages 45–51, Nottingham, UK, Apr. 19, 2013.

[9] F. Casella, F. Donida, and M. Lovera. Beyond simulation: Computer-aided control system design using equation-based object-oriented modelling for the next decade. *Simulation News Europe*, 19(1):29–41, 2009.

[10] P. Caspi, D. Pilaud, N. Halbwachs, and J. A. Plaice. Lustre: A declarative language for real-time programming. In *Proceedings of the 14th ACM SIGACT-SIGPLAN Symposium on Principles of Programming Languages*, POPL '87, pages 178–188, New York, NY, USA, 1987. ACM.

[11] F. E. Cellier and E. Kofman. *Continuous System Simulation*. Springer-Verlag, 2006.

[12] E. Christen and K. Bakalar. VHDL-AMS-a hardware description language for analog and mixed-signal applications. *Circuits and Systems II: Analog and Digital Signal Processing, IEEE Transactions*, 46(10):1263–1272, 1999.

[13] M. Diehl, H. G. Bock, and J. P. Schlöder. A real-time iteration scheme for nonlinear optimization in optimal feedback control. *SIAM Journal on Control and Optimization*, 43(5):1714–1736, 2005.

[14] M. Diehl, H. G. Bock, H. Diedam, and P.-B. Wieber. Fast direct multiple shooting algorithms for optimal robot control. In M. Diehl and K. Mombaur, editors, *Fast Motions in Biomechanics and Robotics*, volume 340 of *Lecture Notes in Control and Information Sciences*, pages 65–93. Springer Berlin Heidelberg, 2006.

[15] F. Donida, F. Casella, and G. Ferretti. Model order reduction for object-oriented models: A control systems perspective. *Mathematical and Computer Modeling of Dynamical Systems*, 16(3):269–284, June 2010.

[16] H. Elmqvist, M. Otter, and F. Cellier. Inline integration: A new mixed symbolic/numeric approach for solving differential-algebraic equation systems. In *Proc. ESM'95, European Simulation Multiconference*, pages xxiii–xxxiv, Prague, Czech Republic, Jun. 5–8, 1995.

[17] R. Franke, F. Casella, M. Otter, M. Sielemann, H. Elmqvist, S. E. Mattsson, and H. Olsson. Stream connectors: An extension of Modelica for device-oriented modeling of convective transport phenomena. In *Proceedings 7th International*

Modelica Conference, pages 108–121, Como, Italy, Sep. 20–22, 2009. The Modelica Association.

[18] P. Fritzson. *Principles of Object Oriented Modeling and Simulation with Modelica 3.3*. Wiley IEEE Press, 2014.

[19] P.-O. Larsson, F. Casella, F. Magnusson, J. Andersson, M. Diehl, and J. Åkesson. A framework for nonlinear model-predictive control using object-oriented modeling with a case study in power plant start-up. In *Proceedings of the 2013 IEEE Multi-conference on Systems and Control*, pages 346–351, Hyderabad, India, Aug. 28–30, 2013. IEEE.

[20] F. Magnusson and J. Åkesson. Collocation methods for optimization in a Modelica environment. In *Proceedings of the 9th International Modelica Conference*, pages 649–658, Munich, Germany, Sep. 3–5, 2012.

[21] S. E. Mattsson, H. Elmqvist, and M. Otter. Physical system modeling with Modelica. *Control Engineering Practice*, 6(4):501–510, 1998.

[22] S. E. Mattsson and G. Söderlind. Index reduction in differential-algebraic equations using dummy derivatives. *SIAM Journal on Scientific Computing*, 14(3):677–692, 1993.

[23] M. Otter, H. Elmqvist, and S. E. Mattsson. The new Modelica Multi-Body library. In *Proceedings 3rd International Modelica Conference*, pages 311–330, Linköping, Sweden, Nov. 3–4, 2003.

[24] C. C. Pantelides. The consistent initialization of differential-algebraic systems. *SIAM Journal on Scientific and Statistical Computing*, 9(2):213–231, 1988.

[25] A. V. Papadopoulos, J. Åkesson, F. Casella, and A. Leva. Automatic partitioning and simulation of weakly coupled systems. In *Proceedings of the 52nd IEEE Conference on Decision and Control*, pages 3172–3177, Firenze, Italy, Dec. 10–13, 2013. IEEE.

[26] The Modelica Association. Modelica – A unified object-oriented language for physical systems modeling – Language specification version 3.2 revision 2. Online, Jul. 30, 2013.

[27] M. Tiller. *Introduction to Physical Modelling with Modelica*. Kluwer, 2001.

Chapter 3

Projection-based model reduction techniques

Pierre Vuillemin[1]*, Charles Poussot-Vassal*[2] *and Daniel Alazard*[3]

Abstract

This chapter is aimed at providing, from the authors point of view, the most representative techniques of linear time-invariant (LTI) model approximation. For an exhaustive overview of all the existing model approximation methods, the reader should refer to [3] and references therein. With this in mind, two sets of methods are presented: truncation methods, and especially the balanced one and Krylov, or moment matching methods. Balanced truncation is grounded on the singular value decomposition (SVD) and is often considered as the gold standard of model reduction. Its standard form as well as its frequency-limited counterpart is presented. The moment matching methods are aimed at interpolating the initial model and its derivatives at some points through projection on some specific Krylov subspaces. These approaches have known significant development in the past few years. Both truncation methods and moment matching methods can be brought together under the underlying framework of approximation by projection. There exist other approaches which are not based on projection but rather on optimization procedures. Since the theoretical framework is quite different, they are not presented in this chapter and the reader should refer to [4, 21] or [31] for further details.

3.1 Introduction

3.1.1 Motivations

Physical systems or phenomena are generally represented by mathematical models which can be used in simulation, control or analysis. In particular, dynamical systems can be described by differential equations (or difference equations in the discrete case). Modern numerical tools enable us to build highly representative models of

[1]ONERA, Toulouse, France, e-mail: pierre.vuillemin@onera.fr
[2]ONERA, Toulouse, France, e-mail: charles.poussot-vassal@onera.fr
[3]ISAE, Toulouse, France, e-mail: Daniel.Alazard@isae.fr

physical systems. However this accuracy generally comes with a high complexity, i.e., the number of equations and variables is large. Theoretically, this complexity is not an issue, but in practice, limited computational capabilities, storage constraints, and finite precision arithmetic result in significant perturbations of theoretical results. Indeed some calculations are not feasible – or not in an acceptable time – when the number of variables involved becomes too large.

Using simplified models instead of the complex original ones is a way to alleviate the problems related to large dimension. The construction of these simplified models is the topic of this chapter. In particular, the approximation of large-scale linear time invariant (LTI) models is addressed. Linear models arise in many industrial applications as they provide a powerful theoretical background for control and analysis, enabling us to perform relevant simulations of physical systems. For this kind of models, a large dimension makes simulation and control synthesis very time-consuming or even infeasible in some cases.

The several examples that illustrate the reduction methods presented in this chapter are based on different medium-scale models available in [16]:

- The model of the International Space Station's (ISS) Russian Service Module 1R has 3 inputs, 3 outputs, and 270 states. Its eigenvalues are represented in Figure 3.1 and the magnitude of its frequency responses is plotted in Figure 3.2.

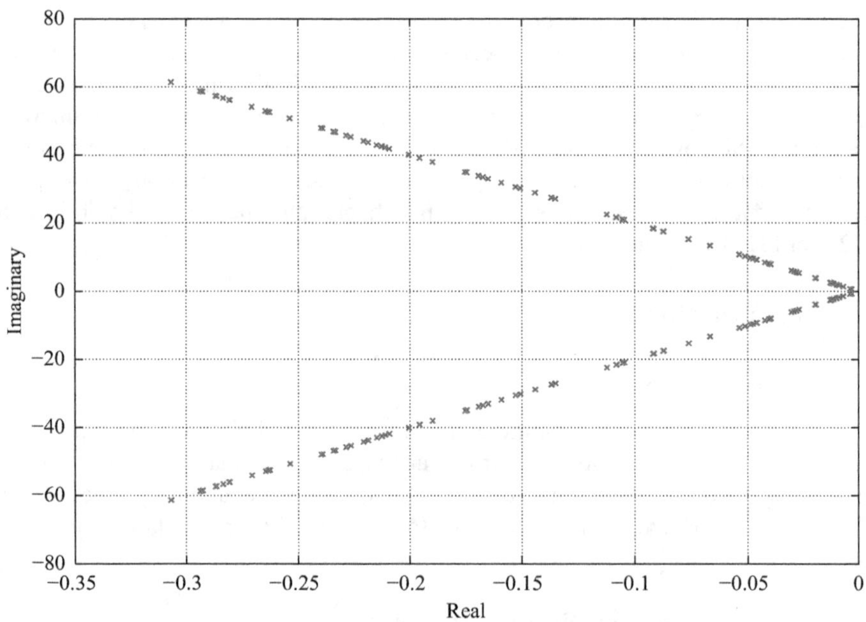

Figure 3.1 Poles of the ISS model

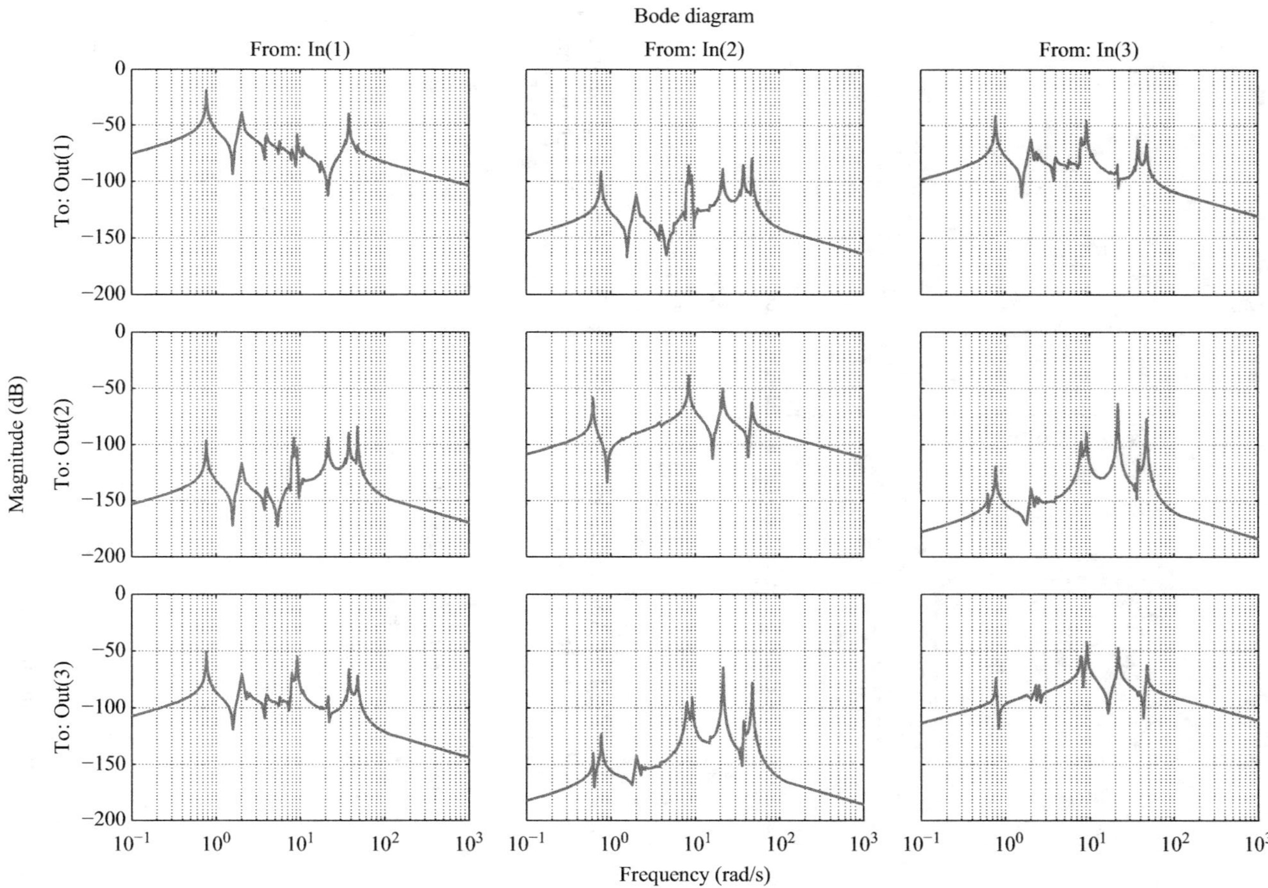

Figure 3.2 Magnitude of ISS model frequency responses

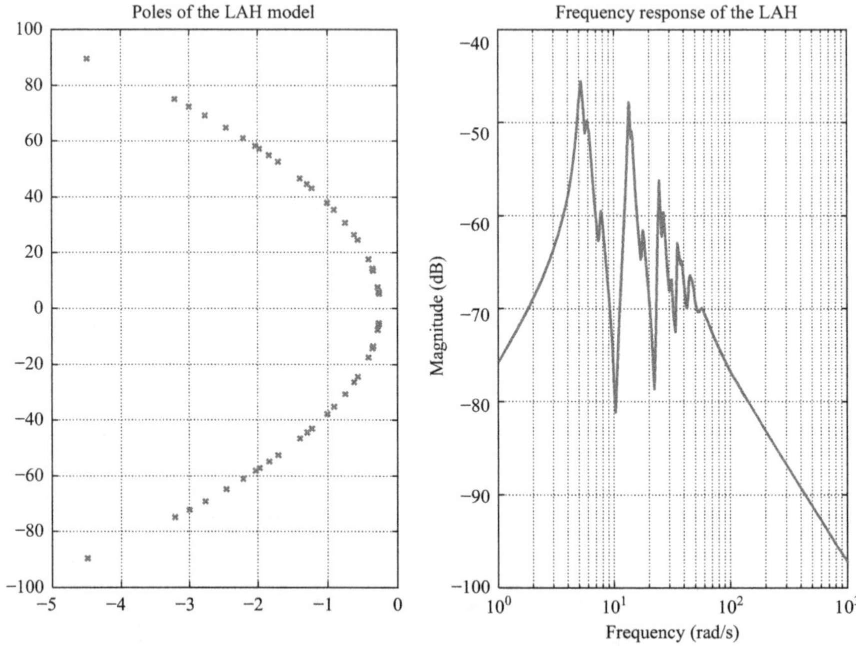

Figure 3.3 Poles and frequency response of the LAH model

- The model of the Los Angeles Hospital (LAH) is a single-input single-output (SISO) model with 48 states. Its eigenvalues and its frequency response are plotted in Figure 3.3.

3.1.2 Model reduction by projection

3.1.2.1 Problem statement

Consider a stable continuous MIMO LTI dynamical model $\mathbf{H}$ described either by its state-space representation:

$$\mathbf{H} := \begin{cases} \dot{\mathbf{x}}(t) = A\mathbf{x}(t) + B\mathbf{u}(t) \\ \mathbf{y}(t) = C\mathbf{x}(t) + D\mathbf{u}(t) \end{cases} \tag{3.1}$$

or by its associated transfer function $H(s)$:

$$\begin{aligned} H : \mathbb{C} &\rightarrow \mathbb{C}^{n_y \times n_u} \\ s &\rightarrow H(s) = C(sI_n - A)^{-1}B + D \end{aligned} \tag{3.2}$$

In many cases, the trajectories $\mathbf{x}(t)$ of $\mathbf{H}$ are contained in a low dimensional subspace $\mathscr{S}$, that is to say $\mathbf{x}(t)$ can be expressed as

$$\mathbf{x}(t) = V_{\mathscr{S}}\mathbf{x}_\mathbf{r}(t) \tag{3.3}$$

where $V_{\mathscr{S}}$ a basis of $\mathscr{S}$ and $\mathbf{x_r}(t)$ is the reduced state vector. The exact basis of $\mathscr{S}$ is in general not known, so an approximation $V \in \mathbb{R}^{n \times r}$ is used instead, thus,

$$\mathbf{x}(t) \approx V\hat{\mathbf{x}}(t) \tag{3.4}$$

Then the ordinary differential equation in (3.1) becomes:

$$V\dot{\hat{\mathbf{x}}}(t) = AV\hat{\mathbf{x}}(t) + B\mathbf{u}(t) + \mathbf{res}(t) \tag{3.5}$$

where $\mathbf{res}(t)$ is a nonzero residue which comes from the approximation of the state $\mathbf{x}(t)$. The *Petrov-Galerkin* conditions [26] are then enforced by forcing the residue $\mathbf{res}(t)$ to be orthogonal to a subspace $\mathscr{W}$. By denoting $W \in \mathbb{R}^{n \times r}$ a basis of $\mathscr{W}$ and left-multiplying (3.5) by W^T, it comes that,

$$W^T V\dot{\hat{\mathbf{x}}}(t) = W^T AV\hat{\mathbf{x}}(t) + W^T B\mathbf{u}(t) + \underbrace{W^T\mathbf{res}(t)}_{=0} \tag{3.6}$$

If $W^T V$ is nonsingular, the reduced differential equation becomes:

$$\dot{\hat{\mathbf{x}}}(t) = \left(W^T V\right)^{-1} W^T AV\hat{\mathbf{x}}(t) + \left(W^T V\right)^{-1} W^T B\mathbf{u}(t) \tag{3.7}$$

and the associated output $\hat{\mathbf{y}}(t)$ is given by

$$\hat{\mathbf{y}}(t) = CV\hat{\mathbf{x}}(t) + D\mathbf{u}(t) \tag{3.8}$$

The problem of model reduction by projection is summarized below in Problem 3.1.

Problem 3.1 (Reduction by projection). Given a continuous MIMO LTI dynamical model $\mathbf{H}$ defined as

$$H := \begin{cases} \dot{\mathbf{x}}(t) &= A\mathbf{x}(t) + B\mathbf{u}(t) \\ y(t) &= C\mathbf{x}(t) + D\mathbf{u}(t) \end{cases} \tag{3.9}$$

where $A \in \mathbb{R}^{n \times n}$, $B \in \mathbb{R}^{n \times n_u}$, $C \in \mathbb{R}^{n_y \times n}$, and $D \in \mathbb{R}^{n_y \times n_u}$, the projection-based model order reduction problem consists in finding two biorthogonal projectors $V, W \in \mathbb{R}^{n \times r}$ such that the reduced-order model $\hat{\mathbf{H}}$ of order $r \ll n$ described by

$$\hat{\mathbf{H}} := \begin{cases} \dot{\hat{\mathbf{x}}}(t) &= \hat{A}\hat{\mathbf{x}}(t) + \hat{B}\mathbf{u}(t) \\ y(t) &= \hat{C}\hat{\mathbf{x}}(t) + \hat{D}\mathbf{u}(t) \end{cases} \tag{3.10}$$

with $\hat{A} = W^T AV \in \mathbb{R}^{r \times r}$, $\hat{B} = W^T B \in \mathbb{R}^{r \times n_u}$, $\hat{C} = CV \in \mathbb{R}^{n_y \times r}$, and $\hat{D} \in \mathbb{R}^{n_y \times n_u}$, accurately reproduces the behaviour of the full-order system $\mathbf{H}$.

3.1.2.2 Measuring the approximation quality

To measure the accuracy of the reduced-order model, several metrics are available:

- The $\mathcal{H}_\infty$-norm (see Definition 3.1) of the error $\|H - \hat{H}\|_{\mathcal{H}_\infty}$ represents the worst error between the full-order model and the reduced-order one.
- The $\mathcal{H}_2$-norm (see Definition 3.2) of the error $\|H - \hat{H}\|_{\mathcal{H}_2}$ represents the average error over the whole frequency range between the full-order model and the reduced-order one. This measurement is useful when the global mismatch error must be considered.
- The $\mathcal{H}_{2,\Omega}$-norm (see Definition 3.3) of the error $\|H - \hat{H}\|_{\mathcal{H}_{2,\Omega}}$ represents the average error over a bounded frequency range $\Omega = [0, \omega]$ between the full-order model and the reduced-order one. This measurement is useful whenever the whole frequency response of the initial system is not accurately known or not needed [32, 31].

Definition 3.1 ($\mathcal{H}_\infty$-norm). The $\mathcal{H}_\infty$-norm of a stable LTI dynamical system $\mathbf{H}$, denoted $\|H\|_{\mathcal{H}_\infty}$ is defined as

$$\begin{aligned} \|H\|_{\mathcal{H}_\infty} &= \max_{\omega \in \mathbb{R}} \|H(j\omega)\|_2 \\ &= \max_{\omega \in \mathbb{R}} \sigma_{\max}(H(j\omega)) \end{aligned} \tag{3.11}$$

It represents the maximum gain in any direction and at any frequency (for SISO systems it is indeed the maximum gain of the frequency response).

Definition 3.2 ($\mathcal{H}_2$-norm). The $\mathcal{H}_2$-norm of a stable LTI dynamical system $\mathbf{H}$ denoted by $\|H\|_{\mathcal{H}_2}$ is defined as

$$\|H\|_{\mathcal{H}_2} = \sqrt{\frac{1}{2\pi} \int_{-\infty}^{\infty} \mathbf{trace}(H(j\nu)H(-j\nu)^T) d\nu} \tag{3.12}$$

where $H(s)$ is the transfer function associated with $\mathbf{H}$. The $\mathcal{H}_2$-norm represents the output energy of the system when its input is white noise.

Definition 3.3 ($\mathcal{H}_{2,\Omega}$-norm [1]). The $\mathcal{H}_{2,\Omega}$-norm of an LTI dynamical system $\mathbf{H}$, denoted $\|H\|_{\mathcal{H}_{2,\Omega}}$, is defined as the restriction of the $\mathcal{H}_2$-norm over $\Omega = [0, \omega]$, $\omega \in \mathbb{R}_+^*$, i.e.,

$$\|H\|_{\mathcal{H}_{2,\Omega}} = \sqrt{\frac{1}{2\pi} \int_{-\omega}^{\omega} \mathbf{trace}(H(j\nu)H(-j\nu)^T) d\nu} \tag{3.13}$$

The $\mathcal{H}_\infty$, $\mathcal{H}_2$, and $\mathcal{H}_{2,\Omega}$ norms are not always finite[1] as stated in Property 3.1, 3.2, and 3.3.

[1] For further information on the norms of linear systems, the reader should refer to [5] or [35].

Property 3.1. The $\mathcal{H}_2$-norm of **H** is finite iff **H** is stable and strictly proper.

Property 3.2. The $\mathcal{H}_\infty$-norm of **H** is finite iff **H** is stable and proper.

Property 3.3. For a finite ω, the $\mathcal{H}_{2,\Omega}$-norm of **H** is finite iff $\omega < |\underline{\lambda}^{im}|$ where $\underline{\lambda}^{im}$ is the purely imaginary pole of **H** with the smallest modulus (if **H** has no poles on the imaginary axis then its $\mathcal{H}_{2,\Omega}$-norm is finite).

Remark 3.1. Computing the norms presented in Definitions 3.1, 3.2, and 3.3 is, especially for large-scale systems, a nontrivial matter that is not covered in this chapter. In particular, the $\mathcal{H}_\infty$-norm is computed with iterative procedures that tends toward its true value up to a given tolerance. The reader can refer to [35] and references therein for the computation of the $\mathcal{H}_2$ and $\mathcal{H}_\infty$ norms.

In the sequel, rather than considering directly the $\mathcal{H}_i$-norm ($\mathcal{H}_i$ being $\mathcal{H}_\infty$, $\mathcal{H}_2$, or $\mathcal{H}_{2,\Omega}$) of the error $\|H - \hat{H}\|_{\mathcal{H}_i}$, the relative error is considered, i.e.,

$$E_{\mathcal{H}_i} = 100\frac{\|H - \hat{H}\|_{\mathcal{H}_i}}{\|H\|_{\mathcal{H}_i}} \tag{3.14}$$

The relative error $E_{\mathcal{H}_i}$ is a more convenient metric since it represents a percentage of error instead of an absolute value.

3.2 Model reduction by truncation

The first set of model reduction methods presented in this chapter is the set of approaches that consists in discarding some states of the full-order model state-space representation. From a projection point of view it consists in constructing an orthogonal projector $V \in \mathbb{R}^{n \times r}$ such that some elements of the state $\mathbf{x}(t)$ are retained after projection. However, discarding some states must not be done in any basis. In fact this procedure has an interest only in the *modal basis* and in the *balanced basis* where the states have a particular meaning.

In Section 3.2.1, the concepts of state-space truncation and residualization are introduced. Then, in Section 3.2.1.3, the state-space truncation in the modal basis, also called *modal approximation*, is presented and illustrated. Finally, the well known *balanced truncation* (BT) is described in Section 3.2.2 as well as its frequency-limited version.

3.2.1 State-space truncation and residualization

3.2.1.1 State-space truncation

Given a n-th order linear time-invariant (LTI) dynamical system **H** described by the following state-space representation:

$$\mathbf{H} := \begin{cases} \dot{\mathbf{x}}(t) = A\mathbf{x}(t) + B\mathbf{u}(t) \\ \mathbf{y}(t) = C\mathbf{x}(t) + D\mathbf{u}(t) \end{cases} \tag{3.15}$$

where $A \in \mathbb{R}^{n\times n}$, $B \in \mathbb{R}^{n\times n_u}$, $C \in \mathbb{R}^{n_y\times n}$, and $D \in \mathbb{R}^{n_y\times n_u}$, the state vector $\mathbf{x}(t)$ can be divided between the states which must be retained $\mathbf{x}_1(t)$ and those which must be discarded $\mathbf{x}_2(t)$, i.e.,

$$\mathbf{x}(t) = \begin{bmatrix} \mathbf{x}_1(t) \\ \mathbf{x}_2(t) \end{bmatrix} \tag{3.16}$$

The state-space representation (3.15) can be structured according to the partitioning of $\mathbf{x}(t)$,

$$A = \begin{bmatrix} A_{11} & A_{12} \\ A_{21} & A_{22} \end{bmatrix}, \quad B = \begin{bmatrix} B_1 \\ B_2 \end{bmatrix}, \quad C = \begin{bmatrix} C_1 & C_2 \end{bmatrix} \tag{3.17}$$

Then, the truncated state-space realization $\hat{\mathbf{H}}$ of $\mathbf{H}$ is obtained by keeping only the sub-matrices associated to the states $\mathbf{x}_1(t)$, i.e., (A_{11}, B_1, C_1, D). The order of $\hat{\mathbf{H}}$ is given by the number of states that are retained.

In general, the truncated model $\hat{\mathbf{H}}$ has no particular property. The only guarantee is that

$$H(\infty) = \hat{H}(\infty) \tag{3.18}$$

because the direct feedthrough is not affected by truncation. In fact, state-space truncation is interesting only when performed on particular realizations such as the modal or balanced ones.

3.2.1.2 State-space residualization

The state-space truncation procedure guarantees a perfect matching of $\mathbf{H}$ and $\hat{\mathbf{H}}$ at infinity. However, the steady-state error is not equal to zero, indeed,

$$H(0) - \hat{H}(0) = C_1 {A_{11}}^{-1} B_1 - CA^{-1}B \tag{3.19}$$

If a perfect matching is required at low frequencies, then state residualization (or singular perturbation approximation) should be used instead of state truncation. It consists in saying that the dynamics of $\mathbf{x}_2(t)$ are fast compared to those of $\mathbf{x}_1(t)$, i.e., $\dot{\mathbf{x}}_2(t) = 0$. This leads to the equation:

$$A_{22}\mathbf{x}_2(t) = A_{21}\mathbf{x}_1(t) + B_2\mathbf{u}(t) \tag{3.20}$$

If A_{22} is nonsingular, then the reduced order model $\hat{\mathbf{H}}$ is then given by

$$\hat{\mathbf{H}} = \left(\begin{array}{c|c} A_{11} - A_{12}{A_{22}}^{-1}A_{21} & B_1 - A_{12}{A_{22}}^{-1}B_2 \\ \hline C_1 - C_2{A_{22}}^{-1}A_{21} & D - C_2{A_{22}}^{-1}B_2 \end{array} \right) \tag{3.21}$$

and with this method,

$$H(0) = \hat{H}(0) \tag{3.22}$$

This result can be explained by the fact that the state residualization is related to the state truncation by the bilinear transformation $s \to \frac{1}{s}$. Indeed a residualization of $H(s)$ can be obtained by performing a state-truncation to $G(\omega) = H(\omega^{-1})$ and setting $\hat{H}(s) = \hat{G}(s^{-1})$. That is why, in the sequel, no distinction is made between truncation and residualization. The reader should refer to [14] or [17] for further information on this topic.

3.2.1.3 Modal approximation

Truncation of modal realizations is widely used by engineers since it is conceptually simple and enables to keep the meaningful modes of the initial system. For instance, high frequency modes are often discarded because they play a secondary role in the system's dominant dynamics and they might also be out of the actuators' bandwidths.

Consider the LTI dynamical system $\mathbf{H}$ described by the state-space representation (3.15). For simplicity, assume that the matrix A is diagonalizable with simple eigenvalues denoted by λ_i, $i = 1, \ldots, n$. By projection onto the eigenspace of A, the realization of $\mathbf{H}$ becomes

$$A_\Delta = \begin{bmatrix} \lambda_1 & & \\ & \ddots & \\ & & \lambda_n \end{bmatrix}, \quad B_\Delta = \begin{bmatrix} \mathbf{b_1} \\ \vdots \\ \mathbf{b_n} \end{bmatrix}, \quad C_\Delta = \begin{bmatrix} \mathbf{c_1}^T & \ldots & \mathbf{c_n}^T \end{bmatrix}, \quad D_\Delta = D \tag{3.23}$$

The reduced order model $\hat{\mathbf{H}}$ is then obtained by retaining only some modes in (3.23). If the first r modes of $\mathbf{H}$ are retained, then the error between the two models can be expressed as[2]

$$H(s) - \hat{H}(s) = \sum_{i=r+1}^{n} \frac{\mathbf{c_i}^T \mathbf{b_i}}{s - \lambda_i} \tag{3.24}$$

where H and $\hat{H}$ are the transfer functions associated to $\mathbf{H}$ and $\hat{\mathbf{H}}$. In particular, the $\mathscr{H}_\infty$-norm of the error is given by

$$\|H - \hat{H}\|_{\mathscr{H}_\infty} = \max_{\omega \in \mathbb{R}} \left\| \sum_{i=r+1}^{n} \frac{\mathbf{c_i}^T \mathbf{b_i}}{j\omega - \lambda_i} \right\|_2 \tag{3.25}$$

$$\leq \max_{\omega \in \mathbb{R}} \sum_{i=r+1}^{n} \frac{\left\| \mathbf{c_i}^T \mathbf{b_i} \right\|_2}{|j\omega - \lambda_i|} \tag{3.26}$$

$$\leq \sum_{i=r+1}^{n} \frac{\left\| \mathbf{c_i}^T \mathbf{b_i} \right\|_2}{|\mathbf{Re}(\lambda_i)|} \tag{3.27}$$

[2] Since A has, by hypothesis, only simple eigenvalues, the transfer function $H(s)$ of $\mathbf{H}$ can be put in its Gilbert form.

Remark 3.2. The bound of the approximation error (3.27) can be very pessimistic (see Example 3.1). This comes from the use of the triangular inequality (from (3.25) to (3.26)) and the transition from (3.26) to (3.27) which consists in saying that the maximum of a sum of functions is smaller than the sum of each maximum.

There are several ways to select the modes that should be retained. Engineers often select those which have a particular physical meaning. This approach implies a good knowledge of the physical system behind the equations. For more systematic selection methods, some criteria can be formulated to:

- discard modes which have a fast decay rate, the modes with the smallest real part $\mathbf{Re}(\lambda_i)$ should be retained;
- discard modes which have a high natural frequency, the modes with the smallest absolute value $|\lambda_i|$ should be retained;
- minimize the $\mathcal{H}_\infty$ bound of the error (3.27), the dominant modes, i.e., those with the largest $\frac{\|\mathbf{c_i}^T\mathbf{b_i}\|_2}{|\mathbf{Re}(\lambda_i)|}$ ratio, should be retained.

Remark 3.3. In large-scale settings, the computation of the eigenvalues and corresponding eigenvectors can become a tedious task. Specific iterative methods have been developed to tackle this issue (see for instance [26]). In [25], the author uses those algorithm to efficiently find the dominant modes of large-scale systems.

Example 3.1. Consider the ISS model described in [16]. This model is reduced to several orders r going from 2 to 40 with modal truncation by considering the last selection method presented above. The relative $\mathcal{H}_\infty$-norm of the approximation error and its bound, i.e.,

$$100\frac{\|H-\hat{H}\|_{\mathcal{H}_\infty}}{\|H\|_{\mathcal{H}_\infty}} \quad \text{and} \quad \frac{100}{\|H\|_{\mathcal{H}_\infty}}\sum_{i=r+1}^{n}\frac{\|\mathbf{c_i}^T\mathbf{b_i}\|_2}{|\mathbf{Re}(\lambda_i)|} \tag{3.28}$$

are computed for each value of r and represented in Figure 3.4. As expected the upper bound is very pessimistic and, obviously, the smaller the reduced-order model is, the worse the bound is. On this example, the approximation error is reasonably small that shows that modal approximation may be sufficient in some cases. In Figure 3.5, the magnitude of the frequency responses of the full-order and reduced-order models is plotted. Due to the modes selection criterion, the input–output transfers with the highest magnitude (like the 1st input to 1st output) are quite well matched while other transfers with smaller magnitude (such as the 2nd input to 3rd output) are not as well reproduced.

Modal approximation is a simple reduction technique that enables us to preserve some modes of the initial system. Its implementation is straightforward and does not require any particular precaution. This technique may be sufficient for some simple applications, but it is clearly limited when quantitative metrics, such as the $\mathcal{H}_2$ or $\mathcal{H}_\infty$

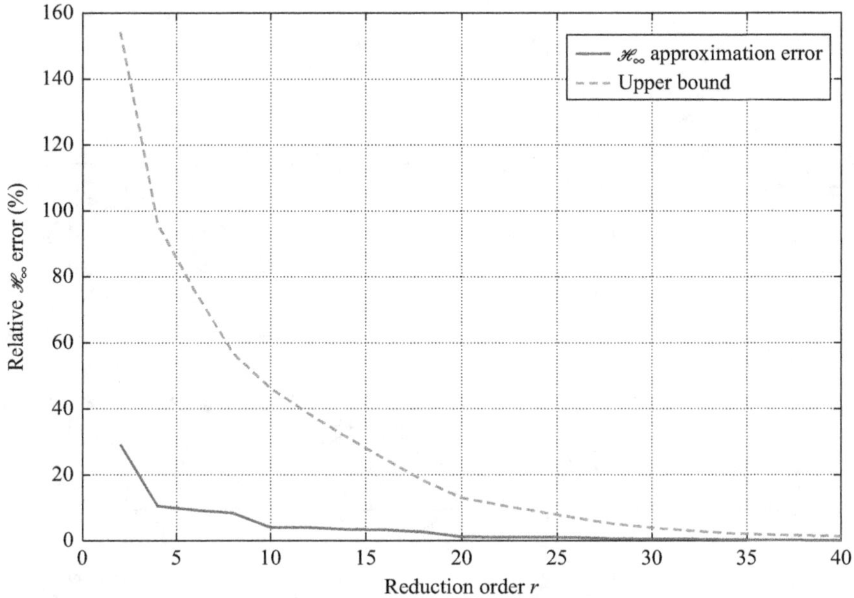

Figure 3.4 Relative $\mathscr{H}_\infty$ approximation error and its bound obtained after reduction of the ISS model with the modal truncation for r going from 2 to 40

norms of the error, are considered. *Balanced truncation* is a more efficient method and is described in the following section.

3.2.2 Balanced truncation

Considering an LTI dynamical system **H** and its approximation $\hat{\mathbf{H}}$. The $\mathscr{H}_\infty$-norm of the error between **H** and $\hat{\mathbf{H}}$ can be expressed in the time domain as

$$\|H - \hat{H}\|_{\mathscr{H}_\infty} = \sup_{u \in \mathscr{L}_2} \frac{\|y - \hat{y}\|_2}{\|u\|_2} \tag{3.29}$$

that is to say as the maximum absolute error between the outputs $y(t)$ and $\hat{y}(t)$ of **H** and $\hat{\mathbf{H}}$, respectively, when both systems are driven by the same input, as illustrated in Figure 3.6. The underlying idea behind the balanced truncation is to discard the states which have the smallest impact on this input to output energy transfer.

3.2.2.1 Singular value decomposition

Among the six matrix decompositions described in [28], the *Singular Value Decomposition* (SVD) is of particular importance, especially for balanced truncation. Its definition is recalled in Definition 3.4.

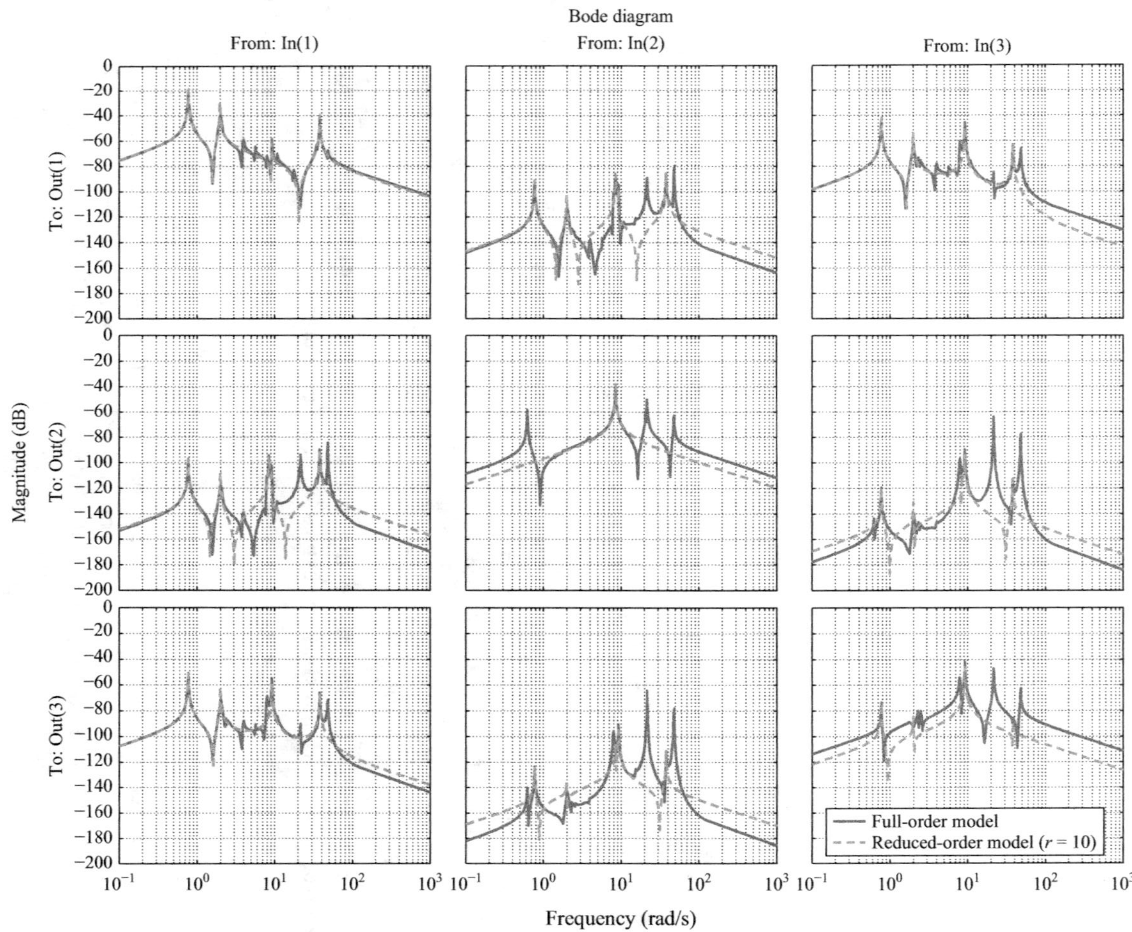

Figure 3.5 Magnitude of the frequency responses of the full-order and reduced-order models for r =10

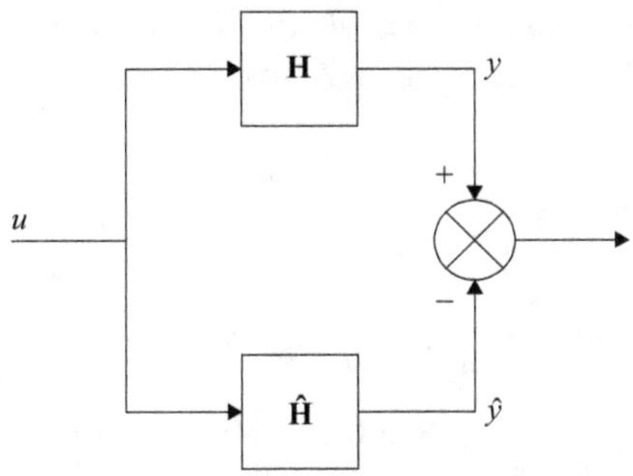

Figure 3.6 Block diagram representation of the error between a full-order model **H** *and its approximation* $\hat{\mathbf{H}}$

Definition 3.4 (The Singular Value Decomposition). Given a matrix $A \in \mathbb{C}^{m\times n}$ with $m \leq n$, the singular value decomposition of A is defined as

$$A = U\Sigma V^H \tag{3.30}$$

where $U \in \mathbb{C}^{m\times n}$ and $V \in \mathbb{C}^{n\times n}$ are unitary matrices, i.e., $UU^H = I_m$, $VV^H = I_n$, and $\Sigma \in \mathbb{R}^{m\times n}$ is given by $\Sigma_{ii} = \sigma_i$, $i = 1, \ldots, n$ and zero elsewhere. The σ_i, $i = 1, \ldots, m$ are nonnegative numbers called the *signular values* of A. They correspond to the eigenvalues of AA^H. Similarly, the columns of U and V are called the left and right singular vectors of A. They correspond to the eigenvectors of AA^H.

The SVD of a matrix A has several interesting properties among which:

- The rank r of A is given by the number of its nonzero singular values.
- Considering that the singular values of A are ordered by decreasing value, i.e., $\sigma_i > \sigma_{i+1}$, A can be written as

  $$A = \sigma_1\mathbf{u}_1\mathbf{v}_1^H + \cdots + \sigma_r\mathbf{u}_r\mathbf{v}_r^H \tag{3.31}$$

  where $\mathbf{u}_i$ and $\mathbf{v}_i$, $i = 1, \ldots, r$ are the left and right singular vectors of A. This decomposition is called the *dyadic decomposition* of A.
- The 2-induced norm of A is given by the largest singular value σ_1 of A,

  $$\|A\|_2 = \sigma_1(A) \tag{3.32}$$

- The Frobenius norm of A is given by $\|A\|_F = \sqrt{\sigma_1^2 + \cdots + \sigma_r^2}$

Theorem 3.1 (Schmidt-Eckart-Young-Mirsky). *Given a complex matrix $A \in \mathbb{C}^{m\times n}$ of rank r, if the singular values of A are ordered such that $\sigma_i > \sigma_i + 1$, then,*

$$\min_{X,\,\mathbf{rank}(X)=k} \|A - X\|_2 = \sigma_{k+1}(A) \tag{3.33}$$

A non-unique minimizer X_* of (3.33) *is obtained by truncating the dyadic decomposition* (3.31) *to retain only the first $k < r$ terms, i.e.,*

$$X_* = \sum_{i=1}^{k} \sigma_i \mathbf{u_i} \mathbf{v_i}^H \tag{3.34}$$

Theorem 3.1 can be directly applied for image compression (see Example 3.2).

Example 3.2. A grayscale image like the one at top left of Figure 3.7 (408×700 pixels) can be represented by a matrix $A \in \mathbb{R}^{m \times n}$ (here, $m = 408$ and $n = 700$). The singular values of A are computed and plotted at the top right of Figure 3.7. We can see that the singular values decrease quickly, so it seems natural to keep the largest ones only. In particular, the matrix A, by using the Theorem 3.1, can be approximated by a rank k matrix X_k (with $k = 100$ and $k = 20$) by truncating its dyadic decomposition. The reconstructed images obtained from X_{100} and X_{20} are displayed at the bottom of Figure 3.7. The image constructed with the 20 largest singular values of A is slightly blurred but it is still of good quality.

To measure the compression efficiency, Table 3.1 shows the image compression ratio, defined as

$$\frac{\text{memory needed for } A}{\text{memory needed for } X_k} = \frac{mn}{k\,(m+n+1)} \tag{3.35}$$

obtained with the compressed image X_k. It also shows the Peak Signal to Noise Ratio (PSNR), defined as

$$PSNR = 10\,\mathbf{log}_{10}\left(\frac{b^2}{mse}\right) \tag{3.36}$$

where b is the maximal value taken by the signal (typically 255 for a grayscale image) and mse is the mean square difference between the two images. The PSNR is widely used to measure the quality of reconstructed images after a lossy compression (such as SVD). In lossy image compression, typical values of the PSNR are between 30 and 50 dB, a higher value implying a better reconstruction.

3.2.2.2 Observability and controllability gramians

The controllability and observability gramians (see Definition 3.5) are useful matrices in system theory, they are related to the concepts of state reachability and observability (see [35] for more details).

Definition 3.5 (Infinite Gramians). Given a stable LTI dynamical system $\mathbf{H} := (A, B, C, D)$, the infinite controllability and observability gramians associated with $\mathbf{H}$, denoted $\mathscr{P}$ and $\mathscr{Q}$ respectively, are defined as

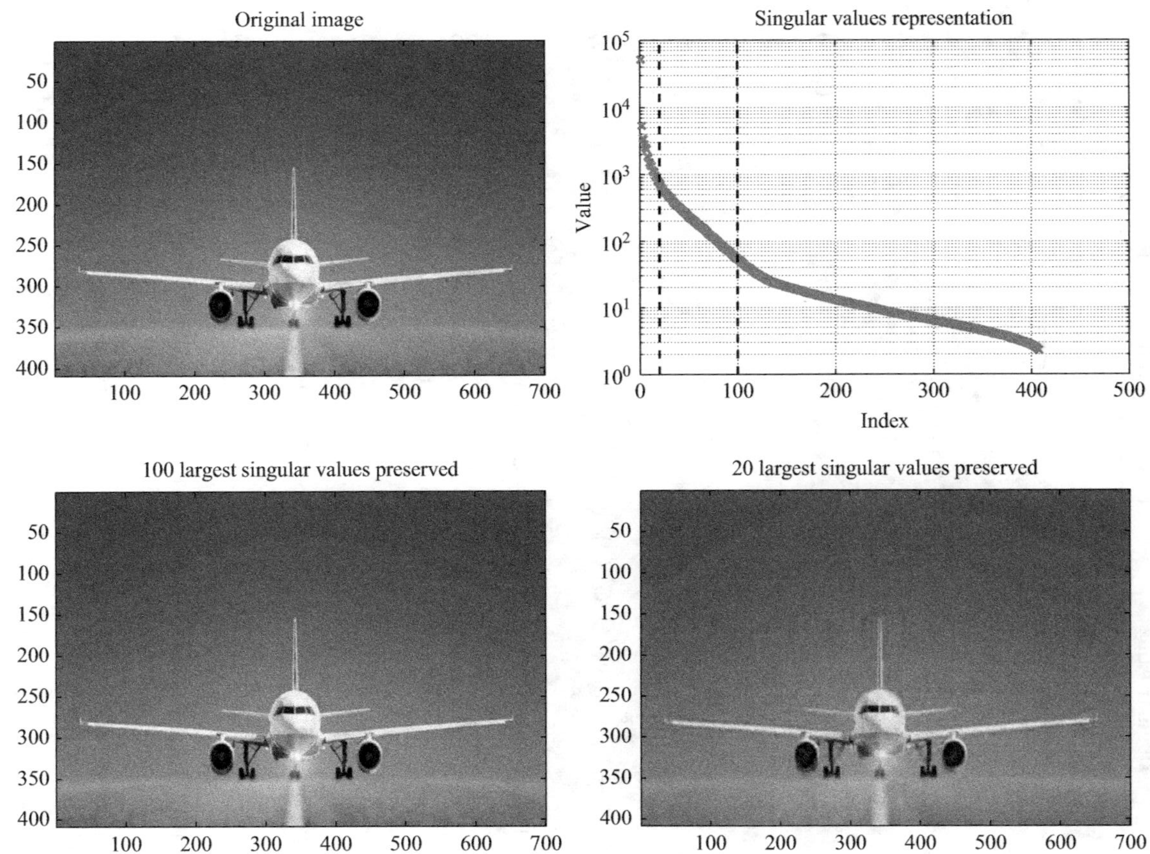

Figure 3.7 *Example of the use of the SVD for image compression. The singular values are plotted at the top right, the original image is at the top left, and the approximated images are at the bottom*

Table 3.1 Compression ratio and Peak Signal to Noise Ratio (PSNR)

Number k of retained singular values	Compression ratio	PSNR (dB)
100	2.6	54.9
20	12.9	34.7

$$\mathcal{P} = \int_0^{\infty} e^{At} BB^T e^{A^T t} dt \tag{3.37}$$

$$\mathcal{Q} = \int_0^{\infty} e^{A^T t} C^T C e^{At} dt \tag{3.38}$$

in the time domain and as

$$\mathcal{P} = \frac{1}{2\pi} \int_{-\infty}^{\infty} T(\nu)^H BB^T T(\nu) d\nu \tag{3.39}$$

$$\mathcal{Q} = \frac{1}{2\pi} \int_{-\infty}^{\infty} T(\nu)^H C^T C T(\nu) d\nu \tag{3.40}$$

with $T(\nu) = (j\nu I - A)^{-1}$, in the frequency domain.

The controllability and observability gramians can be computed by solving two Lyapunov equations. Indeed the controllability gramian $\mathcal{P}$ is the solution of

$$A\mathcal{P} + \mathcal{P}A^T + BB^T = 0 \tag{3.41}$$

while the observability gramian $\mathcal{Q}$ is the solution of

$$A^T \mathcal{Q} + \mathcal{Q}A + C^T C = 0 \tag{3.42}$$

One of the main interest of the gramians relies in the fact that they give information about the input–output energy of the system. In particular,

- The minimal energy ε_r required to make the system go from the state 0 to the state $\mathbf{x}$ is given by

$$\varepsilon_r = \mathbf{x}^T \mathcal{P}^{-1} \mathbf{x} \tag{3.43}$$

- The maximal observation energy ε_o obtained by releasing the system from an initial state $\mathbf{x}$ and no input excitation is given by

$$\varepsilon_o = \mathbf{x}^T \mathcal{Q} \mathbf{x} \tag{3.44}$$

The idea behind balanced truncation is to discard the states that require a lot of energy to be reached and/or the states that release only little energy. In other words, balanced truncation consists in discarding the states that are "hard to reach" and/or "hard to observe", because those states have a minor impact in the input to output energy transfer.

However, the gramians depend of the basis in which the system is expressed. Hence a state which is hard to observe might be easy to reach and conversely in another basis. To overcome this problem, the system must be expressed in a basis where the states hard to observe are also hard to reach. This basis is called the balanced one. In this basis, both gramians are equal and diagonal, i.e.,

$$\mathscr{P} = \mathscr{Q} = \Sigma \tag{3.45}$$

where Σ is a diagonal matrix containing the *Hankel singular values* σ_i of the system $\mathbf{H}$.

Suppose that the Hankel singular values σ_i are ordered such that $\sigma_1 \geq \ldots \geq \sigma_n > 0$ and consider the state $\mathbf{x_0} = \mathbf{e_i} = \begin{bmatrix} 0 & \ldots & 1 & \ldots & 0 \end{bmatrix}^T$, then

$$\varepsilon_r = \mathbf{x_0}^T \mathscr{P}^{-1} \mathbf{x_0} = \mathbf{e_i}^T \Sigma^{-1} \mathbf{e_i} = \sigma_i^{-1} \tag{3.46}$$

and

$$\varepsilon_o = \mathbf{x_0}^T \mathscr{Q} \mathbf{x_0} = \sigma_i \tag{3.47}$$

In the balanced realization, $\varepsilon_r = \varepsilon_0$, hence the states that are easily reachable are also easily observable.

Remark 3.4. The Hankel singular values σ_i are the square roots of the eigenvalues of the product $\mathscr{P}\mathscr{Q}$,

$$\sigma_i = \sqrt{\lambda_i(\mathscr{P}\mathscr{Q})} \tag{3.48}$$

Unlike the gramians $\mathscr{P}$ and $\mathscr{Q}$, the eigenvalues of $\mathscr{P}\mathscr{Q}$ are invariant under state-space transformations.

Remark 3.5 (Hankel operator). The Hankel operator, denoted here as $\mathscr{H}$, is a restriction of the convolution operator $\mathscr{S}$ which maps the past inputs of the system u_- to its future outputs y_+, i.e.,

$$\begin{array}{llll} \mathscr{H}: & \mathscr{L}_2(-\infty, 0) & \rightarrow \mathscr{L}_2(0, \infty) & \\ & u_- & \rightarrow y_+ & \text{where } y_+ = \int_{-\infty}^{0} h(t-\tau)u(\tau)d\tau, \quad t \in \mathbb{R}_+ \end{array} \tag{3.49}$$

where h is the impulse response of the system $\mathbf{H}$. For discrete systems, the Hankel operator can be represented as a rank n (the order of the system) infinite matrix which can be optimally approximated in the 2-norm, thus leading to the optimal approximation in the Hankel norm (see [8] for further information). This method is not presented in this chapter because it is more complex than balanced truncation and does not necessarily gives better results.

3.2.2.3 Truncation of the balanced realization

The balanced realization $(A_{bal}, B_{bal}, C_{bal}, D_{bal})$ of an LTI dynamical system $\mathbf{H} := (A, B, C, D)$ is obtained through a state transformation $\mathbf{x}(t) = T\mathbf{x_{ball}}(t)$ where $T \in \mathbb{R}^{n\times n}$ is a nonsingular transformation matrix, as

$$\begin{aligned} A_{bal} &= T^{-1}AT \\ B_{bal} &= T^{-1}B \\ C_{bal} &= CT \\ D_{bal} &= D \end{aligned} \tag{3.50}$$

In this basis, the gramians are given by

$$\mathscr{P}_{bal} = T^{-1}\mathscr{P}T^{-T} \quad \text{and} \quad \mathscr{Q}_{bal} = T^T\mathscr{Q}T \tag{3.51}$$

If both $\mathscr{P}_{bal}$ and $\mathscr{Q}_{bal}$ are diagonal, the transformation T is called a *contragradient* transformation. In particular, if $\mathscr{P}_{bal} = \mathscr{Q}_{bal} = \Sigma$ the coordinates are said to be internally balanced and T is called the *balanced transformation* [19, 15].

Different approaches exist to find the balanced transformation [19, 15], two popular methods are presented in Algorithm 3.1 and Algorithm 3.2.

We can verify that the transformation T constructed in Algorithm 3.1 is indeed a contragradient transformation by projecting the gramians,

$$\begin{aligned} T^{-1}\mathscr{P}T^{-T} &= \Sigma^{-\frac{1}{2}}U^T L_o^T \mathscr{P} L_o U \Sigma^{-\frac{1}{2}} \\ &= \Sigma^{-\frac{1}{2}}U^T \underbrace{L_o^T L_c}_{U\Sigma V^T} \underbrace{L_c^T L_o}_{V\Sigma U^T} U\Sigma^{-\frac{1}{2}} = \Sigma \end{aligned} \tag{3.52}$$

Algorithm 3.1 Balanced Transformation

1: Compute the lower Cholesky factorizations of the gramians [13]:

$$\mathscr{P} = L_c L_c^T \quad \text{and} \quad \mathscr{Q} = L_o L_o^T$$

2: Compute the singular value decomposition of $L_o^T L_r$,

$$L_o^T L_c = U\Sigma V^T$$

where Σ is the diagonal matrix containing the Hankel singular values of $\mathbf{H}$

3: The balanced transformation is given by

$$T = L_c V\Sigma^{-\frac{1}{2}}, \quad \text{and} \quad T^{-1} = \Sigma^{-\frac{1}{2}}U^T L_o^T$$

Similarly,

$$
\begin{aligned}
T^T \mathscr{Q} T &= \Sigma^{-\frac{1}{2}} V^T L_c^T \mathscr{Q} L_c V \Sigma^{-\frac{1}{2}} \\
&= \Sigma^{-\frac{1}{2}} V^T \underbrace{L_c^T L_o}_{V\Sigma U^T} \underbrace{L_o^T L_c}_{U\Sigma V^T} V \Sigma^{-\frac{1}{2}} = \Sigma
\end{aligned} \tag{3.53}
$$

Algorithm 3.2 Balanced Transformation

1: Compute the gramians $\mathscr{P}$ and $\mathscr{Q}$ with (3.41) and (3.42).

2: Compute the lower Cholesky factorization of $\mathscr{P}$,

$$\mathscr{P} = L_c L_c^T$$

3: Solve the following symmetric eigenvalue problem:

$$L_c^T \mathscr{Q} L_c = V \Sigma^2 V^T$$

where Σ contains the Hankel singular values of the system.

4: The balanced transformation is given by

$$T = L_c V \Sigma^{-\frac{1}{2}} \quad \text{and} \quad T^{-1} = \Sigma^{\frac{1}{2}} V^T L_c^{-1}$$

Again, the transformation T given by Algorithm 3.2 is a contragradient transformation,

$$T^{-1} \mathscr{P} T^{-T} = \Sigma^{\frac{1}{2}} V^T L_c^{-1} \underbrace{\mathscr{P}}_{L_c L_c^T} L_c^{-T} V \Sigma^{\frac{1}{2}} = \Sigma \tag{3.54}$$

and

$$T^T \mathscr{Q} T = \Sigma^{-\frac{1}{2}} V^T \underbrace{L_c^T \mathscr{Q} L_c}_{V\Sigma^2 V^T} V \Sigma^{-\frac{1}{2}} = \Sigma \tag{3.55}$$

Remark 3.6. In Algorithm 3.2, the gramians are explicitly computed whereas in Algorithm 3.1, only their Cholesky factorizations are required [13]. The latter method is more reliable numerically because it does not involve the explicit computation of the products BB^T and $C^T C$ arising in the Lyapunov equations (3.41) and (3.42), which can lead to discard some small singular values. An illustration of the difference between the two algorithms is shown in Example 3.3. The counterpart is that Algorithm 3.2 requires less operations than Algorithm 3.1 (see [15] for a detailed description of the number of operations required for both algorithms).

Once the system **H** is in the balanced basis, its state-space representation can be partitioned accordingly to the partitioning of the Hankel singular values, i.e.,

$$A_{bal} = \begin{bmatrix} A_{11} & A_{12} \\ A_{21} & A_{22} \end{bmatrix}, \; B_{bal} = \begin{bmatrix} B_1 \\ B_2 \end{bmatrix}, \; C_{bal} = \begin{bmatrix} C_1 & C_2 \end{bmatrix}, \; \Sigma = \begin{bmatrix} \Sigma_1 & 0 \\ 0 & \Sigma_2 \end{bmatrix} \tag{3.56}$$

The reduced-order models obtained by balanced truncation are then given by

$$\hat{\mathbf{H}}_i = \left(\begin{array}{c|c} A_{ii} & B_i \\ \hline C_i & D \end{array} \right) \tag{3.57}$$

The states that should be retained are those which correspond to the largest Hankel singular values because they have a larger impact on the input–output energy transfer. Balanced truncation has some interesting properties that are recalled in Theorem 3.2.

Theorem 3.2. *Given the stable, reachable and observable continuous system* $\mathbf{H}$, *the reduced-order models* $\hat{\mathbf{H}}_i$, $i = 1,2$ *(3.57) obtained by balanced truncation have the following properties:*

1. $\hat{\mathbf{H}}_i$ *is balanced and has no poles in the open right half plane.*
2. *If the singular values contained in* Σ_1 *and* Σ_2 *(3.56) are different, then* $\hat{\mathbf{H}}_i$ *are reachable, observable and have no poles on the imaginary axis.*
3. *The* $\mathcal{H}_\infty$*-norm of the error between* $\mathbf{H}$ *and* $\hat{\mathbf{H}}_i$ *is bounded by twice the sum of the neglected Hankel singular values without their multiplicities [20], i.e., if* $\hat{\mathbf{H}}_1$ *has an order r, then*

$$\|H - \hat{H}_1\|_{\mathcal{H}_\infty} \leq 2 \sum_{i=r+1}^{n} \sigma_i \tag{3.58}$$

Remark 3.7. The proof of Theorem 3.2 can be found in [3].

Remark 3.8. Unlike the modal approximation, the poles of the reduced-order model obtained with the balanced truncation are not necessarily a subset of the poles of the initial system.

Example 3.3. In this example, the Hankel singular values of the ISS model are computed with the two algorithms presented above. The results are plotted in Figure 3.8.

The largest singular values are not affected by the algorithm used, but we can see that differences appear for small singular values. Indeed for the singular values smaller than 10^{-10}, the algorithm that explicitly computes the gramians $\mathcal{P}$ and $\mathcal{Q}$ gives erroneous results.

Example 3.4. In this example, the ISS model is reduced to several orders r going from 2 to 40 with the modal approximation described in Section 3.2.1.3 and with the balanced truncation presented above. For each order r, the $\mathcal{H}_\infty$-norm of the relative error is computed together with the corresponding upper bounds of each method. The results are plotted in Figure 3.9.

On this example, the two methods lead to similar results, the balanced truncation being slightly more efficient. It should be mentioned, however, that in most problems balanced truncation outperforms modal approximation.

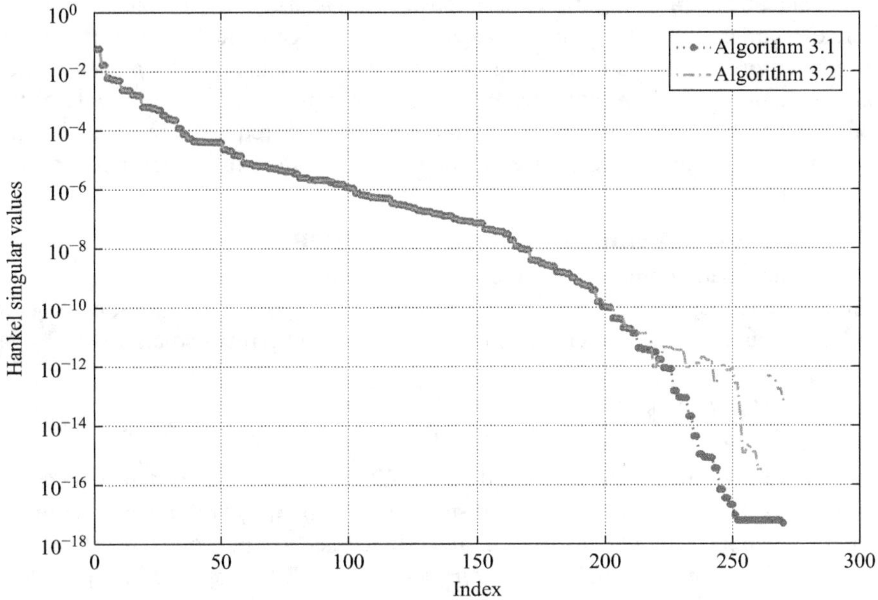

Figure 3.8 *Hankel singular values of the ISS model obtained with Algorithm 3.1 and Algorithm 3.2*

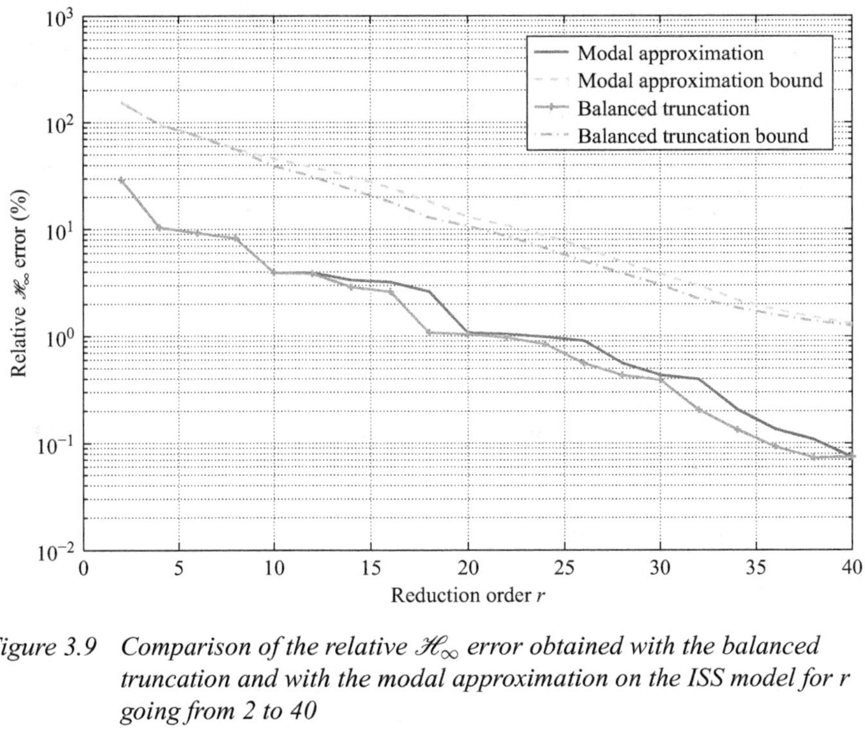

Figure 3.9 *Comparison of the relative $\mathscr{H}_\infty$ error obtained with the balanced truncation and with the modal approximation on the ISS model for r going from 2 to 40*

The balanced transformation presented above is also called *Lyapunov balancing*. There exist other ways of balancing[3] a system depending on the desired properties of the reduced-order model obtained after truncation. In particular, the *frequency-limited balanced truncation* represents, from the authors' point of view, a very interesting method since it enables to focus only on some bounded frequency intervals instead of taking into account the whole frequency behaviour of the full-order model.

3.2.2.4 Frequency-limited balanced truncation

A commonly used approach to tackle the issue of approximating a system **H** over a bounded frequency interval $\Omega = [\omega_1, \omega_2]$, where $\omega_1, \omega_2 \in \mathbb{R}_+$, $\omega_1 < \omega_2$, consists in finding a reduced-order model $\hat{\mathbf{H}}$ such that the following error is small:

$$\|W_o(H - \hat{H})W_i\|_{\mathscr{H}_\infty} \tag{3.59}$$

where W_i and W_o are suitable input and output filters. Despite the existence of several strategies to choose the filters (see for instance [3, Chap. 7] and references therein), it remains a tedious task which can be avoided by considering a weight-free approach such as the *frequency-limited balanced truncation* (FL-BT) proposed in [7].

This approach relies on the use of the frequency-limited controllability and observability gramians, denoted $\mathscr{P}_\omega$ and $\mathscr{Q}_\omega$ respectively (see Definition 3.6), instead of the infinite gramians (3.39) and (3.40) to achieve the balanced truncation.

Definition 3.6 (Frequency-limited gramians). Given an LTI dynamical system **H** admitting a state-space realization is (A, B, C, D), its frequency-limited controllability and observability gramians, denoted $\mathscr{P}_\omega$ and $\mathscr{Q}_\omega$ respectively, are defined as the restriction of the infinite gramians (3.39) and (3.40) over $[-\omega, \omega]$, i.e.,

$$\mathscr{P}_\omega = \frac{1}{2\pi}\int_{-\omega}^{\omega} T(\nu)^H BB^T T(\nu)d\nu \tag{3.60}$$

$$\mathscr{Q}_\omega = \frac{1}{2\pi}\int_{-\omega}^{\omega} T(\nu)^H C^T CT(\nu)d\nu \tag{3.61}$$

with $T(\nu) = (j\nu I - A)^{-1}$.

Similarly to the infinite gramians, $\mathscr{P}_\omega$ and $\mathscr{Q}_\omega$ can be obtained by solving the Lyapunov equations

$$A\mathscr{P}_\omega + \mathscr{P}_\omega A^T + W_c(\omega) = 0 \tag{3.62}$$

$$A^T\mathscr{Q}_\omega + \mathscr{Q}_\omega A + W_o(\omega) = 0 \tag{3.63}$$

[3] The reader should refer to [3, Chap. 7] or [11] for a complete description of the other types of balancing.

where

$$W_c(\omega) = S(\omega)BB^T + BB^TS^*(\omega) \tag{3.64}$$

$$W_o(\omega) = S^*(\omega)C^TC + C^TCS(\omega) \tag{3.65}$$

and

$$S(\omega) = \frac{j}{2\pi}\mathbf{log\,m}((A + j\omega I_n)(A - j\omega I_n)^{-1}) \tag{3.66}$$

Remark 3.9. The frequency-limited gramians can easily be expressed on a frequency interval $\Omega = [\omega_1, \omega_2]$, indeed $\mathscr{P}_\Omega = \mathscr{P}_{\omega_2} - \mathscr{P}_{\omega_1}$ and $\mathscr{Q}_\Omega = \mathscr{Q}_{\omega_2} - \mathscr{Q}_{\omega_1}$.

Unlike the standard balanced truncation, the frequency-limited balanced truncation does not offer any guarantee concerning the stability of the reduced-order model and there is no error bound available with this method. A modification has been proposed in [11] that enables to preserve stability and offer a bound on the error. However, this modification has sometimes a major impact on the quality of the approximation (see Example 3.5).

Example 3.5. In this example, the LAH model is reduced to an order $r = 14$ with balanced truncation (BT), frequency-limited balanced truncation (FL-BT) and the same method with the modification proposed in [11]. The frequency interval considered is set to $\Omega = [0, 7]$. The frequency responses of the different models together with the frequency responses of the error models are plotted on Figure 3.10.

The efficiency of the FL-BT clearly appears on the frequency responses of the error models. Indeed, it leads to a very small error in the considered frequency interval Ω. This can be verified more quantitatively by computing the $\mathscr{H}_{2,\Omega}$ norm of the relative error, i.e.,

$$E_{\mathscr{H}_{2,\Omega}} = 100\frac{\|H - \hat{H}\|_{\mathscr{H}_{2,\Omega}}}{\|H\|_{\mathscr{H}_{2,\Omega}}} \tag{3.67}$$

for each reduced-order model. The results are reported in Table 3.2. All the methods lead to a good approximation, the worst relative error being of 6% (BT), but the best result is clearly given by the FL-BT on this example.

3.2.3 Conclusion

Model truncation is the most direct approach to address the problem of model reduction. When performed in a particular basis, such as the modal or balanced one, the reduced-order models obtained by truncation can have some interesting properties. Indeed, both modal and balanced truncation preserve the stability of the full-order model and offer a bound on the $\mathscr{H}_\infty$-norm of the error. The advantage of the modal approximation lies in the fact that some poles of the initial model are preserved

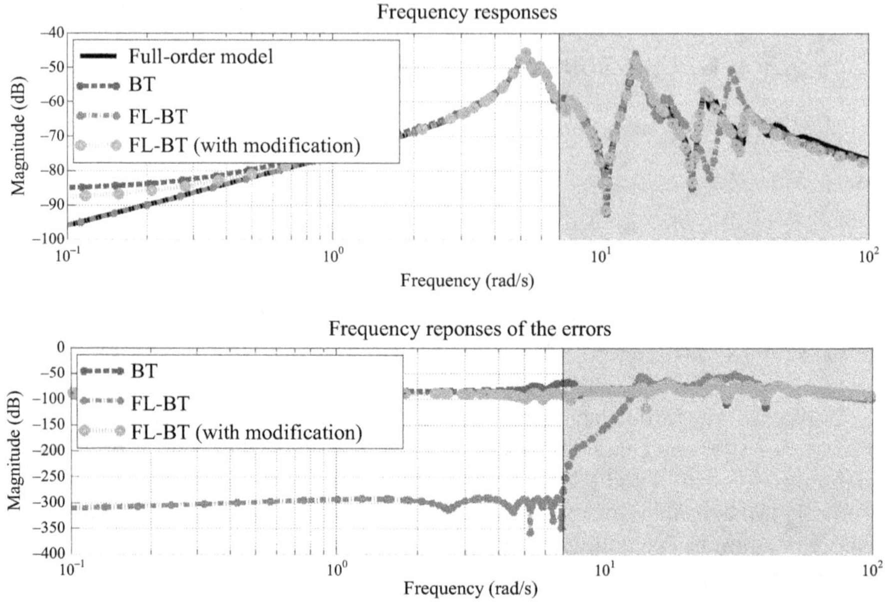

Figure 3.10 Frequency responses of the initial and reduced-order models obtained after approximation of the LAH model to an order $r = 14$ over $\Omega = [0, 7]$ with the BT, the FL-BT and the modifier FL-BT

Table 3.2 $\mathcal{H}_{2,\Omega}$-norm of the relative error $E_{\mathcal{H}_{2,\Omega}}$ achieved by the BT, the FL-BT and the modified FL-BT

Method	**$\mathcal{H}_{2,\Omega}$-norm of the relative error $E_{\mathcal{H}_{2,\Omega}}$ (%)**
BT	6.43
FL-BT	2.4×10^{-7}
FL-BT with modification	1.74

whereas this is not the case with balanced truncation. However, balanced truncation tends to give better results in terms of $\mathcal{H}_\infty$ or $\mathcal{H}_2$ norm of the error than modal approximation. From a computational point of view, balanced truncation requires to compute both gramians. For large-scale systems, solving Lyapunov equations is a complex and ill-conditioned task, hence the method cannot be applied in those cases. Similarly, the eigenvalue decomposition of a matrix is hard to find in large-scale settings.

Some alternative model approximation methods based on interpolation have known a significant development in the past few years. These methods can be implemented through iterative procedures that make them tractable even in (very) large-scale settings.

3.3 Moment matching methods

Krylov subspaces, denoted $\mathcal{K}_k$ and defined as

$$\mathcal{K}_k(A, v) = \mathbf{span}\big(v, Av, \ldots, A^{k-1}v\big) \tag{3.68}$$

arise in many areas of research and are of particular interest for large-scale problems since they can be constructed iteratively. In particular, they are used to

- solve systems of linear equations $Ax = b$ [26];
- compute the eigenvalues of matrices [27];
- reduce large-scale dynamical models [9].

The last point is described in this section. It consists in using the Krylov subspaces to perform an implicit matching of the full-order model's moments at some points.

In Section 3.3.1, the moments of a transfer function are defined and the problem of moment matching is introduced. The problem is then addressed for SISO systems within a *rational interpolation* framework. Even if this framework can theoretically be directly generalized to MIMO systems, the associated algorithms become very complex. Instead, the *tangential interpolation* framework is used to address the MIMO case. Section 3.3.2 concerns the optimal model approximation problem in the $\mathcal{H}_2$-norm. The problem is introduced and the first-order necessary optimality conditions are described. Then, grounded on the tangential interpolation framework, two algorithms are presented:

- *The Iterative Tangential Interpolation Algorithm* (ITIA), proposed in [29, 12], enables to efficiently construct a reduced-order model that satisfies the first-order optimality conditions.
- *The SVD-Iterative Tangential Interpolation Algorithm* (ISTIA), proposed in [10, 22], is a modified version of the ITIA that leads to a sub-optimal approximation but guarantees the stability of the reduced-order model.

A frequency-limited extension of the ISTIA, called *Frequency-Limited ISTIA* is also presented.

Remark 3.10. All the methods presented in this chapter (and others) have been implemented in the MORE Toolbox presented in [24].

3.3.1 Moment matching through Krylov subspaces

3.3.1.1 Moment matching problem

The moments of a transfer function are presented in Definition 3.7 and the problem of model reduction by moment matching is introduced in Problem 3.2.

Definition 3.7 (Moments of a transfer function). Consider a system $\mathbf{H}$, its transfer function $H(s) = C(sI_n - A)^{-1}B + D \in \mathbb{C}^{n_y \times n_u}$ can be decomposed through a Laurent series expansion around a given shift point $\sigma \in \mathbb{C}$ as

$$H(s)|_\sigma = \sum_{i=0}^{\infty} \eta_i(\sigma)\,(s-\sigma)^i \tag{3.69}$$

where $\eta_i(\sigma) \in \mathbb{C}^{n_y \times n_u}$, $i \in \mathbb{N}$ are the i-th moments of $H(s)$ at σ associated with the system, defined as

$$\eta_i(\sigma) = (-1)^i \left.\frac{d^i H(s)}{ds^i}\right|_{s=\sigma} \tag{3.70}$$

Remark 3.11. For $\sigma = \infty$, the moments (3.70) are called the *Markov parameters* of $\mathbf{H}$ and are given by

$$\eta_0(\infty) = D \quad \text{and} \quad \eta_i(\infty) = CA^{i-1}B, \quad i > 0 \tag{3.71}$$

Their computation implies the evaluation of the successive powers of the matrix A.

Problem 3.2 (Moment matching problem). Given a n-th order system $\mathbf{H}$ the transfer function $H(s)$ of which is decomposed at $\sigma \in \mathbb{C}$ as

$$H(s)|_\sigma = \sum_{i=0}^{\infty} \eta_i(\sigma)\,(s-\sigma)^i \tag{3.72}$$

the moment matching problem consists in finding a reduced-order model $\hat{\mathbf{H}}$ the first r moments $\hat{\eta}_i(\sigma)$ of which are such that at σ

$$\hat{\eta}_i(\sigma) = \eta_i(\sigma), \quad i = 1, \ldots, r \tag{3.73}$$

Remark 3.12. The moment matching problem presented above has different names depending on the point σ:

- If $\sigma = 0$, it is called a *Padé approximation.*
- If $\sigma = \infty$, then it is a *partial realization* problem.
- Otherwise it is a *rational interpolation* problem.

The first approach which could be used to address Problem 3.2 consists in computing the moments of the initial system $\mathbf{H}$ and finding the reduced-order model that matches the first r ones [9]. However, computing the moments is ill-conditioned, thus this method is not reliable. A more suitable approach consists in implicitly matching the moments of the full-order model by projecting it onto particular Krylov subspaces.

3.3.1.2 Implicit moment matching: the SISO case

The use of Krylov subspaces in model reduction has been developed by Grimme in [9]. The first approach consists in projecting the full-order model **H** on the Krylov subspaces $\mathcal{K}_r\left((\sigma I_n - A)^{-1}, (\sigma I_n - A)^{-1}B\right)$ as stated in Theorem 3.3. This enables the reduced-order model to match the first r moments of **H** at $\sigma \in \mathbb{C}$.

Theorem 3.3 (One-sided moment matching). *Consider a SISO LTI dynamical system* $\mathbf{H} := (A, B, C, D)$ *and an interpolation point* $\sigma \in \mathbb{C}$ *such that* $(\sigma I_n - A)$ *is nonsingular. If* $V \in \mathbb{C}^{n\times r}$ *is a full-rank matrix the columns of which span the Krylov subspace* $\mathcal{K}_r\left((\sigma I_n - A)^{-1}, (\sigma I_n - A)^{-1}B\right)$*, i.e.,*

$$\mathcal{K}_r\left((\sigma I_n - A)^{-1}, (\sigma I_n - A)^{-1}B\right) \subseteq \mathcal{V} = \mathbf{span}(V) \tag{3.74}$$

then the reduced-order model $\hat{\mathbf{H}} = ((V^TV)^{-1}V^TAV, (V^TV)^{-1}V^TB, CV, D)$ *obtained by projection matches the first r moments of* **H** *at* σ*, i.e.,*

$$\hat{\eta}_i(\sigma) = \eta_i(\sigma), \quad \text{for} \quad i = 1, \ldots, r \tag{3.75}$$

To efficiently create a basis $V \in \mathbb{R}^{n\times r}$ for the Krylov subspace used in Theorem 3.3, the *Arnoldi algorithm* can be used. Its basic version is presented in Algorithm 3.3. This procedure enables to construct an orthonormal basis for a Krylov subspace $\mathcal{K}_r(A, \mathbf{v}_1)$ without explicitly computing the successive powers of the matrix A.

Algorithm 3.3 Arnoldi Algorithm

Require: $A \in \mathbb{C}^{n\times n}$, $\mathbf{v_1} \in \mathbb{C}^{n\times 1}$ with $\|\mathbf{v_1}\|_2 = 1$, and $r \in \mathbb{N}_+^*$.
for $i = 1, \ldots, r$ **do**
  Compute $\mathbf{w} = A\mathbf{v_i}$
  for $j = 1, \ldots, i$ **do**
    $h_{j,i} = \mathbf{v_j}^H\mathbf{w}$
  end for
  Set $\mathbf{w}_\perp = \mathbf{w} - \sum_{j=1}^{i} h_{ji}\mathbf{v_j}$
  Set $h_{i+1,i} = \|\mathbf{w}_\perp\|_2$
  if $h_{i+1,i} = 0$ **then**
    Stop
  end if
  Set $\mathbf{v_{i+1}} = \mathbf{w}_\perp / h_{i+1,i}$
end for

At each iteration of Algorithm 3.3 a new vector $\mathbf{w}$ is created by multiplying the matrix A with the Arnoldi vector $\mathbf{v_i}$, and by applying a simple Gram-Schmidt procedure, this vector $\mathbf{w}$ is then orthogonalized with respect to the previous Arnoldi vectors $\mathbf{v_j}, j = 1, \ldots, i$. The new Arnoldi vector $\mathbf{v_{i+1}}$ if finally obtained by normalizing the orthogonal vector $\mathbf{w}_\perp$. The orthogonalization step is very important, so a more evolved orthogonalization procedure than the simple Gram-Schmidt one can be used. For instance, the modified Gram-Schmidt procedure [26] may be more suitable.

Remark 3.13. A more detailed presentation of the Arnoldi procedure and its properties can be found in [26].

By using the Arnoldi procedure with $A \leftarrow (\sigma I_n - A)^{-1}$ and $\mathbf{v}_1 \leftarrow (\sigma I_n - A)^{-1}B$ then the resulting matrix $V \in \mathbb{C}^{n\times r}$ is an orthonormal basis for the Krylov subspace $\mathcal{K}_r\left((\sigma I_n - A)^{-1}, (\sigma I_n - A)^{-1}B\right)$. Note that since the Arnoldi procedure directly creates an orthonormal basis V, the reduced-order model is obtained by $\hat{\mathbf{H}} = \left(V^T AV, V^T B, CV, D\right)$.

In Theorem 3.3, an orthogonal projection is performed, but better results can be obtained by using an oblique projection as presented in Theorem 3.4 (see [26] for further informations on projections). Indeed, twice the number of moments of $\mathbf{H}$ are matched with a two-sided approach compared to the one-sided approach.

Theorem 3.4 (Two-sided moment matching). *Consider a SISO LTI dynamical system $\mathbf{H} := (A, B, C, D)$ and an interpolation point $\sigma \in \mathbb{C}$ such that $(\sigma I_n - A)$ is nonsingular. If $V \in \mathbb{C}^{n\times r}$ and $W \in \mathbb{C}^{n\times r}$ are full-rank matrices the columns of which span the Krylov subspaces:*

$$\begin{aligned} \mathcal{K}_r\left((\sigma I_n - A)^{-1}, (\sigma I_n - A)^{-1}B\right) &\subseteq \mathcal{V} = \mathbf{span}(V) \\ \mathcal{K}_r\left((\sigma I_n - A)^{-T}, (\sigma I_n - A)^{-T}C^T\right) &\subseteq \mathcal{W} = \mathbf{span}(W) \end{aligned} \tag{3.76}$$

then the reduced-order model $\hat{\mathbf{H}} = ((W^T V)^{-1} V^T AV, (W^T V)^{-1} V^T B, CV, D)$ obtained by projection matches the first $2r$ moments of $\mathbf{H}$ at σ, i.e.,

$$\hat{\eta}_i(\sigma) = \eta_i(\sigma), \quad \textit{for} \quad i = 1, \ldots, 2r \tag{3.77}$$

Similarly to the basis V, $W \in \mathbb{C}^{n\times r}$ can be efficiently computed by using the Arnoldi procedure with $A \leftarrow (\sigma I_n - A)^{-T}$ and $\mathbf{v}_1 \leftarrow (\sigma I_n - A)^{-T} C^T$. The advantage of the two-sided moment matching approach over the one-sided one is illustrated in Example 3.6.

Remark 3.14. The *Lanczos procedure*, which can be found in [26, 3], can be used to directly compute V and W at the same time. It is computationally cheaper than the Arnoldi procedure but more complex to implement since numerical issues can arise more easily than with the Arnoldi procedure.

Example 3.6. To compare the one-sided and two-sided moment matching approaches, the LAH model is reduced to an order 2 at $\sigma = 0$ with both techniques. The frequency responses of the error models are plotted in Figure 3.11. As expected, the two-sided method leads to better results than the one-sided method. Since the interpolation is done at $\sigma = 0$, the error is smaller in low frequencies than in high frequencies.

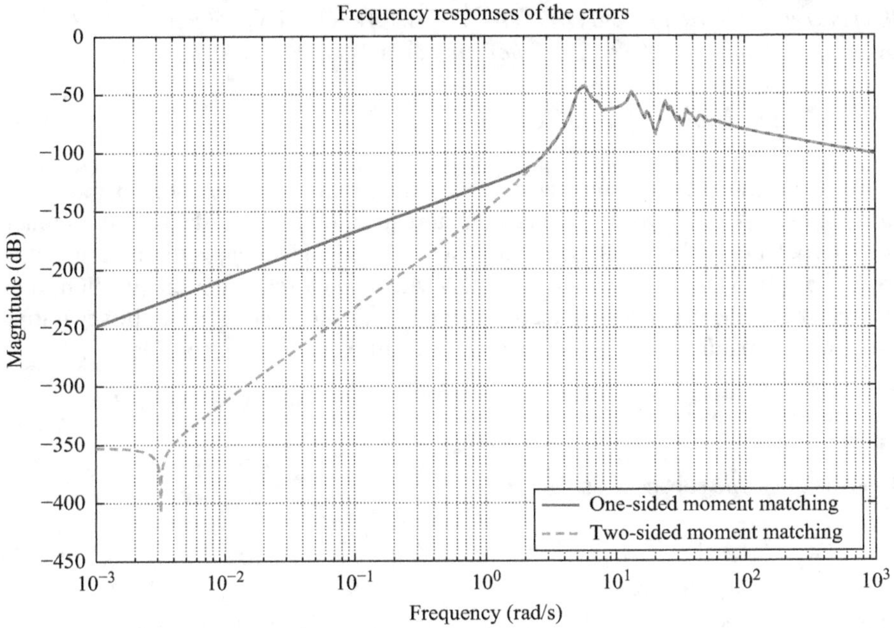

Figure 3.11 Comparison of one-sided and two-sided moment matching on the LAH model which is approximated to an order 2 at $\sigma = 0$

Note that by explicitly computing the moments of the different models, we can check that the one-sided technique matches the first two moments whereas the two-sided approach matches the first four moments.

Both Theorem 3.3 and Theorem 3.4 only enable us to match the moments of the full-order model $\mathbf{H}$ at one single point $\sigma \in \mathbb{C}$. For a moment matching at several points, the generalized Krylov subspaces, which are the union of different Krylov subspaces at different points $\sigma_i \in \mathbb{C}$, $i = 1, \ldots, n_\sigma$, must be considered. The result is presented in Theorem 3.5.

Theorem 3.5 (Multiple points two-sided moment matching). *Consider a SISO LTI dynamical system $\mathbf{H} := (A, B, C, D)$ and n_σ interpolation points $\{\sigma_1, \ldots, \sigma_{n_\sigma}\} \in \mathbb{C}$ such that each $(\sigma_k I_n - A)$, $k = 1, \ldots, n_\sigma$ is nonsingular. If $V \in \mathbb{C}^{n \times r}$ and $W \in \mathbb{C}^{n \times r}$ are full-rank matrices the columns of which span the generalized Krylov subspaces:*

$$\begin{aligned} \bigcup_{k=1}^{n_\sigma} \mathcal{K}_{r_k}\left((\sigma_k I_n - A)^{-1}, (\sigma_k I_n - A)^{-1} B\right) &\subseteq \mathcal{V} = \mathbf{span}(V) \\ \bigcup_{k=1}^{n_\sigma} \mathcal{K}_{r_k}\left((\sigma_k I_n - A)^{-T}, (\sigma_k I_n - A)^{-T} C^T\right) &\subseteq \mathcal{W} = \mathbf{span}(W) \end{aligned} \tag{3.78}$$

then the reduced-order model $\hat{\mathbf{H}} = ((W^T V)^{-1} V^T AV, (W^T V)^{-1} V^T B, CV, D)$ *obtained by projection matches the first* $2r_k$ *moments of* $\mathbf{H}$ *in each* σ_k, *i.e., for* $k = 1, \ldots, n_\sigma$

$$\hat{\eta}_i(\sigma_k) = \eta_i(\sigma_k), \quad i = 1, \ldots, 2r_k \tag{3.79}$$

All the moment matching theorems presented above consider complex interpolation points $\sigma_i \in \mathbb{C}$. However, to obtain a real realization for the reduced-order model, the interpolation points must be either real or closed under conjugation. In the latter case, the columns of the matrices V and W are also closed under conjugation, thus they can be replaced by two real matrices which span the same subspace by separating the real and imaginary parts of each complex pair. For instance, consider

$$V = \begin{bmatrix} \mathbf{v}_1 & \bar{\mathbf{v}}_1 \end{bmatrix} \in \mathbb{C}^{n \times 2} \tag{3.80}$$

V spans the same subspace as

$$V_{\mathbb{R}} = \begin{bmatrix} \mathbf{Re}(\mathbf{v}_1) & \mathbf{Im}\,(\mathbf{v}_1) \end{bmatrix} \in \mathbb{R}^{n \times 2} \tag{3.81}$$

This is exploited in Example 3.7 where the multiple points two-sided moment matching approach is compared to the single point one.

Example 3.7. This example illustrates the rational interpolation at multiple points. The LAH model is reduced to an order 3 by a two-sided moment matching at $\sigma = \{0\}$ and at $\sigma = \{0, 5j, -5j\}$. The reduced model obtained with a Padé approximation, denoted $\hat{\mathbf{H}}_{\text{padé}}$, matches the first six moments of the full-order model $\mathbf{H}$ at 0 while the one obtained with a rational interpolation at $\sigma = \{0, 5j, -5j\}$, denoted $\hat{\mathbf{H}}_{\text{rat}}$, matches the first two moments of $\mathbf{H}$ at each interpolation points. This can be seen in Figure 3.12 where the frequency responses of the error are plotted. Indeed, for low frequencies, $\hat{\mathbf{H}}_{\text{padé}}$ leads to a better approximation than $\hat{\mathbf{H}}_{\text{rat}}$ whereas at $\omega = 5$ rad/s (which corresponds to the complex point $5j$), the error $\mathbf{H} - \hat{\mathbf{H}}_{\text{rat}}$ falls drastically.

All the results presented in this section can theoretically be generalized to MIMO systems. However, in the latter case, the available algorithms, such as the block Arnoldi procedure [26], are very tricky to implement and do not necessarily lead to satisfactory results. Instead, the *tangential interpolation* framework proposed in [6] enables to elegantly and naturally address the MIMO case.

3.3.1.3 Tangential interpolation

An in-depth study of the tangential interpolation framework is done in [6]. Here, only the result involved in the optimal model approximation is presented. Theorem 3.6 shows how a bi-tangential interpolation of an LTI model can be achieved by projection onto a particular subspace. It is very similar to the previous generalized Krylov subspaces except that in this case, tangential directions are involved.

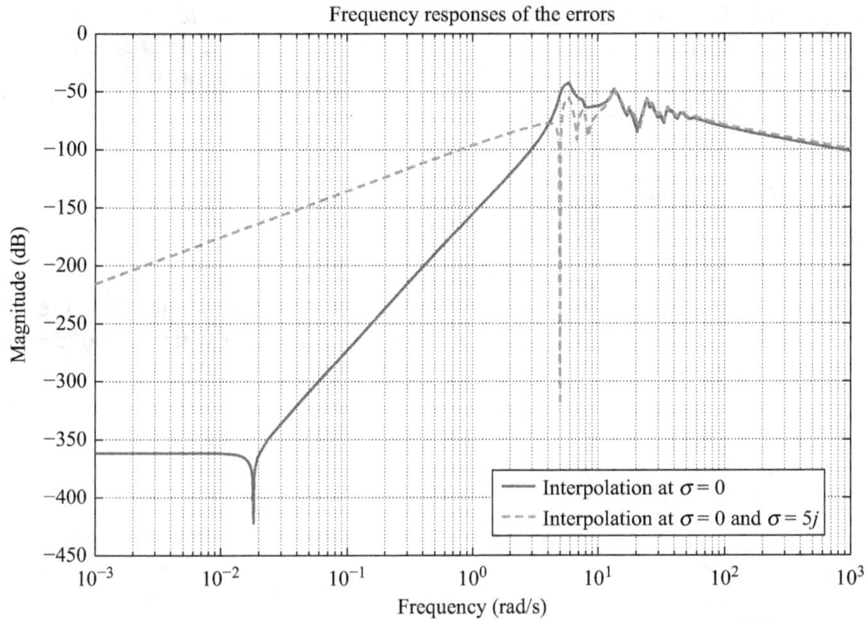

Figure 3.12 Comparison of the third-order approximations obtained by interpolation of the LAH model at $\sigma = \{0\}$ and at $\sigma = \{0, 5j, -5j\}$

Theorem 3.6 (Bi-tangential interpolation). *Consider a MIMO LTI dynamical system* $\mathbf{H}$ *with* n_u *inputs and* n_y *outputs. Let* $V,\ W \in \mathbb{C}^{n\times r}$ *be full rank matrices and let* $\sigma_i \in \mathbb{C}$, $\hat{\mathbf{b}}_\mathbf{i} \in \mathbb{C}^{1\times n_u}$, *and* $\hat{\mathbf{c}}_\mathbf{i} \in \mathbb{C}^{1\times n_y}$, $i = 1,\ldots,n_\sigma$ *be a given set of interpolation points and associated right and left tangential directions, respectively. Suppose that the points* σ_i *are such that* $(\sigma_i I_n - A)$ *is nonsingular. If for* $i = 1,\ldots,n_\sigma$,

$$\begin{aligned} (\sigma_i I_n - A)^{-1} B\hat{\mathbf{b}}_\mathbf{i}^T &\in \mathbf{span}(V) \\ (\sigma_i I_n - A)^{-T}\, C^T\hat{\mathbf{c}}_\mathbf{i}^T &\in \mathbf{span}(W) \end{aligned} \tag{3.82}$$

then the reduced order model $\hat{\mathbf{H}} = ((W^TV)^{-1}V^TAV, (W^TV)^{-1}V^TB, CV, D)$ *satisfies the following bi-tangential interpolation equations,*

$$\begin{aligned} \hat{\mathbf{c}}_\mathbf{i} H(\sigma_i) &= \hat{\mathbf{c}}_\mathbf{i}\hat{H}(\sigma_i), \\ H(\sigma_i)\hat{\mathbf{b}}_\mathbf{i}^T &= \hat{H}(\sigma_i)\hat{\mathbf{b}}_\mathbf{i}^T, \\ \hat{\mathbf{c}}_\mathbf{i}\left.\frac{dH(s)}{ds}\right|_{s=\sigma_i}\hat{\mathbf{b}}_\mathbf{i}^T &= \hat{\mathbf{c}}_\mathbf{i}\left.\frac{d\hat{H}(s)}{ds}\right|_{s=\sigma_i}\hat{\mathbf{b}}_\mathbf{i}^T, \end{aligned} \tag{3.83}$$

for $i = 1,\ldots,n_\sigma$.

The interpolation framework presented throughout this section enables to implicitly match several moments of a linear model at several complex points. The choice of these interpolations point will determine the quality of the approximation. In the next section, a particular set of interpolation points (and associated tangential directions) is used so that the reduced-order model is an optimal approximation, in the $\mathcal{H}_2$ sense, of the initial full-order model.

3.3.2 $\mathcal{H}_2$ optimal model reduction

3.3.2.1 First-order optimality conditions

Problem 3.3 ($\mathcal{H}_2$ optimal model reduction problem). Given $\mathbf{H}$ a n-th order continuous LTI dynamical system, the problem of optimal model reduction in the $\mathcal{H}_2$ norm consists in finding a r-th order ($r \ll n$) reduced model $\hat{\mathbf{H}}$ which minimizes the $\mathcal{H}_2$-norm of the error

$$\mathcal{J}_{\mathcal{H}_2} = \|H - \hat{H}\|^2_{\mathcal{H}_2} \tag{3.84}$$

Minimizing $\mathcal{J}_{\mathcal{H}_2}$ is a non-convex problem (see Example 3.8), thus finding a global minimum is a complex task. Instead, finding a local minimizer is way more tractable. To this goal, first-order necessary optimality conditions have been derived: first in [18], for SISO systems, as interpolation conditions between the transfer functions of the full-order and reduced-order models. Then in a state-space form that involves the gramians of the error model $\mathbf{H} - \hat{\mathbf{H}}$ in [33]. More recently in [6], tangential interpolation conditions have been formulated thus generalizing the interpolation conditions of [18] to MIMO systems. These tangential interpolation conditions have finally been revisited in a very comprehensive way in [29] and are recalled in Theorem 3.7.

Example 3.8. To illustrate the non-convexity of Problem 3.3, the LAH model is approximated by a second order model $\hat{\mathbf{H}}$ of transfer function:

$$\hat{H}(s) = \frac{\phi}{s-\lambda} + \frac{\phi^*}{s-\lambda^*} \tag{3.85}$$

where the pole $\lambda \in \mathbb{C}$ is varying and the residue $\phi \in \mathbb{C}$ is taken constant. The resulting $\mathcal{H}_2$-norm is plotted in Figure 3.13.

We can see that there exist several local minima of different amplitude. By looking at the frequency response and the poles of the LAH model (Figure 3.3), it appears that the smallest minima are located near the biggest resonances of the system $\mathbf{H}$. This comes from the fact that modes with a high gain have more impact on the $\mathcal{H}_2$-norm than modes with a smaller gain.

Theorem 3.7 (First-order optimality conditions for the $\mathcal{H}_2$ model reduction problem). *Consider a n-th order LTI dynamical system* $\mathbf{H}$ *the transfer function is given by* $H(s) = C(sI_n - A)^{-1}B + D \in \mathbb{C}^{n_y \times n_u}$ *and* $\hat{\mathbf{H}}$ *a r-th order model with transfer*

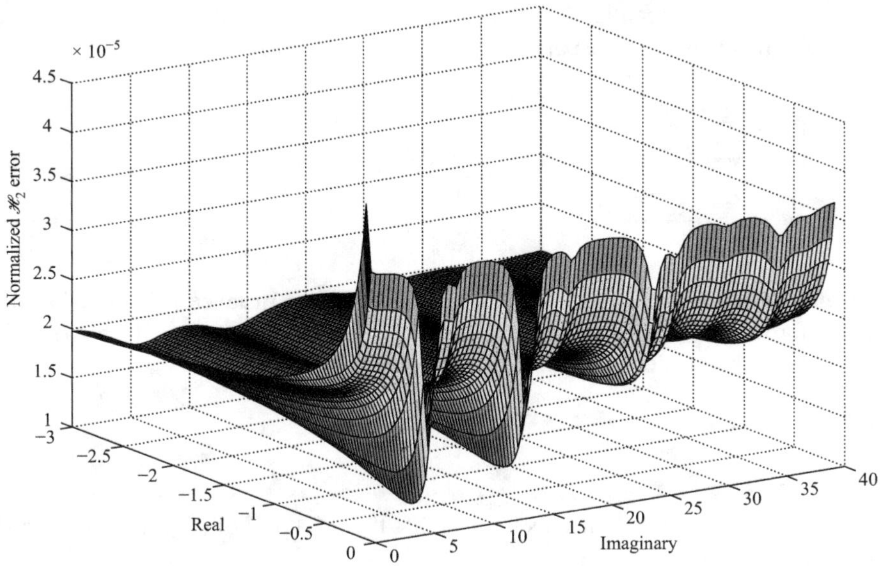

Figure 3.13 $\mathcal{H}_2$*-norm of the error for an order 2 approximation of the LAH model*

function $\hat{H}(s) = \hat{C}(sI_r - \hat{A})^{-1}B_r + Dr \in \mathbb{C}^{n_y \times n_u}$. *If* $\hat{\mathbf{H}}$ *solves Problem 3.3 then, for* $i = 1, \ldots, r$,

$$\begin{aligned} \hat{\mathbf{c}}_\mathbf{i} H(-\hat{\lambda}_i) &= \hat{\mathbf{c}}_\mathbf{i} \hat{H}(-\hat{\lambda}_i), \\ H(-\hat{\lambda}_i)\hat{\mathbf{b}}_\mathbf{i}^T &= \hat{H}(-\hat{\lambda}_i)\hat{\mathbf{b}}_\mathbf{i}^T, \\ \hat{\mathbf{c}}_\mathbf{i} \left.\frac{dH(s)}{ds}\right|_{s=-\hat{\lambda}_i} \hat{\mathbf{b}}_\mathbf{i}^T &= \hat{\mathbf{c}}_\mathbf{i} \left.\frac{d\hat{H}(s)}{ds}\right|_{s=-\hat{\lambda}_i} \hat{\mathbf{b}}_\mathbf{i}^T \end{aligned} \tag{3.86}$$

where $\hat{\mathbf{c}}_\mathbf{i} \in \mathbb{C}^{1 \times n_y}$ *and* $\hat{\mathbf{b}}_\mathbf{i} \in \mathbb{C}^{1 \times n_u}$, $i = 1, \ldots, r$ *are tangential directions given as*

$$\begin{bmatrix} \hat{\mathbf{c}}_\mathbf{1}^T & \ldots & \hat{\mathbf{c}}_\mathbf{r}^T \end{bmatrix} = \hat{C}\hat{X} \tag{3.87}$$

and

$$\begin{bmatrix} \hat{\mathbf{b}}_\mathbf{1} \\ \vdots \\ \hat{\mathbf{b}}_\mathbf{r} \end{bmatrix} = \hat{X}^{-1}\hat{B} \tag{3.88}$$

where $\hat{X}$ *is the matrix which columns are the right eigenvectors of* $\hat{A}$.

Remark 3.15. The tangential directions (3.87) and (3.88) are the outer product of the residues ϕ_i that appear in the partial fraction expansion of $\hat{H}(s)$,

$$\hat{H}(s) = \sum_{i=1}^{r} \frac{\phi_i}{s - \hat{\lambda}_i} \tag{3.89}$$

$$= \sum_{i=1}^{r} \frac{\hat{\mathbf{c}}_i^T \hat{\mathbf{b}}_\mathbf{i}}{s - \hat{\lambda}_i} \tag{3.90}$$

They naturally appear in $\mathcal{J}_{\mathcal{H}_2}$ and its gradient (see for instance [4]) by using the residue formulation of the $\mathcal{H}_2$-norm, i.e., the $\mathcal{H}_2$-norm of a system $\mathbf{H} := (A, B, C)$ which have a diagonalizable matrix A can be expressed as

$$\|H\|_{\mathcal{H}_2}^2 = \sum_{i=1}^{n} \mathbf{trace}(\phi_i H(-\lambda_i)^T) \tag{3.91}$$

with $\phi_i = \mathbf{c_i}^T \mathbf{b_i}$, $i = 1, \ldots, n$. This implies that the $\mathcal{H}_2$-norm of the error between $\mathbf{H}$ and $\hat{\mathbf{H}}$ can be expressed as

$$\mathcal{J}_{\mathcal{H}_2} = \|\tilde{H}\|_{\mathcal{H}_2} = \|H - \hat{H}\|_{\mathcal{H}_2}^2 = \sum_{i=1}^{n+r} \mathbf{trace}(\tilde{\phi}_i \tilde{H}(-\tilde{\lambda}_i)^T) \tag{3.92}$$

where (by assuming that the elements are ordered),

$$\tilde{\phi}_i = \begin{cases} \phi_i & i = 1, \ldots, n \\ -\hat{\phi}_i & i = n+1, \ldots, n+r \end{cases} \tag{3.93}$$

and

$$\tilde{\lambda}_i = \begin{cases} \lambda_i & i = 1, \ldots, n \\ \hat{\lambda}_i & i = n+1, \ldots, n+r \end{cases} \tag{3.94}$$

By separating the sums in (3.92), it comes that

$$\mathcal{J}_{\mathcal{H}_2} = \|H - \hat{H}\|_{\mathcal{H}_2}^2 = \|H\|_{\mathcal{H}_2}^2 + \|\hat{H}\|_{\mathcal{H}_2}^2 - 2\sum_{i=1}^{r} \hat{\mathbf{c}}_\mathbf{i} H(-\hat{\lambda}_i)\hat{\mathbf{b}}_\mathbf{i}^T \tag{3.95}$$

In the SISO case, the first-order optimality conditions consist in saying that if $\hat{\mathbf{H}}$ is a stationary point of 3.3 then it has to match the first two moments of the full-order model $\mathbf{H}$ at the mirror image of its poles $\hat{\lambda}_i$, $i = 1, \ldots, r$. In the MIMO case, the moment matching is replaced with bi-tangential interpolation conditions. Those tangential conditions can be satisfied through projection of the full-order model on a particular subspace as stated in Theorem 3.6. The problem is that the interpolation points (and associated tangential directions) depend on the reduced-order model. Hence, the first-order optimality conditions cannot be directly satisfied, therefore an iterative procedure has to be considered.

3.3.2.2 Iterative tangential interpolation algorithm

In [12], the authors proposed an iterative algorithm called *Iterative Rational Krylov Algorithm* (IRKA) that enables to construct, for SISO systems, a reduced order model that satisfies the first-order optimality conditions. This algorithm has then been generalized to MIMO systems in [29] with the tangential interpolation framework introduced in [6]. Here, this procedure is called *Iterative Tangential Interpolation Algorithm* (ITIA) and is recalled in Algorithm 3.4.

Algorithm 3.4 Iterative Tangential Interpolation Algorithm (ITIA or MIMO IRKA)

Require: $A \in \mathbb{R}^{n\times n}$, $B \in \mathbb{R}^{n\times n_u}$, $C \in \mathbb{R}^{n_y\times n}$, $D \in \mathbb{R}^{n_y\times n_u}$, $\{\sigma_1^{(0)},\dots,\sigma_r^{(0)}\} \in \mathbb{C}^r$, $[\hat{\mathbf{b}}_\mathbf{1}^T,\dots,\hat{\mathbf{b}}_\mathbf{r}^T]^T \in \mathbb{C}^{r\times n_u}$, $[\hat{\mathbf{c}}_\mathbf{1}^T,\dots,\hat{\mathbf{c}}_\mathbf{r}^T] \in \mathbb{C}^{n_y\times r}$, $\varepsilon > 0$

1: Construct,

$$V = [(\sigma_1^{(0)}I_n - A)^{-1}B\hat{\mathbf{b}}_\mathbf{1}^T \dots (\sigma_r^{(0)}I_n - A)^{-1}B\hat{\mathbf{b}}_\mathbf{r}^T]$$
$$W = [(\sigma_1^{(0)}I_n - A)^{-T}C^T\hat{\mathbf{c}}_\mathbf{1}^T \dots (\sigma_r^{(0)}I_n - A)^{-T}C^T\hat{\mathbf{c}}_\mathbf{r}^T]$$

2: Compute $W \leftarrow W(V^TW)^{-1}$

3: **while** $|\sigma^{(i)} - \sigma^{(i-1)}| > \varepsilon$ **do**

4: $\quad i \leftarrow i+1$

5: $\quad \hat{A} = W^TAV$, $\hat{B} = W^TB$ and $\hat{C} = CV$

6: $\quad$ Solve the eigenvalue problem $\hat{A}\hat{X} = \Delta\hat{X}$ where $\Delta = \mathbf{diag}(\hat{\lambda}_1,\dots,\hat{\lambda}_r)$.

7: $\quad$ Compute $[\hat{\mathbf{b}}_\mathbf{1}^T \dots \hat{\mathbf{b}}_\mathbf{r}^T]^T = \hat{X}^{-1}\hat{B}$, $[\hat{\mathbf{c}}_\mathbf{1}^T \dots \hat{\mathbf{c}}_\mathbf{r}^T] = \hat{C}\hat{X}$

8: $\quad$ Set $\sigma^{(i)} = -\lambda(\hat{A})$

9: $\quad$ Construct,

$$V = [(\sigma_1^{(i)}I_n - A)^{-1}B\hat{\mathbf{b}}_\mathbf{1}^T \dots (\sigma_r^{(i)}I_n - A)^{-1}B\hat{\mathbf{b}}_\mathbf{r}^T]$$
$$W = [(\sigma_1^{(i)}I_n - A)^{-T}C^T\hat{\mathbf{c}}_\mathbf{1}^T \dots (\sigma_r^{(i)}I_n - A)^{-T}C^T\hat{\mathbf{c}}_\mathbf{r}^T]$$

10: $\quad$ Compute $W \leftarrow W(W^TV)^{-T}$

11: **end while**

12: Construct $\hat{\mathbf{H}} = (W^TAV, W^TB, CV, D)$

Ensure: $V, W \in \mathbb{R}^{n\times r}$, $W^TV = I_r$

At each iteration, a reduced-order model is constructed (step 5) by tangential interpolation of the full-order model at the mirror images of the previous reduced-order model's poles (step 8). To perform the interpolation, two projectors $V, W \in \mathbb{C}^{n \times r}$ are constructed (step 9) accordingly to Theorem 3.6. Note that the interpolation points σ_i and the associated tangential directions are complex under conjugation. Hence, the matrices V and W can be transformed into real matrices which span the same subspace so that the reduced-order realization $(\hat{A}, \hat{B}, \hat{C})$ is real (see Section 3.3.1).

Some remarks can be made about the ITIA algorithm:

- It does not offer any guarantee concerning the stability of the reduced-order model. However, in practice, unstable models are rarely obtained.
- At each iteration, the algorithm requires to solve r large-scale linear systems. These steps are the most expensive in terms of computational costs. However, efficient methods that exploit sparsity exist to address this problem even in very large-scale settings ([26]). Hence, the ITIA is suitable even for (very) large-scale model approximation.
- The final reduced-order model satisfies the first-order optimality conditions and in practice, it is often a local minimum of the $\mathcal{H}_2$ approximation Problem 3.3.
- The quality of the resulting reduced-order model depends on the choice of the initial interpolation points. However, no systematic selection procedure has proven to be more effective in all the cases yet. An intuitive approach consists in taking the mirror images of the initial model's dominant poles. A restarting procedure can also be added to Algorithm 3.4 so that several initializations are considered.

As presented in the next section, a modification can be made on the ITIA so that it guarantees the stability of the reduced-order model at the cost of a lower numerical efficiency.

Remark 3.16. A tangential interpolation can also be constructed by solving particular Sylvester equations as presented in [34].

3.3.2.3 Iterative SVD-tangential interpolation algorithm

In [10], a modified version of the IRKA has been proposed to bring together the numerical efficiency of the IRKA and the stability preservation of the balanced truncation. The resulting procedure is called the *Iterative SVD Rational Krylov Algorithm* (ISRKA) and enables, for SIMO or MISO systems, to construct a reduced-order model that satisfies a subset of the first-order optimality conditions and preserves the stability of the full-order model. The ISRKA has then been generalized to MIMO systems in [22] leading to the *Iterative SVD Tangential Interpolation Algorithm* (ISTIA) recalled in Algorithm 3.5.

Algorithm 3.5 Iterative SVD-Tangential Interpolation Algorithm (ISTIA)

Require: $A \in \mathbb{R}^{n\times n}$, $B \in \mathbb{R}^{n\times n_u}$, $C \in \mathbb{R}^{n_y\times n}$, $D \in \mathbb{R}^{n_y\times n_u}$ $\{\sigma_1^{(0)}, \ldots, \sigma_r^{(0)}\} \in \mathbb{C}^{n\times r}$, $[\hat{\mathbf{b}}_1^T, \ldots, \hat{\mathbf{b}}_r^T]^T \in \mathbb{C}^{r\times n_u}$, $\varepsilon > 0$

Construct, $V = [(\sigma_1^{(0)} I_n - A)^{-1} B\hat{\mathbf{b}}_1^T \ldots (\sigma_r^{(0)} I_n - A)^{-1} B\hat{\mathbf{b}}_r^T]$

Solve $\mathcal{Q}A + A^T\mathcal{Q} + C^TC = 0$ in $\mathcal{Q}$

Compute $W = \mathcal{Q}V(V^T\mathcal{Q}V)^{-1}$

while $|\sigma^{(i)} - \sigma^{(i-1)}| > \varepsilon$ **do**

  $i \leftarrow i + 1, \hat{A} = W^TAV, \hat{B} = W^TB$

  Compute $\hat{A}X = \mathbf{diag}(\lambda(\hat{A}))X$

  Compute $[\hat{\mathbf{b}}_1^T \ldots \hat{\mathbf{b}}_r^T]^T = \hat{X}^{-1}\hat{B}$

  Set $\sigma^{(i)} = -\lambda(\hat{A})$

  Construct, $V = [(\sigma_1^{(i)} I_n - A)^{-1} B\hat{\mathbf{b}}_1^T \ldots (\sigma_r^{(i)} I_n - A)^{-1} B\hat{\mathbf{b}}_r^T]$

  Compute $W = \mathcal{Q}V(V^T\mathcal{Q}V)^{-1}$

end while

Construct $\hat{\mathbf{H}} = (W^TAV, W^TB, CV, D)$

Ensure: $V, W \in \mathbb{R}^{n\times r}$, $W^TV = I_r$

The ISTIA is very similar to the ITIA. The only difference lies in the use of the observability gramian to construct the left projector W. Note that the controllability gramian could be used instead of the observability one as mentioned in [22]. The properties of the ISTIA are the following:

- Unlike the ITIA, the ISTIA preserves the stability of the initial model [10].
- Since a gramian is required by the ISTIA, this method might not be suitable for very large-scale problems. It remains computationally cheaper than the balanced truncation which requires both gramians.
- In practice, the ISTIA seems to be less dependent of the initial point selection than the ITIA.

The ITIA and the ISTIA are compared to balanced truncation in Example 3.9. This example illustrates the fact that both the ITIA and the ISTIA are efficient alternatives to balanced truncation. There is no frequency-limited version of the ITIA[4] yet, but the ISTIA can easily be extended to the frequency-limited case as presented in the next section.

[4] For a frequency-weighted version of the ITIA, see [2].

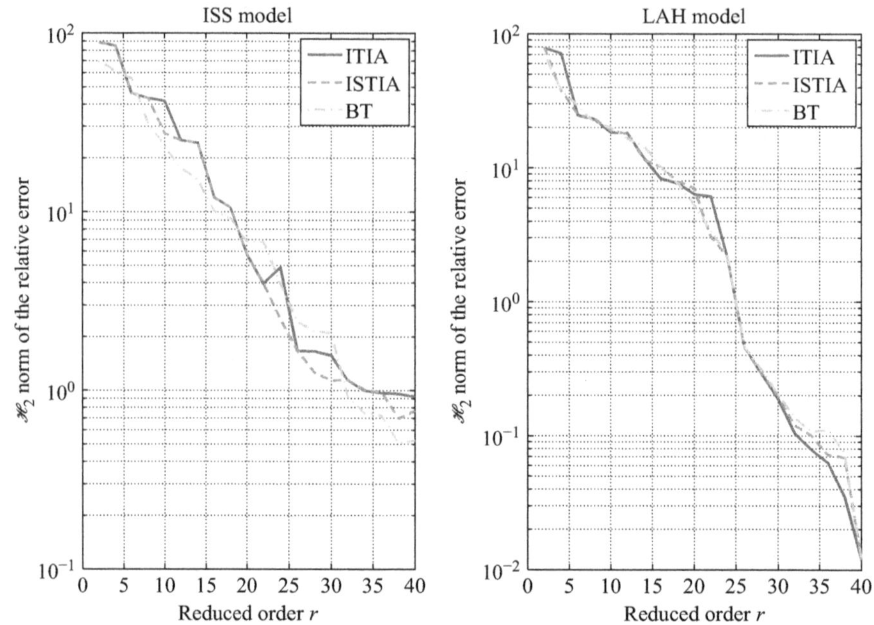

Figure 3.14 *$\mathcal{H}_2$-norms of the relative errors obtained by approximation of the ISS and LAH models with the ITIA, the ISTIA and the BT to several order r going grom 2 to 40*

Example 3.9. In this example the LAH and the ISS models are reduced for several orders r going from $r = 2$ to $r = 40$ with the ITIA, the ISTIA and the BT. The resulting $\mathcal{H}_2$-norms of the relative errors are plotted in Figure 3.14.

We can see that the all the approaches lead to very close results. In spite of the optimality of the ITIA, it does not necessarily leads to the best result in terms of $\mathcal{H}_2$-norm. This can be explained by the numerous local minima that characterizes the $\mathcal{H}_2$ model approximation problem.

3.3.2.4 Frequency-limited SVD-tangential interpolation algorithm

Similarly to the FL-BT, the use of a frequency-limited gramian (see Definition 3.6) in the ISTIA instead of the infinite gramian makes the method more efficient in terms of $\mathcal{H}_{2,\Omega}$-norm. This modification has been proposed in [30] and is based on an experimental approach, which means that there is no optimality consideration. Yet the approach has revealed to be efficient in practice and has successfully been used on and industrial use case [23]. The properties of the frequency-limited ISTIA (FL-ISTIA) are as follows:

- Due to the use of a frequency-limited gramian instead of the infinite one, the reduced-order model is no longer guaranteed to be stable. However, in practice, unstable models are only obtained when considering extremely tight frequency intervals.

- If the considered frequency interval is $[0, \omega]$ then the FL-ISTIA becomes equivalent to the ISTIA when $\omega \to \infty$. This comes from the fact that the frequency-limited gramians tends toward the infinite gramians when $\omega \to \infty$.
- As illustrated in [30], the FL-ISTIA seems to be numerically more robust than the FL-BT when the large-scale models are ill-conditioned.

The efficiency of the FL-ISTIA is illustrated in Example 3.10.

Example 3.10. In this example, the ISS model is reduced to a 10-th order model with the ISTIA, the BT, the FL-ISTIA, and the FL-BT. The approximation is done over $[0, \omega]$ with ω ranging from 5 rad/s to 60 rad/s. The $\mathcal{H}_{2,\Omega}$-norms of the relative errors are plotted in Figure 3.15.

We can see that the FL-ISTIA leads to better results that the ISTIA in terms of $\mathcal{H}_{2,\Omega}$-norm. The difference is large for small ω and decreases when ω increases. In particular, for frequencies larger than 40 rad/s, the FL-ISTIA becomes equivalent to the ISTIA. For small ω, the FL-ISTIA gives similar results to those obtained with the FL-BT.

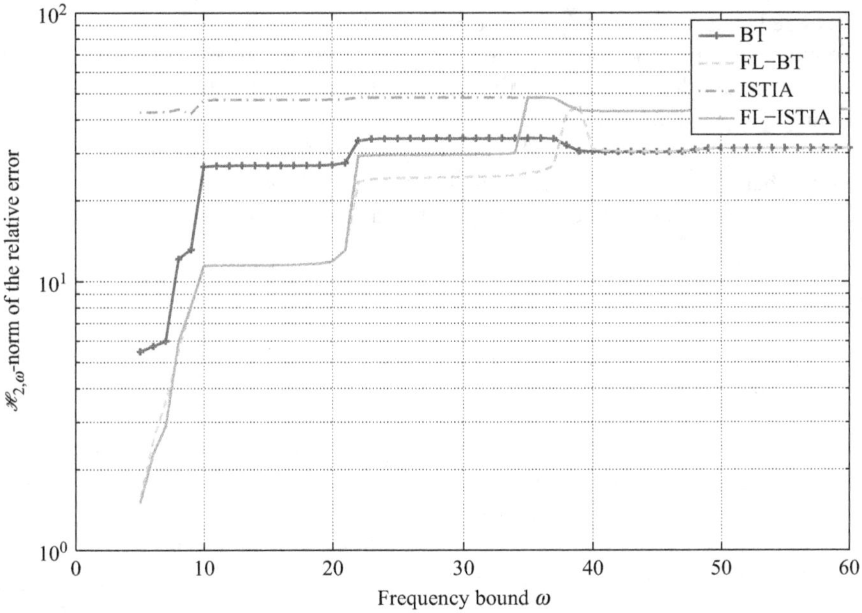

Figure 3.15 *$\mathcal{H}_{2,\Omega}$-norms of the relative errors obtained by approximation of the ISS model with the ISTIA, the FL-ISTIA, the BT, and the FL-BT to an order 10 for ω going from 5 rad/s to 60 rad/s*

3.3.3 Conclusion

In this section, moment maching model approximation has been introduced. The methods are based on rational interpolation for the SISO case and on tangential interpolation in the MIMO case. Such interpolations can be achieved by projection of the initial large-scale model onto particular Krylov subspaces. The projectors can be efficiently computed through iterative procedures such as the Arnoldi (or Lanczos) one. If the interpolation is done at some particular points, then an optimal reduced-order model for the $\mathcal{H}_2$ model approximation problem can be obtained. The moment matching methods offer an interesting alternative to the truncations methods since some of them are suitable for very large-scale systems and are as efficient.

3.4 Conclusion

In this chapter, different model approximation methods have been presented. On one hand, truncation methods consist in discarding some states of the initial large-scale model's state-space realization after it has been transformed into an interesting basis, such as the modal or balanced one. These methods preserve the stability of the initial model and also offer a bound on the $\mathcal{H}_\infty$-norm of the error. However, they are computationally expensive and might not be suitable for very large-scale applications. On the other hand, moment matching methods interpolate the large-scale initial model at some complex points. If these points are selected in a particular manner then an optimal reduced-order model, with respect to the $\mathcal{H}_2$ approximation problem, can be found. These methods do not offer any guarantee concerning the stability of the reduced-order model or bounds on the error but they are computationally cheaper and can thus be applied on much larger models than the truncation ones.

In the literature, these model approximation methods are often compared, and it seems that there is no best method. Each approach has its own advantages and drawbacks and is more suited for a given kind of application.

References

[1] Anderson, M.R., Emami-Naeni, A., Vincent, J.: Measures of merit for multivariable flight control. Tech. rep., *Systems Control Technology Inc*, Palo Alto, California, USA (1991)

[2] Anic, B., Beattie, C.A., Gugercin, S., Antoulas, A.C.: *Interpolatory Weighted-$\mathcal{H}_2$ Model Reduction*. ArXiv e-prints (2012)

[3] Antoulas, A.C.: Approximation of large-scale dynamical systems. *Society for Industrial and Applied Mathematics*, Philadelphia, USA (2005)

[4] Beattie, C., Gugercin, S.: A trust region method for optimal $\mathcal{H}_2$ model reduction. In: *Proceedings of the Joint 48th IEEE Conference on Decision and Control and 28th Chinese Control Conference*, pp. 5370–5375 (2009)

[5] Doyle, J., Francis, B., Tannenbaum, A.: *Feedback Control Theory*, Macmillan (1990)

[6] Gallivan, K.A., Vandendorpe, A., Van Dooren, P.: Model reduction of MIMO systems via tangential interpolation. *SIAM J. Matrix Anal. Appl.* **26**(2), 328–349 (2004)

[7] Gawronski, W., Juang, J.: Model reduction in limited time and frequency intervals. *International Journal of Systems Science* **21**(2), 349–376 (1990)

[8] Glover, K.: All optimal hankel-norm approximations of linear multivariable systems and their $\mathscr{L}_\infty$-error bounds. *International Journal of Control* **39**(6), 1115–1193 (1984)

[9] Grimme, E.J.: Krylov projection methods for model reduction. Ph.D. thesis, University of Illinois at Ubarna-Champaign (1997)

[10] Gugercin, S.: An iterative SVD-Krylov based method for model reduction of large-scale dynamical systems. *Linear Algebra and its Applications* **428**(89), 1964–1986 (2007)

[11] Gugercin, S., Antoulas, A.C.: A survey of model reduction by balanced truncation and some new results. *International Journal of Control* **77**(8), 748–766 (2004)

[12] Gugercin, S., Antoulas, A.C., Beattie, C.: $\mathscr{H}_2$ model reduction for large-scale linear dynamical systems. *SIAM Journal on Matrix Analysis and Applications* **30**(2), 609–638 (2008)

[13] Hammarling, S.J.: Numerical solution of the stable, non-negative definite Lyapunov equation. *IMA Journal of Numerical Analysis* **2**(3), 303–323 (1982). DOI 10.1093/imanum/2.3.303. Available from http://imajna.oxfordjournals.org/content/2/3/303.abstract

[14] Kokotovic, P., Khalil, H., O'Reilly, J.: Singular Perturbation Methods in Control: Analysis and Design. *Society for Industrial and Applied Mathematics* (1986)

[15] Laub, A., Heath, M., Paige, C., Ward, R.: Computation of system balancing transformations and other applications of simultaneous diagonalization algorithms. *IEEE Transactions on Automatic Control* **32**(2), 115–122 (1987)

[16] Leibfritz, F., Lipinski, W.: Description of the benchmark examples in *COMPl$_e$ib* 1.0. Tech. rep., University of Trier (2003)

[17] Liu, Y., Anderson, B.D.: Singular perturbation approximation of balanced systems. In: *Proceedings of the 28th Conference on Decision and Control*, pp. 1355–1360 (1989)

[18] Meier, L., Luenberger, D.: Approximation of linear constant systems. *IEEE Transactions on Automatic Control* **12**(5), 585–588 (1967)

[19] Moore, B.: Principal component analysis in linear systems: Controllability, observability, and model reduction. *IEEE Transactions on Automatic Control* **26**(1), 17–32 (1981). DOI 10.1109/TAC.1981.1102568

[20] Pernebo, L., Silverman, L.: Model reduction via balanced state space representations. *IEEE Transactions on Automatic Control* **27**(2), 382–387 (1982). DOI 10.1109/TAC.1982.1102945

[21] Petersson, D., Löfberg, J.: Model reduction using a frequency-limited $\mathscr{H}_2$-cost (2012). Available from http://arxiv.org/abs/1212.1603

[22] Poussot-Vassal, C.: An iterative SVD-tangential interpolation method for medium-scale MIMO systems approximation with application on flexible aircraft. In: *Proceedings 50th IEEE Conference on Decision and Control – European Control Conference (CDC-ECC'11)*, pp. 7117–7122. Florida (2011)

[23] Poussot-Vassal, C., Loquen, T., Vuillemin, P., Cantinaud, O., Lacoste, J.: Business jet large-scale model approximation and vibration control. In: *Proceedings of the 11th IFAC International Workshop on Adaptation Learning in COntrol and Signal Processing (ALCOSP'13)* pp. 199–204. Caen, France (2013)

[24] Poussot-Vassal, C., Vuillemin, P.: Introduction to MORE: a MOdel REduction toolbox. In: *Proceedings of the IEEE Multi Systems Conference (MSC CCA'12)*. Dubrovnik, Croatia (2012)

[25] Rommes, J., Martins, N.: Efficient computation of multivariable transfer function dominant poles using subspace acceleration. *IEEE Transactions on Power Systems* **21**, 1471–1483 (2006)

[26] Saad, Y.: Iterative methods for sparse linear systems, 2nd edn. *Society for Industrial and Applied Mathematics*, Philadelphia, USA (2003)

[27] Sorensen, D.: Implicitly restarted Arnoldi/Lanczos methods for large scale eigenvalue calculations (1996). Technical report

[28] Stewart, G.W.: The decompositional approach to matrix computation. *Computing in Science Engineering* **2**(1), 50–59 (2000). DOI 10.1109/5992.814658

[29] Van Dooren, P., Gallivan, K.A., Absil, P.A.: $\mathcal{H}_2$-optimal model reduction of MIMO systems. *Applied Mathematics Letters* **21**(12), 1267–1273 (2008)

[30] Vuillemin, P., Poussot-Vassal, C., Alazard, D.: $\mathcal{H}_2$ optimal and frequency limited approximation methods for large-scale lti dynamical systems. In: *Proceedings of the 5th Symposium on System Structure and Control*. Grenoble, France (2013)

[31] Vuillemin, P., Poussot-Vassal, C., Alazard, D.: Poles residues descent algorithm for optimal frequency-limited $\mathcal{H}_2$ model approximation (2013). Submitted to CDC 2013

[32] Vuillemin, P., Poussot-Vassal, C., Alazard, D.: A spectral expression for the frequency-limited $\mathcal{H}_2$-norm (2013). Available from http://arxiv.org/abs/1211.1858

[33] Wilson, D.A.: Model reduction for multivariable systems. *International Journal of Control* **20**(1), 57–64 (1974)

[34] Xu, Y., Zeng, T.: Optimal $\mathcal{H}_2$ model reduction for large scale MIMO systems via tangential interpolation. *International Journal of Numerical Analysis and Modeling* **8**(1), 174–188 (2011)

[35] Zhou, K., Doyle, J.C., Glover, K.: *Robust and Optimal Control*. Prentice Hall (1995)

Chapter 4

Integrated modelling and parameter estimation: an LFR–Modelica approach

Marco Lovera[1] *and Francesco Casella*[2]

Abstract

Linear fractional representations (LFRs) are a widely used model description formalism in modern control and system identification theory. Deriving such models from physical first principles is a non-trivial and often tedious and error-prone process, if carried out manually. Tools already exist to transform symbolic transfer functions and symbolic state-space representations into reduced-order LFRs, but these descriptions are still quite far from a natural, physical-based, object-oriented description of physical and technological systems and are moreover hard to integrate with model identification tools. In this chapter a new approach to LFR modelling and identification starting from equation-based, object-oriented descriptions of the plant dynamics (formulated using the Modelica language) and input–output data is presented. This approach allows to reduce the gap between user-friendly model representations, based on object diagrams with physical connections, block diagrams with signal connections, and generic differential-algebraic models, and the use of advanced LFR-based identification and control techniques.

4.1 Introduction

The process of developing control-oriented mathematical models of physical systems is a complex task, which in general calls for a careful combination of prior knowledge about the physics of the system under study with information coming from experimental data (leading to the so-called problem of grey-box modelling; see, e.g. [1]), in view of the application of the model to control systems analysis and design. As recently discussed in [18], the critical issue in the development of an effective approach to control-oriented grey-box modelling lies in the integration of existing methods and tools for physical systems modelling and simulation

[1]Politecnico di Milano, Milano, Italy, e-mail: marco.lovera@polimi.it
[2]Politecnico di Milano, Milano, Italy, e-mail: francesco.casella@polimi.it

with methods and tools for parameter estimation. Furthermore, throughout the modelling exercise, one should always keep in mind that the eventual application of the model is control system analysis and design, so the mathematical structure of the model has to be compatible with currently available methods and tools for such problems. Indeed, integrating information from experimental data at the simulator level to refine the values of uncertain parameters, as currently made possible by some commercial simulation tools, leads to optimisation problems that can turn out to be ill-conditioned and generally hard to keep under control in terms of system identification (i.e. structural and experimental identifiability). An effort in the direction of overcoming such difficulties has been recently presented in [13], where dedicated parameter and state estimation schemes have been proposed for models given in the form of differential-algebraic equations (DAEs), such as generated by modern object-oriented modelling methods and tools. In this chapter an approach is proposed to bridge the gap between physical and control/estimation-oriented system modelling, by automatically deriving standard model structures used in system identification and control starting from object-oriented models of non-linear plants. As far as object-oriented modelling is concerned, the Modelica modelling language is considered here, as it allows to describe the plant dynamics in a very general, natural and user-friendly way; note, however, that the concepts and algorithms introduced here can be applied to any acausal, equation-based modelling language without any substantial change. On the other hand, the linear fractional representation (LFR) model structure has been considered as the natural target for identification and control applications. Indeed, LFRs are a widely used model description formalism both in modern control [26, 19] and in system identification [17, 25, 7, 15]. In this chapter the results obtained in the development of an approach to the derivation of an LFR starting from a system model written with the Modelica language and input–output data are presented. The chapter extends the preliminary results presented in [3, 4], which are limited to linear models, to generic non-linear models with uncertain parameters (see also [9]).

The chapter is organised as follows: in section 4.2 the basic concepts of object-oriented and LFR modelling are briefly reviewed. The main results concerning the algorithm are presented in section 4.3. Finally, section 4.4 presents some examples of application of the algorithm.

4.2 Applicable models and LFRs

As mentioned in section 4.1, in the following we will refer to the Modelica language as a paradigm of an object-oriented approach to mathematical modelling. The basics of the object-oriented approach and Modelica language have been presented in Chapter 2 of this book, so in this section the emphasis will be on the definition of the sub-set of models to which the methods presented in this chapter are applicable, and on the definition of linear fractional representations.

4.2.1 Applicable plant models

In this chapter we consider DAE models obtained from the flattening of systems built by the aggregation of many sub-models through either causal (input–output) or acausal (physical ports carrying effort and flow variables) connectors. Pure block diagrams are included as a special case, but in general it is possible to combine block diagrams with object-oriented descriptions of physical systems of diverse nature. It is assumed that the model is time-invariant (i.e. the *time* variable does not appear explicitly within any model equation) and that all variables and equations are continuous-time (i.e. there are no *when* statements and discrete variables in the model). The model is supplemented by a list of the uncertain parameters, with their minimum and maximum values; this information can be either supplied separately or directly embedded within the Modelica model by means of the *min/max/nominal* attributes and of custom annotations to flag the uncertain parameters.

4.2.2 Linear fractional representations

LFRs are a widely used model description formalism in modern control and system identification theory [26, 22, 19]. The considered LFR of the plant dynamics, shown in Figure 4.1, corresponds to the equations

$$\dot{x} = Ax + B_1 w + B_2 \zeta + B_3 u \tag{4.1}$$

$$z = C_1 x + D_{11} w + D_{12} \zeta + D_{13} u \tag{4.2}$$

$$\omega = C_2 x + D_{21} w + D_{22} \zeta + D_{23} u \tag{4.3}$$

$$y = C_3 x + D_{31} w + D_{32} \zeta + D_{33} u \tag{4.4}$$

$$w = \mathrm{diag}\left\{\delta_1 I_{r_1}, \ldots, \delta_{n_p} I_{r_{np}}\right\} z = \Delta z \tag{4.5}$$

$$\zeta = \Theta(\omega) \tag{4.6}$$

where $x \in \mathbb{R}^{n_x}$ is the vector of state variables, $u \in \mathbb{R}^{n_u}$ is the vector of input variables, $y \in \mathbb{R}^{n_y}$ is the vector of output variables, δ_j, $j = 1, \ldots, n_p$ are uncertain (possibly measurable) parameters and r_j are the sizes of the corresponding identity matrices in the Δ block ($j = 1, \ldots, n_p$). The structure of the vector function $\Theta(\omega)$ is such that its input vector ω and its output vector ζ can each be partitioned into n_f sub-vectors ω_i and ζ_i, such that $\zeta_i = \Theta_i(\omega_i)$.

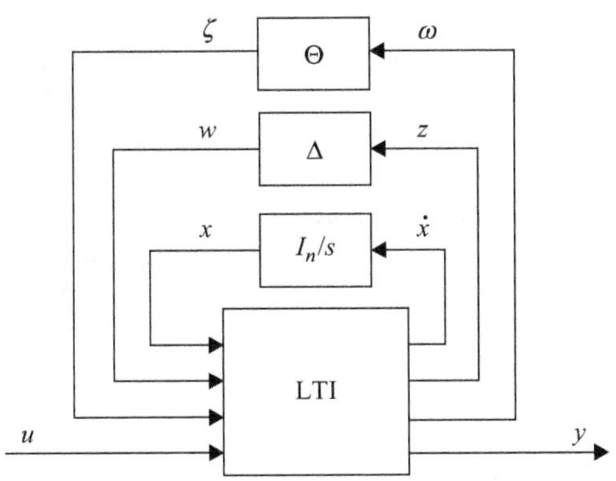

Figure 4.1 Block diagram of the considered LFR

Deriving LFRs from first-principles physical models is a non-trivial, error-prone and tedious process, if carried out manually, but is an essential preliminary step in many problems arising in model identification and control (such as robust stability and performance analysis; see [22]), so there is clearly a strong need of user-friendly tools to perform this task automatically. A first step in this direction is taken by control-oriented tools such as the LFR Toolbox [19]. This tool provides several basic functionalities, which allow to, e.g. define LFRs corresponding to linear models with uncertain parameters and to static non-linear functions; hierarchically build complex models by series, parallel and feedback interconnection, as well as horizontal or vertical vector concatenation; and reduce the size of the Δ block by suitable algorithms, obtaining minimal (or close to minimal) realisations. A fundamental limitation of these tools is that they can only handle causal models, where the interaction between the different sub-models is expressed in terms of input–output interaction. This is not the most natural way to describe the interaction between physical components in a plant. A further step (as first suggested in [4]) is to obtain an LFR starting from a system model written with the Modelica language, which allows to describe the plant dynamics in a much more general, natural and user-friendly way. Preliminary results are presented in [3], limited to the case of linear models. This chapter presents a more general algorithm to deal with generic non-linear models.

4.3 Transformation of non-linear DAE models into LFR

4.3.1 Definitions and assumptions

Definition 4.1. A time-invariant DAE model with parameter-binding equations is defined as

$$F(x, \dot{x}, v, u, p_0) = 0 \tag{4.7}$$

$$p_0 = G(p_0) \tag{4.8}$$

where $F(\cdot)$ is a vector-valued function, $x \in \mathbb{R}^{n_x}$ is the vector of the state variables (i.e. variables which appear together with their time derivative in the model), $v \in \mathbb{R}^{n_v}$ is the vector of the algebraic variables (i.e. variables which appear free of derivatives in the model), $u \in \mathbb{R}^{n_u}$ is the vector of exogenous input variables, $p_0 \in \mathbb{R}^{n_{p0}}$ is the vector of all the system parameters. Finally, the vector of uncertain parameters $p \in \mathbb{R}^{n_p}$ contains a sub-set of the elements of p_0.

Definition 4.2. The set of consistent solutions of (4.7) is the set $\Omega \subset \mathbb{R}^{n_x} \times \mathbb{R}^{n_x} \times \mathbb{R}^{n_v} \times \mathbb{R}^{n_u}$ of 4-tuples $(x, \dot{x}, v, u)$ that satisfy (4.7) for a given choice of p_0.

Assumption 4.1. *The function $F(\cdot)$ in (4.7) is continuosly differentiable in Ω with respect to its first four arguments.*

Assumption 4.2. *Equation (4.8) is such that, by suitable re-ordering, it is possible to solve it for all elements of p_0 explicitly.*

Remark 4.1. Assumption 4.2 corresponds to the way parameter values are defined in the Modelica language, i.e. by explicit parameter-binding equations, the right-hand side of which is an expression possibly involving other parameters:

$$p_i = g_i(p_j, p_h, \ldots), \quad i \neq j, \quad i \neq h, \ldots \tag{4.9}$$

e.g. $p_1 = 10$, $p_2 = 1 + p_2$, $p_3 = p_1/p_2$. The language specification ([23], section 4.4.3) mandates that the parameter-binding equations are acyclic, so they can be solved explicitly by suitable ordering.

Definition 4.3. A semi-explicit, time-invariant DAE model with parameter-binding equations is defined as

$$\dot{x} = f(x, v, u, p_0) \tag{4.10}$$

$$0 = g(x, v, u, p_0) \tag{4.11}$$

$$p_0 = G(p_0) \tag{4.12}$$

Assumption 4.3. *Physical models considered in this chapter, that can be built with Modelica contain dynamic conservation laws (stating that the time derivative of conserved quantities such as mass, momentum, energy, charge, etc., are equal to the flows of such quantities), as well as algebraic constitutive laws and connection equations. The conservation and constitutive laws are time-invariant, while all time-variability is embedded in the input variables u. The conserved quantities Q, which are part of the algebraic variables v, are then expressed as a function of the state variables, $Q = Q(x)$, having a non-singular Jacobian Q_x. The model can thus be formulated as*

$$Q_x(x)\dot{x} = h(x, v, u, p_0) \tag{4.13}$$

$$0 = g(x, v, u, p_0) \tag{4.14}$$

$$p_0 = G(p_0) \tag{4.15}$$

where the DAEs (4.13) and (4.14) are equivalent to semi-explicit DAEs in the form (4.10) and (4.11), since the Jacobian Q_x is non-singular:

$$\dot{x} = Q_x(x)^{-1} h(x, v, u, p_0) = f(x, v, u, p_0) \tag{4.16}$$

$$0 = g(x, v, u, p_0) \tag{4.17}$$

In the remainder of this chapter, the DAEs (4.13) and (4.14) will still be written in the more general form (4.7) for compactness of notation.

Definition 4.4. The semi-explicit, time-invariant DAE system (4.10) and (4.11) is said to have index 1 if only one differentiation of the constraint equations (4.11) is sufficient to explicitly compute the derivatives $\dot{x}, \dot{v}$ as a function of $(x, v, u, \dot{u}, p_0)$, i.e. the Jacobian g_v is non-singular, so it is possible to write

$$\dot{x} = f(x, v, u, p_0) \tag{4.18}$$

$$\dot{v} = -g_v(x, v, u, p_0)^{-1} \left[g_x(x, v, u, p_0)f(x, v, u, p_0) + g_u(x, v, u, p_0)\dot{u}\right] \tag{4.19}$$

(see [2], section 2.2).

Assumption 4.4. *The time-invariant DAE models considered in this chapter have index 1.*

Remark 4.2. Many object-oriented models that are obtained by interconnection of sub-models have index higher than one, i.e. they are characterised by a singular Jacobian g_v and require at least one more differentiation of the constraint equations (4.11) to explicitly solve for all the derivatives. This typically occurs when the connection equations introduce algebraic constraints among two or more state variables. Symbolic index reduction algorithms such as the dummy derivative algorithm [20], which requires symbolic differentiation of the constraint equations (4.17), can usually be applied to such high-index physical models, obtaining an index-1 formulation as the output. These algorithms are of standard use in Modelica tools, so we can assume that the DAE description of the system is equivalent to a semi-explicit index-1 DAE without significant loss of generality.

Definition 4.5. The set of solutions Π of a DAE model is the set of all the triples of continuously differentiable time functions $(x(t), v(t), u(t))$ that satisfy the DAE (4.7) over a time interval $I \subseteq \{t \in \mathbb{R}, t \geq 0\}$.

Assumption 4.5. *The solutions of the index-1 DAE model are unique: if two elements of Π are characterised by the same function $u(t)$ and by the same initial values $x(0), v(0)$, then they are the same element.*

Remark 4.3. Assumptions 4.1, 4.4 and 4.5 hold for well-posed physical models that can be numerically simulated over the time interval I.

4.3.2 Re-ordering of the system equations

At each time instant, the values x, u and p_0 are known, and the goal is to compute $\dot{x}$ and v. To this end, the equations and the variables of the DAE (4.7) are re-ordered so that the incidence matrix (equations on the rows, unknowns on the columns) is brought into block-lower-triangular (BLT) form. This task is accomplished by using the well-known Tarjan algorithm [12], applied to the equations-variables bipartite graph that is equivalent to the incidence matrix of the system. The equations can then be solved one block at a time, possibly considering previously computed values of other variables in the process.

After re-ordering, the DAE (4.7) can be formulated as

$$\Phi(x, u, \Xi, p_0) = 0 \tag{4.20}$$

where $\Phi(\cdot)$ is the set of re-ordered equation residuals and Ξ is the re-ordered set of the system unknowns, containing all the elements of vectors $\dot{x}$ and v. By defining $\Phi_j(\cdot)$ as the sub-set of equations corresponding to the j-th block of the BLT form, Ξ_j as the corresponding sub-set of unknown variables, and q as the number of blocks on the diagonal of the BLT incidence matrix, the re-ordered system (4.20) can be formulated as

$$\Phi_1(x, u, \Xi_1, p_0) = 0 \tag{4.21}$$

$$\Phi_2(x, u, \Xi_1, \Xi_2, p_0) = 0 \tag{4.22}$$

$$\ldots \tag{4.23}$$

$$\Phi_q(x, u, \Xi_1, \ldots, \Xi_{q-1}, \Xi_q, p_0) = 0 \tag{4.24}$$

4.3.3 Elimination of known parameters

The next step of the transformation is to re-formulate the DAE so that it only depends explicitly on the uncertain parameter vector $p \in \mathbb{R}^{n_p}$, the elements of which are a subset of the elements of p_0. This operation is performed on a symbolic representation of the DAE model (4.21)–(4.24), supplemented by the parameter-binding equations (4.8):

1. The parameter-binding equations corresponding to the uncertain parameters p, which give their nominal values, are removed from the system (4.8) and the corresponding variables are kept as symbolic unknowns.
2. The remaining equations in (4.8) are symbolically solved for the remaining parameters in p_0; the results are either known numerical constants or symbolic expressions involving elements of p.
3. The corresponding expressions are symbolically substituted in the system equations (4.21)–(4.24), thus re-defining Φ_j to depend on the uncertain parameters p only:

$$\Phi_1(x, u, \Xi_1, p) = 0 \tag{4.25}$$

$$\Phi_2(x, u, \Xi_1, \Xi_2, p) = 0 \tag{4.26}$$

$$\ldots \tag{4.27}$$

$$\Phi_q(x, u, \Xi_1, \ldots, \Xi_{q-1}, \Xi_q, p) = 0 \tag{4.28}$$

Remark 4.4. Since the parameter-binding equations (4.8) can be solved explicitly by Assumption 4.2, any symbolic manipulation tool can derive the closed-form solutions of step 2 without problems.

4.3.4 Solving the system equations

Proposition 4.1. *Assumptions 4.3 and 4.4 imply that the Jacobian $\partial\Phi/\partial\Xi$ of the re-ordered system of equations is non-singular. Moreover, since the equations are in BLT form, the said Jacobian is block-diagonal, with non-singular blocks on the main diagonal.*

Proposition 4.2. *The property of the DAE model stated in Proposition 4.1 and Assumption 4.1 imply that the functions Φ_j in (4.25)–(4.28) can be locally inverted with respect to Ξ_j in an open neighbourhood of every 4-tuple $(x, \dot{x}, v, u)$ in Ω.*

Proof: Proof immediately follows from the application of the implicit function theorem to each of the equations within (4.25)–(4.28).

Corollary 4.1. *For each consistent solution of (4.25)–(4.28) in Ω, continuously differentiable inverse functions Ξ_j exist in a neighbourhood of (x, u) such that*

$$\Xi_1 = \Psi_1(x, u, p) \tag{4.29}$$

$$\Xi_2 = \Psi_2(x, u, \Xi_1, p) \tag{4.30}$$

$$\Xi_3 = \Psi_3(x, u, \Xi_1, \Xi_2, p) \tag{4.31}$$

$$\dots \tag{4.32}$$

$$\Xi_q = \Psi_q(x, u, \Xi_1, \dots, \Xi_{q-1}, p) = 0 \tag{4.33}$$

Remark 4.5. If the function Φ_j is linear with respect to its argument Ξ_j, since the Jacobian is non-singular a unique solution and thus a unique inverse function Ψ_j exist and can always be obtained in closed form via symbolic manipulation. On the other hand, if Φ_j is non-linear with respect to its argument Ξ_j, then a solution Ψ_j in closed form might or might not be found. In the former situation, one can proceed as for the linear case and a closed form expression for Ψ_j can be obtained. In the latter one, the function $\Phi_j(\Xi_j)$, together with its symbolically computed Jacobian, which is non-singular in view of Assumptions 4.3–4.4, can be used to compute the inverse function Ψ_j numerically, e.g. by using Newton–Raphson's algorithm. Therefore, the function Ψ_j is not known globally in closed form, but can only be computed numerically along a simulation trajectory; note that Assumption 4.5 guarantees that a particular, well-defined inverse can be computed along the entire time interval I, provided that sufficiently short time steps are taken.

4.3.5 Formulation of the system equations as a cascaded connection of LFRs

Collect all the input arguments of each $\Psi_j(\cdot)$ function (except the uncertain parameters p) in a vector X_j for convenience:

$$X_1 = \begin{bmatrix} x \\ u \end{bmatrix}, \quad X_j = \begin{bmatrix} x \\ u \\ \Xi_1 \\ \cdots \\ \Xi_{j-1} \end{bmatrix}, \quad j = 1, \ldots, q \tag{4.34}$$

Equations stemming from object-oriented models are usually characterised by a high sparsity degree, so that each function $\Psi_j(X_j)$ will only depend on a few elements of the vector X_j. In order to obtain a more compact formulation, the functions $\Psi_j(\cdot)$ can be re-defined so that they are a function of the vectors Σ_j, which only contain those elements of X_j that appear explicitly as input arguments of the function

$$\Xi_j = \Psi_j(\Sigma_j, p), \quad j = 1, \ldots, q \tag{4.35}$$

The relationship between Σ_j and X_j is given by suitable selection matrices S_j (i.e. matrices containing just zeros and ones):

$$\Sigma_j = S_j X_j, \quad j = 1, \ldots, q \tag{4.36}$$

It is now possible to cast the DAE (4.29)–(4.33) in the recursive form

$$X_1 = \begin{bmatrix} x \\ u \end{bmatrix}, \quad X_{j+1} = \begin{bmatrix} X_j \\ \Psi_j(S_j X_j, p) \end{bmatrix}, \quad j = 1, \ldots, q-1 \tag{4.37}$$

which can be interpreted as the cascaded connection of sub-systems, as shown by the block diagram in Figure 4.2.

As already noted in Remark 4.5, the functional relationships expressed by the functions $\Psi_j(\Sigma_j, p)$ can be either linear or non-linear with respect to the input argument Σ_j. There are two cases in which such functional relationships can be represented as LFRs, which are summarised in the following propositions.

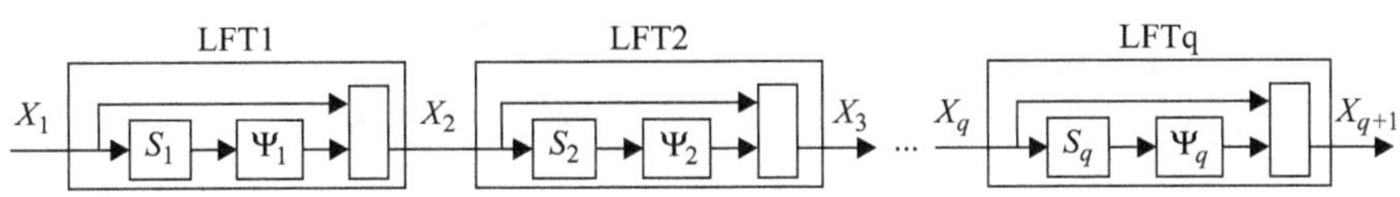

Figure 4.2 Block diagram representation of the recursive LFR definition

Proposition 4.3. *A linear Ψ_j block can be represented as an instance of an LFR such as (4.1)–(4.6) with a suitable Δ block and neither dynamics nor non-linear function feedback blocks, if the coefficients of the Jacobian are rational functions of the uncertain parameters p.*

Proposition 4.4. *A non-linear Ψ_j block can be represented as an instance of the LFR (4.1)–(4.6) with no dynamics and no Δ blocks and with the $\Psi_j(\Sigma_j)$ function as the non-linear function block $\Theta(\cdot)$, and with the trivial equations*

$$\omega = u \tag{4.38}$$

$$y = \zeta \tag{4.39}$$

assuming that it does not involve the uncertain parameters p.

Remark 4.6. If the coefficients of some linear equation block Φ_j are not rational functions of the uncertain parameters, the corresponding linear relationships cannot be represented exactly as an LFR, and therefore the proposed procedure will only be possible in approximate terms by replacing non-rational expressions involving uncertain parameters by means of, e.g. Taylor or Padé expansions around a nominal value.[1]

Remark 4.7. If a a non-linear Φ_j block involving uncertain parameters is encountered, then the above described procedure fails. Though a detailed treatment of this problem is left as future work, a possible approach is sketched here to rewrite the block, separating the equations containing the non-linearities from those containing the uncertain parameters. Each non-linear equation in the block can be written as a composition of n-ary operators (such as $+, -, \times, /, f(\cdot), g(\cdot,\cdot,\cdot)$), so each equation can be represented as a tree, where the leaf nodes represent the literal numerical values, the variables and the parameters appearing in the expression, while the other nodes represent the operators. Therefore, for each equation in the non-linear Φ_j block, the equivalent tree can be constructed and the following algorithm can be executed: for every non-leaf node in the tree associated with every equation a new algebraic variable v_k is added; for every non-leaf node in the graph, a new equation is added, stating that v_k equals the result of the operator corresponding to the node; the original equation is removed from the Φ_j block and a new equation $v_r = 0$ is added, where v_r corresponds to the root node. Finally, the BLT algorithm is re-run on the resulting system of equations. The resulting Φ_j block(s) replaces the original one. Following this re-formulation, every equation involves only one operator, so that it is more likely that it can be represented according to either Proposition 4.3 or Proposition 4.4.

As an example, consider the case of a Φ_j block given by

$$v_2 - p_1 \sin(p_2 v_1) = 0 \tag{4.40}$$

[1] The nominal value could be supplied by the *nominal* attribute of the parameter, as defined in the Modelica language, and the series expansion might be performed by suitable symbolic manipulation.

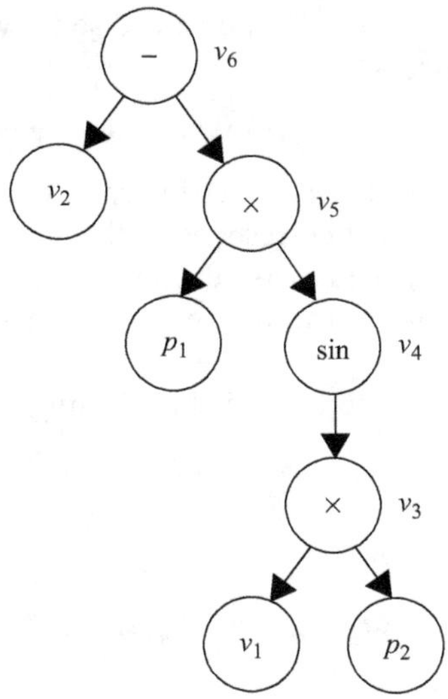

Figure 4.3 Tree representation of (4.40)

which must be solved for v_2, but involves both the non-linear function $\sin(\cdot)$ and two uncertain parameters p_1 and p_2. The expression on the left-hand side of (4.40) is equivalent to the tree shown in Figure 4.3. We then define four new algebraic variables v_3, v_4, v_5 and v_6, corresponding to the non-leaf nodes, add four new equations corresponding to the non-leaf nodes and replace the original equation with $v_6 = 0$. After re-ordering the equations with the BLT algorithm, we get

$$v_3 - p_2 v_1 = 0 \quad \text{(solved for } v_3\text{)} \tag{4.41}$$

$$v_4 - \sin(v_3) = 0 \quad \text{(solved for } v_4\text{)} \tag{4.42}$$

$$v_5 - p_1 v_4 = 0 \quad \text{(solved for } v_5\text{)} \tag{4.43}$$

$$v_6 = 0 \quad \text{(solved for } v_6\text{)} \tag{4.44}$$

$$v_6 - (v_2 - v_5) = 0 \quad \text{(solved for } v_2\text{)} \tag{4.45}$$

Each of these equations will thus form a new Φ_j block that can be handled by means of Propositions 4.3 and 4.4.

4.3.6 Construction of the LFR of the DAE

The LFR from the inputs $X_1 = \begin{bmatrix} x \\ u \end{bmatrix}$ to the outputs $X_q = \begin{bmatrix} x \\ u \\ \Xi \end{bmatrix}$ can then be built by recursive series connection and vertical vector concatenation, as suggested by (4.37) and by Figure 4.2. These operations are supported by tools such as the LFR Toolbox (see [19], section 7 for details). The LFR obtained so far will contain a Δ block and a Θ block, but no dynamics block.

The LFR from the inputs $\tilde{u} = \begin{bmatrix} x \\ u \end{bmatrix}$ to the outputs $\tilde{y} = \begin{bmatrix} \dot{x} \\ y \end{bmatrix}$ is then obtained by selecting and arranging in the proper order the rows of the matrices of the output equation in the previous LFR that correspond to the elements of $\dot{x}$ and y within X_q. Finally, it is possible to apply the order-reduction algorithms discussed in [14] and available in the LFR Toolbox [19] to derive equivalent LFR having a Δ block of minimal dimension.

After these steps have been performed, the resulting LFR will have the structure

$$z = D_{11}w + D_{12}\zeta + D_{13}\tilde{u} \tag{4.46}$$

$$\omega = D_{21}w + D_{22}\zeta + D_{23}\tilde{u} \tag{4.47}$$

$$\tilde{y} = D_{31}w + D_{32}\zeta + D_{33}\tilde{u} \tag{4.48}$$

$$w = \operatorname{diag}\left\{\delta_1 I_{r_1}, \ldots, \delta_{n_p} I_{r_{n_p}}\right\} z \tag{4.49}$$

$$\zeta = \Theta(\omega) \tag{4.50}$$

Remark 4.8. When system (4.20) is brought into BLT form (4.21)–(4.24), all the implicit non-linear algebraic equations in the system are confined within the single Φ_j blocks and then solved either symbolically or numerically to yield the corresponding Ψ_j functions (see Proposition 4.4). The function Θ collects those functions, while the ω and ζ vectors collect the corresponding input and output vectors. As a consequence, D_{22} will be such that there are no algebraic loops going through the Θ block; it is only possible that the output of a certain non-linear Ψ_j function is then used as the input of another one, in an acyclical fashion.

The system (4.46)–(4.50) can now be brought into standard form (4.1)–(4.6) by splitting D_{13} and D_{23} column-wise according to

$$D_{13} = \begin{bmatrix} D_{131} & D_{132} \end{bmatrix}, \quad D_{23} = \begin{bmatrix} D_{231} & D_{232} \end{bmatrix} \tag{4.51}$$

where D_{131} and D_{231} contain, respectively, the first n_x columns (corresponding to x) of D_{13} and D_{23}, and D_{132} and D_{232} are formed with the remaining n_u columns (corresponding to u).

Similarly, the D_{31} and D_{32} matrices are each split as

$$D_{31} = \begin{bmatrix} D_{311} \\ D_{312} \end{bmatrix}, \quad D_{32} = \begin{bmatrix} D_{321} \\ D_{322} \end{bmatrix} \tag{4.52}$$

where D_{311} and D_{321} contain the first n_x rows (corresponding to $\dot{x}$) of D_{31} and D_{32} respectively, while D_{312} and D_{322} are given by the remaining n_y rows (corresponding to y). Finally, matrix D_{33} is split into a 2×2 block matrix, with a partitioning consistent with the previous ones

$$D_{33} = \begin{bmatrix} D_{3311} & D_{3312} \\ D_{3321} & D_{3322} \end{bmatrix} \tag{4.53}$$

The LFR system can thus be written in standard form, including the dynamic feedback block, by re-arranging the matrices according to

$$\dot{x} = D_{3311}x + D_{311}w + D_{321}\zeta + D_{3312}u \tag{4.54}$$

$$z = D_{131}x + D_{11}w + D_{12}\zeta + D_{132}u \tag{4.55}$$

$$\omega = D_{231}x + D_{21}w + D_{22}\zeta + D_{232}u \tag{4.56}$$

$$y = D_{3321}x + D_{312}w + D_{322}\zeta + D_{3322}u \tag{4.57}$$

$$w = \mathrm{diag}\left\{\delta_1 I_{r_1}, \ldots, \delta_{n_p} I_{r_{np}}\right\} z \tag{4.58}$$

$$\zeta = \Theta(\omega) \tag{4.59}$$

4.3.7 Implementation of the algorithm

The transformation of an object-oriented Modelica model into a flattened form is performed by the open-source OpenModelica tool [6], which outputs the model equations (4.21)–(4.24) after flattening, index reduction, and BLT transformation in the form of an XML file, for further processing by other tools.

The XML file is then read from an ad hoc implemented Java routine, where it is possible to perform symbolic computations by means of the Java–Maxima interface (see [8] for further details), and to manipulate LFR objects by exploiting the features of the LFR Toolbox [19]. The LFR Toolbox provides basic functionalities for LFR manipulation, such as building the LFR of a static non-linear function and of linear systems starting from symbolic formulations of their parameter-dependent matrices (via the `lfr()` function). An alternative implementation is possible using MATLAB®/Symbolic Toolbox or Scilab/Maxima and the Scilab version of the LFR toolbox. This option has also been implemented, even if it turned out to be somewhat less efficient than the Java-based version.

The implementation closely follows the procedure outlined in the previous sections and has been integrated within the open-source SimForge environment [21]. In the following the implementation of the algorithm is briefly described, referring neither to the Java sources architecture nor to any coding details; the interested reader can refer to the SimForge documentation (and to the JavaDoc or directly to the comments in the source code) for more details.

Linear LFRs are built by defining symbolic objects for the variables and parameters of the corresponding equations in the $\Phi_j(\cdot)$ block. Initially the binding expressions of parameters, variables, and constants are considered and the corresponding system of equations is solved, thus associating a value to each identifier. The state and algebraic equations are then symbolically solved for the variables within Ξ_j, and the coefficient matrix of the solution, which contains symbolic expressions in the uncertain parameters, is used to build the LFR object.

LFRs containing the non-linear functions Ψ_j are also built using the `lfr()` function; for example, a 2×3 non-linear function corresponding to block Ψ_3 can be defined as `lfr('Psi_3','nlmfr',[2,3])`. Note that the LFR Toolbox per se cannot operate on the non-linear function block, which is just a black box identified by a tag `Psi_3` in this case. The tag is associated with the corresponding block of non-linear equations Φ_j. Depending on the particular application of the LFR (e.g. for control system analysis and design or for system identification), by means of further symbolic processing, it is possible to try to explicitly solve the equations of the block in the form $\Xi_j = \Psi_j(\Sigma_j)$, and/or to linearise them, and so on. Alternatively, the non-linear blocks Φ_j might be scanned by a human reader, who will recognise their content and tag each block with the relevant properties (e.g. min/max slope and min/max bounds).

The core of the implementation is the recursive construction of the LFR, corresponding to (4.37). The input–output relationship between X_1 and X_q is easily built step by step by using the above-mentioned functionalities of the LFR Toolbox. For numerical reasons, the LFR is also normalised, so that a value of -1 in the Δ block corresponds to the lower bound of the parameter, and a value of $+1$ corresponds to the upper bound. The corresponding MATLAB code reads:

```
% The first LFR is just a linear relation such as:
% inputs of the system (state variables and inputs) == outputs.
LFR = lfr(eye(n+m_u));

% Construction of the LFR X_1->X_j+1
for j = 1:max(size(Psi))
  LFR = [lfr(eye(size(S{j},2))); normalizelfr(Psi{j}*S{j})]*LFR;
end
```

4.3.8 *Simulation of the LFR*

In order to assess the correctness of the obtained results, a 'simulation' module has also been implemented within the MATLAB environment to simulate the automatically obtained LFR model. In the following a short overview of this module is provided (see, again, [8] for further details).

The LFR model obtained with the previous method is processed associating a specific value to each element of the Δ block and then used to calculate the state derivative values during the simulation. To that purpose, a specific MATLAB function has been developed, which computes the values of the state derivative using the A, B, C and D matrices of the LFR and, if the LFR has non-linear blocks, some auxiliary functions (in the number of the non-linear blocks). These auxiliary functions are generally written as explicit functions, thanks to the Java–Maxima solving functionality. If the non-linear system cannot be solved, thus is implicit, the MATLAB `fsolve()` function is used, also specifying as input the values of the variables to solve for calculated at the previous step.

To complete the 'simulation' module, a main file is also generated, where the initial values of the states as well as the start and stop simulation times are specified. The `ode45()` function is finally used (within the main file) to obtain the simulation results.

4.4 Application example: identification of LFR models

The problem of model identification formulated over LFR model structures has been a subject of active research for more than 10 years, see for example [17, 25, 7, 15]. More precisely, while in this chapter we consider only the problem of dealing with uncertain parameters, more general formulations of the problem taking also into account, e.g. static non-linearities have been proposed. In the following, we will focus on the case of linear LFRs, for which the the considered identification problem can be formulated as follows. Consider the LFR

$$\dot{x} = Ax + B_1 w + B_3 u \tag{4.60}$$

$$z = C_1 x + D_{11} w + D_{13} u \tag{4.61}$$

$$y = C_3 x + D_{31} w + D_{33} u \tag{4.62}$$

$$w = \text{diag}\left\{\delta_1 I_{r_1}, \ldots, \delta_{n_p} I_{r_{n_p}}\right\} z = \Delta z \tag{4.63}$$

and assume that a data set $\{u(t_k), y(t_k)\}$, $k \in [1, N]$ of sampled input–output data obtained from system (4.60)–(4.63) is available. Then, the problem is to provide a consistent estimate of the unknown parameters $\delta_j, j = 1, \ldots, n_p$. Assume further that only white Gaussian measurement noise affects the system; then it is well known that the identification problem can be formulated as an output error maximum likelihood estimation one, in terms of the cost function

$$J(\delta) = \frac{1}{2}\langle e(\delta), e(\delta)\rangle \tag{4.64}$$

where e is the simulation error associated with the LFR model (see, e.g. [16] for a recent, detailed account of theoretical and computational issues in maximum likelihood identification). In view of the fact that the model class is non-linearly parameterised, the minimisation of the cost function in (4.64) can be faced via gradient-based methods, using suitable expressions for the gradient $g : \mathbb{R}^{n_p} \to \mathbb{R}^{n_p}$ and for a positive semi-definite approximation $\hat{H} : \mathbb{R}^{n_p} \to \mathbb{R}^{n_p \times n_p}$ of the Hessian of the cost function, which can be derived thanks to the LFR structure of the model. In particular, note that the problem can be formulated both in continuous time, as outlined above, or in discrete time, as in [25, 15], on the basis of a suitable discretisation of the system (obtained by, e.g. one of the methods described in [24]).

Identifiability is an important issue in system identification problems; a number of results are available in the literature, in particular, as far as non-linearly parameterised models are concerned, as is the case here (see, e.g. [11] and the more recent [10]).

Consider $\hat{\delta}$, the vector of the parameters minimising (4.64). Local identifiability in $\hat{\delta}$ is verified testing if the Hessian $\hat{H}$ is positive definite in $\hat{\delta}$, i.e. if the rank of $\frac{\partial^2 J(\delta)}{\partial \delta^2}$ at $\hat{\delta}$ is equal to q. If the Hessian $\hat{H}$ in $\hat{\delta}$ is not full rank then the estimated parameters are not unique and the system is not identifiable. The matrix $\hat{H}$ computed at $\hat{\delta}$ is strongly related to the variance of $\hat{\delta}$, since for the Gaussian case we have

$$\mathrm{var}(\hat{\delta}) = V^{-1}$$

where

$$V = E\left[\left. \frac{\partial^2 J(\delta)}{\partial \delta^2} \right|_{\hat{\delta}} \right]$$

Then, to select the identifiable parameter space we consider the singular value decomposition (SVD) of the Hessian

$$\hat{H} = [U_1 \; U_2] \begin{bmatrix} \Sigma_1 & 0 \\ 0 & \Sigma_2 \end{bmatrix} \begin{bmatrix} V_1^T \\ V_2^T \end{bmatrix}$$

where the separation between the singular value matrix Σ_1 and Σ_2 is chosen in such a way that the singular values in Σ_2 are considerably smaller than those in Σ_1. Then the column space of U_1 represents the sub-space of the parameter space that is identifiable.

Example 4.1. A mechanical system

The first example is a simple mechanical system with two rotational degrees of freedom, represented in Figure 4.4 using components from the standard Modelica library. The system is built by the connection of an ideal torque generator, a first rotating body with coefficient of inertia $J = p_1$, a non-linear spring, a second rotating body with coefficient of inertia $J = p_2$, and an angle sensor. The non-linear spring is described by the equations

$$\tau = \tau_l(1 + \alpha) \tag{4.65}$$

$$\tau_l = -K\varphi \tag{4.66}$$

$$\alpha = \frac{\varphi^2}{\varphi_0^2} \tag{4.67}$$

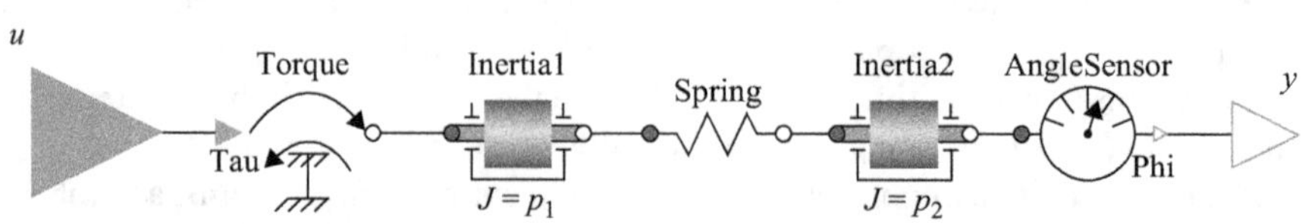

Figure 4.4 Object diagram of Example 4.1

where τ is the torque applied by the end flanges, τ_l is the linear torque component, φ is the relative angle between the two flanges, K is the elastic coefficient, and φ_0 is the characteristic coefficient of the non-linear effect. The model has four state variables, one input, one output, and two uncertain parameters p_1 and p_2 with bounds $10 < p_1 < 15$ and $2 < p_2 < 4$, while $K = 100$ and $\varphi_0^2 = 0.5$.

After applying the algorithm, the model is brought in standard form (4.54)–(4.59), with a 2×2 normalised block $\Delta = \mathrm{diag}(p_1, p_2)$ and with a 2×4 block Θ, which is partitioned into two scalar non-linear functions, both having a two-element vector as input:

$$\zeta_1 = (\omega_1 - \omega_2)^2/0.5 \tag{4.68}$$

$$\zeta_2 = \omega_3(1 + \omega_4) \tag{4.69}$$

The structure of the LFR object as reported by the LFR toolbox is shown in Table 4.1, while the state-space matrices of the LFR are given in Table 4.2. The correctness of the obtained state-space model has been checked both analytically and by numerical validation (details omitted for brevity).

Table 4.1 Structure of the LFR object of Example 4.1

```
LFR-object with 1 output(s), 1 input(s) and 4 state(s).
Uncertainty blocks (globally (4 x 6)):
```

Name	Dims	Type	Real/Cplx	Full/Scal	Bounds
Psi1	1x2	NLM	r	f	[-1,1]
Psi2	1x2	NLM	r	f	[-1,1]
p1	1x1	LTI	r	s	[-1,1]
p2	1x1	LTI	r	s	[-1,1]

Table 4.2 State-space matrices of the LFR of Example 4.1

$$D_{3311} = \begin{bmatrix} 0 & 1 & 0 & 0 \\ 0 & 0 & 0 & 0 \\ 0 & 0 & 0 & 1 \\ 0 & 0 & 0 & 0 \end{bmatrix}, \quad D_{311} = \begin{bmatrix} 0 & 0 \\ 0.11 & 0 \\ 0 & 0 \\ 0 & 0.33 \end{bmatrix}, \quad D_{321} = \begin{bmatrix} 0 & 0 \\ 0 & 0.08 \\ 0 & 0 \\ 0 & -0.33 \end{bmatrix}, \quad D_{3312} = \begin{bmatrix} 0 \\ 0.08 \\ 0 \\ 0 \end{bmatrix}$$

$$D_{131} = 0_{2\times4}, \quad D_{11} = \begin{bmatrix} -0.2 & 0 \\ 0 & -0.33 \end{bmatrix}, \quad D_{12} = \begin{bmatrix} 0 & -0.14 \\ 0 & 0.33 \end{bmatrix}, \quad D_{132} = \begin{bmatrix} -0.14 \\ 0 \end{bmatrix}$$

$$D_{231} = \begin{bmatrix} 1 & 0 & 0 & 0 \\ 0 & 0 & 1 & 0 \\ -100 & 0 & 100 & 0 \\ 0 & 0 & 0 & 0 \end{bmatrix}, \quad D_{21} = 0_{4\times2}, \quad D_{22} = \begin{bmatrix} 0 & 0 \\ 0 & 0 \\ 0 & 0 \\ 1 & 0 \end{bmatrix}, \quad D_{232} = 0_{4\times2}$$

$$D_{3321} = \begin{bmatrix} 0 & 0 & 1 & 0 \end{bmatrix}, \quad D_{312} = 0_{1\times2}, \quad D_{322} = 0_{1\times2}, \quad D_{3322} = 0$$

Example 4.2. Modelling and identification of a thermal plant

The considered plant is loosely based on the experimental facility described in [5] and consists of a cylindrical metal tube, filled with a dense vapour. The tube is covered by an insulating blanket, whose inner layer can generate heat by Joule effect when a current passes through it. The gas temperature is measured by a high-precision sensor.

The system can be modelled by a connection of lumped thermal resistors and thermal capacitors, as shown in Figure 4.5. Starting from the left-hand side of the diagram in the figure, the thermal blanket is modelled as a thermal resistor connected to a fixed temperature source, representing the insulation layer in contact with the external air, and a prescribed heat flow source, which represents the heat generation by the Joule effect. The tube wall is modelled by a finite volume approximation with two layers: the lumped thermal capacitors describe the heat storage of a wall layer, and the lumped resistors the heat conduction between adjacent layers and to the inner and outer surfaces. A thin wall is assumed (thickness much smaller than the radius) for simplicity. The next components describe the convective heat transfer between the wall and the gas, the gas heat capacitance, the heat transfer between the vapour and the sensor, and the sensor heat capacitance. The system input u is the control signal driving the power actuator, which is converted to the prescribed thermal power flow by a suitable gain. The system output y is the sensor temperature.

The control problem in this case is to keep the gas at slightly superheated conditions (around 1 K above the saturation temperature) with a high precision, in order not to destroy the exotic gasdynamic effects under study. The design of the controller is a non-trivial problem as the sensor output alone does not provide sufficiently accurate information on the gas temperature because of its low-pass nature. Therefore, in order to estimate the gas temperature it is important to have an accurate physical model of the system. Some of the parameters of the model can be derived from first principles and geometrical data: the heater gain, the thermal capacitance and conductance of the wall, and the thermal capacitance of the gas. Conversely, the convective heat transfer coefficients and the sensor thermal capacitance cannot be determined precisely, so a direct determination of those coefficient on the specific lab equipment through a grey-box identification procedure is considered.

The system model has been built by connecting components from the standard Modelica library (see again the object-oriented diagram in Figure 4.5). Note that

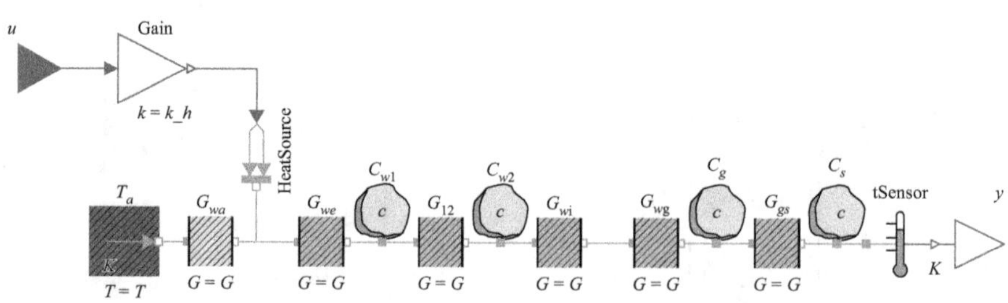

Figure 4.5 Example 4.2: object-oriented diagram of the model

this diagram includes a causal signal processing part (the gain of the power actuator) with signal connectors, as well as acausal physical connections between the various thermal elements. The semantics of such connections is that the temperature is the same and that the sum of the heat flows is equal to zero.

The flattened model corresponds to a DAE system defined by the differential equations

$$\begin{aligned}
\frac{C_w}{2}\frac{dT_{w1}}{dt} &= Q_{hw} - Q_{12} \\
\frac{C_w}{2}\frac{dT_{w2}}{dt} &= Q_{12} - Q_{wg} \\
C_g\frac{dT_g}{dt} &= Q_{wg} - Q_{vs} \\
C_s\frac{dT_s}{dt} &= Q_{vs}
\end{aligned}$$

and by the algebraic ones

$$\begin{aligned}
Q_h &= k_h u \\
Q_h &= Q_{hw} + Q_{ha} \\
Q_{ha} &= G_{wa}(T_{we} - T_a) \\
Q_{hw} &= 4G_w(T_{we} - T_{w1}) \\
Q_{12} &= 2G_w(T_{w1} - T_{w2}) \\
Q_{wg} &= 4G_w(T_{w2} - T_{wi}) \\
Q_{wg} &= G_{wg}(T_{wi} - T_g) \\
Q_{gs} &= G_{gs}(T_g - T_s)
\end{aligned}$$

where Q is the heat flow, G the thermal conductance, T the temperature, C the heat capacitances of the different components, and k_h the gain of the power actuator. Finally, the following binding equations hold among the model parameters

$$\begin{aligned}
G_{wa} &= 2\pi rlh_{wa}, \quad G_w = \frac{\lambda 2\pi rl}{s} \\
G_{wg} &= 2\pi rlh_{wg}, \quad G_{gs} = 2\pi r_s l_s h_{gs} \\
C_w &= \rho_m c_m 2\pi rls, \quad C_g = \rho_g c_g \pi r2l \\
C_s &= \rho_s c_s \pi r_s 2l_s k_{cs}
\end{aligned}$$

where r, l, s, ρ_m, c_m, λ are, respectively, the tube radius, length, thickness, density, specific heat capacity and thermal conductivity; r_s, l_s, ρ_s, c_s are the sensor radius, length, density, and specific heat capacity; k_{cs} is a fudge factor for the sensor heat

Table 4.3 Structure of the LFR object of Example 4.2

```
LFR-object with 5 output(s), 5 input(s) and 0 state(s).
Uncertainty blocks (globally (5 x 5)):
```

Name	Dims	Type	Real/Cplx	Full/Scal	Bounds
h_gs	2x2	LTI	r	s	[-1,1]
h_wa	1x1	LTI	r	s	[-1,1]
h_wg	1x1	LTI	r	s	[-1,1]
k_cs	1x1	LTI	r	s	[-1,1]

capacity; and h_{wa}, h_{wg}, and h_{gs} are the heat transfer coefficients between wall and external air, wall and gas, gas and sensor respectively. The last four mentioned parameters are uncertain, while all others are assumed to be perfectly known.

The sought after LFR form for the model is summarised in the LFR Toolbox output as in Table 4.3 and has been used to formulate the parameter estimation problem for the system. It is interesting to point out that the parameter estimation problem which will be considered in the following, is particularly challenging due to the relatively low sensitivity of the system's dynamics to variations in the parameters of interest. As an example, in Figure 4.6 the Bode plots of the frequency response for the system obtained by varying the parameter k_{cs} over its admissible range are shown; it is apparent from the figure that the system is affected by the variation of the parameter only in a frequency range where the amplitude if its frequency response is very low, so it can be expected that proper tuning of the parameter on the basis of measured data will be a difficult problem. Finally, the discrete-time LFR model, which serves as input for the parameter estimation step, is obtained by discretising the above continuous-time LFR with a sampling period $T = 5$ s, according to the discussion in section 4.4.

An input–output data set has been generated by simulating the Modelica model (with $h_{gs} = 5$, $h_{wa} = 1$, $h_{wg} = 5$, and $k_{cs} = 1.3$) under an excitation of the input variable of the system given by a Schroeder-phased multisine and measuring the corresponding output time history. A similar data set, characterised by a different input spectrum, has been used to validate the identified model.

The uncertain model parameters have been initialised to the starting values $h_{gs} = 4$, $h_{wa} = 1.5$, $h_{wg} = 6$ and $k_{cs} = 0.3$ and the discretised LFR has been optimised by minimising the cost function (4.64) on the basis of the above-described input–output data, using a Gauss–Newton method. The performance of the identified model has been assessed in simulation, using the validation data set. The results of the validation experiment are depicted in Figure 4.7, where the simulation error for the optimised LFR model (bottom of the figure) is compared with the corresponding error for the initial model (top of the figure). As can be seen, the performance of the model has been significantly improved by the optimisation. Finally, the optimal values of the parameters, given by $h^o_{gs} = 6.4579$, $h^o_{wa} = 1.0028$, $h^o_{wg} = 5.0176$ and

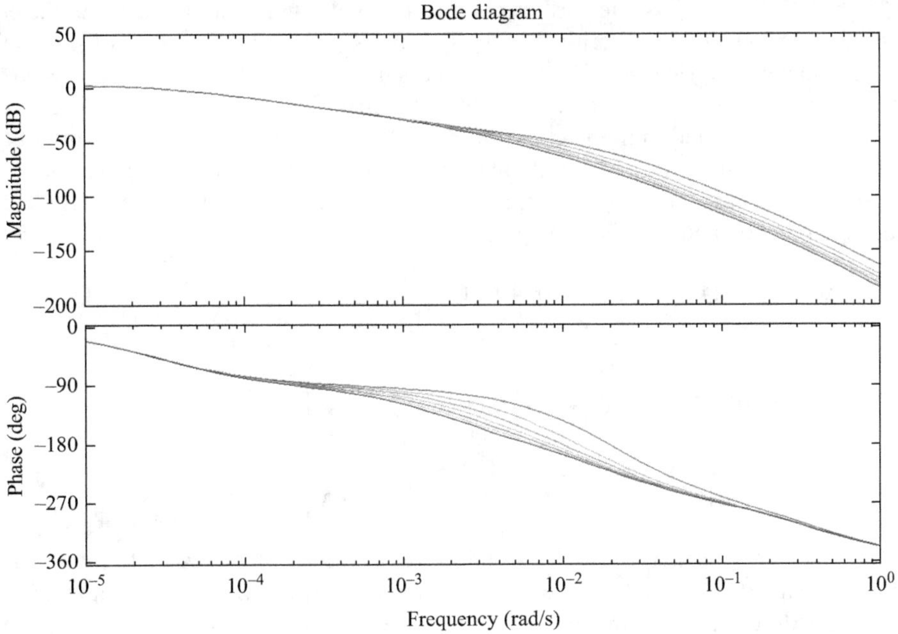

Figure 4.6 Bode plot of the model's frequency response, for k_{cs} ranging between 0.3 and 3

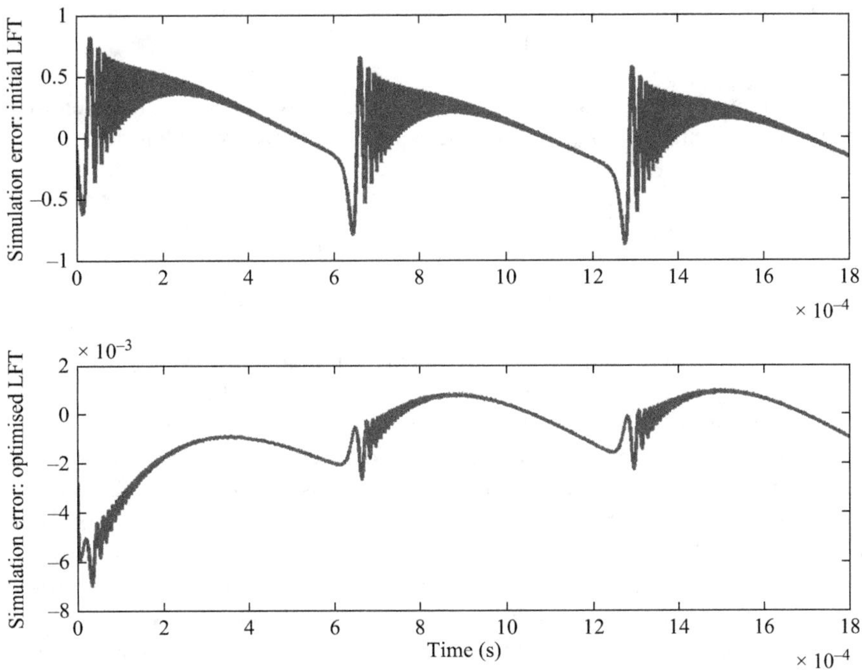

Figure 4.7 Simulation error on validation data: initial (top) and optimised (bottom) LFR models

$k^o_{cs} = 1.6831$, deserve some further comments: as discussed above, the parameter estimation problem is very critical as the system's sensitivity to the value of some of the parameters is very low. This situation occurs frequently in the practice of grey-box modelling, and it is interesting to point out that such difficulties can be easily detected, analysed and handled in the framework of LFR identification, while they would be much harder to deal with when attempting to optimise the values of the parameters directly at the simulator level. In order to demonstrate this, computing the Hessian of the cost function and its SVD, one gets

$$\Sigma = [4.2224 \quad 3.0332 \quad 0.0249 \quad 0]$$

and

$$U = \begin{bmatrix} -0.4590 & 0.3697 & 0.0906 & 0.8028 \\ -0.6276 & -0.7785 & -0.0024 & 0 \\ -0.1168 & 0.0972 & -0.9884 & 0 \\ 0.6179 & -0.4977 & -0.1220 & 0.5963 \end{bmatrix}$$

We consider the matrix U_1 with the first three columns of U associated with the non-zero singular values. In particular the columns of U_1 determine the linear combinations of the original parameters δ_i that will be identifiable from the measurements. To evidence the identifiable parameters we can reparameterise the model structure introducing a parameter ρ defined as $\delta = U_1\rho$, leading to an identifiable model structure in the new parameter. Then, computing the residual

$$v = U_1\rho - \hat{\delta} = [0.2852 \quad 0 \quad 0 \quad 0.2119]^T$$

from the value of v we can state that the parameters associated with the identifiable part are $\delta_2 = h_{wa}$ and $\delta_3 = h_{wg}$; in fact these parameters are well identified.

4.5 Conclusions

The problem of obtaining reduced-order LFR models starting from generic equation-based, object-oriented descriptions of non-linear plant dynamics and experimental data is considered. In particular, an approach to the problem of deriving an equivalent LFR form for a given mathematical representation of a Modelica model is proposed and its capability and limitations are analysed. Integrated modelling and identification based on the combination of the Modelica object-oriented modelling language with LFR identification techniques is presented and discussed. Finally, simulation examples are used to illustrate both the capability for integrated modelling and parameter estimation offered by the proposed approach. Future work will aim at further exploring the algorithm outlined in Remark 4.7 to deal with non-linear, parameter-dependent blocks in the construction of the LFR and investigating the potential of this approach in the framework of robust stability and performance analysis for feedback control systems.

References

[1] Bohlin, T.: *Practical Grey-box Process Identification: Theory and Applications*. Springer-Verlag, London, UK (2006)

[2] Brenan, K.E., Campbell, S.L., Petzold, L.R.: *Numerical Solution of Initial-Value Problems in Differential-Algebraic Equations*. SIAM, Philadelphia, USA (1996)

[3] Casella, F., Donida, F., Lovera, M.: Automatic generation of LFTs from object-oriented Modelica models. In: *Proceedings of the 2nd IEEE Multi-conference on Systems and Control* (Late News Paper). San Antonio, USA (2008)

[4] Casella, F., Donida, F., Lovera, M.: Beyond simulation: Computer-aided control systems design using equation-based object-oriented modelling for the next decade. In: *Proceedings of the 2nd International Workshop on Equation-Based Object-Oriented Languages and Tools*. Paphos, Cyprus (2008). URL: http://www.ep.liu.se/ecp/029/005/

[5] Colonna, P., Guardone, A., Nannan, N., Zamfirescu, C.: Design of the dense gas flexible asymmetric shock tube. *Journal of Fluids Engineering* **130**(3), 034,501 (6 pages) (2008)

[6] Consortium, T.O.: OpenModelica: the open source Modelica compiler. URL: http://openmodelica.org

[7] Demourant, F., Ferreres, G.: Closed loop identification of a LFT model. *Journal Européen des Systémes Automatisés* **36**(3), 449–464 (2002)

[8] Donida, F.: Object-oriented modelling, simulation and control of engineering systems: an integrated approach. Ph.D. thesis, Politecnico di Milano (2010)

[9] Donida, F., Romani, C., Casella, F., Lovera, M.: Integrated modelling and parameter estimation: an LFT–Modelica approach. In: *48th IEEE Conference on Decision and Control*. Shanghai, China (2009)

[10] Doren, J.V., den Hof, P.V., Hansen, J.D., Bosgra, O.: Determining identifiable parameterisations for large-scale physical models in reservoir engineering. In: *Proceedings of the 2008 IFAC World Congress*. Seoul, South Korea (2008)

[11] Dotsch, H., Van den Hof, P.: Test for local structural identifiability of high-order non-linearly parameterised state space models. *Automatica* **32**(6), 875–883 (1996)

[12] Duff, I.S., Reid, J.K.: An implementation of Tarjan's algorithm for the block triangularization of a matrix. *ACM Transactions on Mathematical Software* **4**(2), 137–147 (1978)

[13] Gerdin, M., Schon, T., Glad, T., Gustafsson, F., Ljung, L.: On parameter and state estimation for linear differential-algebraic equations. *Automatica* **43**(3), 416–425 (2007)

[14] Hecker, S., Varga, A.: Symbolic manipulation techniques for low order LFT-based parametric uncertainty modelling. *International Journal of Control* **79**(11), 1485–1494 (2006)

[15] Hsu, K., Vincent, T., Wolodkin, G., Rangan, S., Poolla, K.: An LFT approach to parameter estimation. *Automatica* **44**(12), 3087–3092 (2008)

[16] Klein, V., Morelli, E.A.: *Aircraft System Identification: Theory and Practice*. AIAA, Reston, Virginia, USA (2006)

[17] Lee, L., Poolla, K.: Identification of linear parameter-varying systems using nonlinear programming. *Journal of Dynamic Systems, Measurement and Control – Transactions of the ASME* **121**(1), 71–78 (1999)

[18] Ljung, L.: Perspectives on system identification. In: *2008 IFAC World Congress*. Seoul, South Korea (2008)

[19] Magni, J.F.: *User Manual of the Linear Fractional Representation Toolbox, Version 2.0*. Tech. Rep. TR 5/10403.01F DCSD, ONERA (2006). URL: http://www.cert.fr/dcsd/idco/perso/Magni/download/lfrt_manual_v20.pdf

[20] Mattsson, S.E., Söderlind, G.: Index reduction in differential-algebraic equations using dummy derivatives. *SIAM Journal on Scientific Computing* **14**(3), 677–692 (1993)

[21] Politecnico di Milano, S.d.C.: SimForge: a graphical Modelica environment. URL: https://trac.wd.dei.polimi.it/simforge. https://trac.wd.dei.polimi.it/simforge

[22] Skogestad, S., Postlethwaite, I.: *Multivariable Feedback Control: Analysis and Design*, 2nd edn. Wiley-Interscience, Chichester, UK (2005)

[23] The Modelica Association: Modelica – a unified object-oriented language for physical systems modeling – Language specification version 3.2. Online (2010). URL: https://www.modelica.org/news_items/documents/ModelicaSpec32.pdf. URL: https://www.modelica.org/news_items/documents/ModelicaSpec32.pdf

[24] Toth, R., Lovera, M., Heuberger, P., den Hof, P.V.: Discretisation of linear fractional representations of LPV systems. In: *48th IEEE Conference on Decision and Control*. Shanghai, China (2009)

[25] Wolodkin, G., Rangan, S., Poolla, K.: An LFT approach to parameter estimation. In: *Proceedings of the 1997 American Control Conference*. Albuquerque, USA (1997)

[26] Zhou, K., Doyle, J.U., Glover, K.: *Robust and Optimal Control*. Prentice-Hall, New Jersey (1996)

Chapter 5
Identification for robust control of complex systems: algorithm and motion application

Tom Oomen[1] *and Maarten Steinbuch*[2]

Abstract

Increasing performance demands in control applications necessitate accurate modeling of complex systems for control. The aim of this chapter is to develop a new system identification algorithm that delivers models that are suitable for subsequent robust control design and can be reliably applied to complex systems. To achieve this, an identification algorithm is developed that delivers a system model in terms of recently developed coprime factorizations and thereby extends classical iterative procedures to the closed-loop case. These coprime factorizations have important advantages for uncertainty modeling and robust controller synthesis of complex systems. A numerically optimal implementation is presented that relies on orthonormal polynomials with respect to a data-dependent discrete inner product. Experimental results on a nanometer-accurate positioning system confirm that the algorithm is capable of delivering the required coprime factorizations and the implementation is numerically reliable, which is essential for complex systems as common implementations suffer from severe ill-conditioning.

5.1 Introduction

Increasing performance demands in control applications necessitate accurate models, and hence the development of system identification algorithms for complex systems. Important examples are nanometer-accurate positioning systems, which are for instance used in the production of integrated circuits (ICs). Here, increased performance requirements necessitate taking into account high frequency flexible dynamical behavior [19]. Furthermore, for such motion systems, six inputs and six outputs are traditionally used for control in six degrees-of-freedom. However, more

[1]Eindhoven University of Technology, Faculty of Mechanical Engineering, Control Systems Technology Group, Eindhoven, The Netherlands, e-mail: t.a.e.oomen@tue.nl
[2]Eindhoven University of Technology, Faculty of Mechanical Engineering, Control Systems Technology Group, Eindhoven, The Netherlands, e-mail: m.steinbuch@tue.nl

actuators and sensors will become necessary in the near future to actively compensate for flexible dynamical behavior, as is argued in [28]. Hence, models for controlling next-generation positioning systems are envisaged to be of high order and to have large input and output dimensionality.

Several approaches have been developed to connect system identification and robust controller design, see [23], [17], [32], [13]. Especially, frequency domain approaches, see [31] for a general overview and [7], [26] for robust control related approaches, have led to successful results in traditional motion control, see [41], [8], [36], [28].

Although identification and robust control have been able to enhance robust performance in traditional motion systems, existing techniques are inadequate for complex systems. These complex systems include motion systems with high-order dynamics and large input and output dimensionality. At least three interrelated aspects are highly relevant for identifying complex systems for control:

1. numerical aspects for identifying a nominal model,
2. algorithmic efficiency for identifying a nominal model, and
3. algorithmic aspects related to the model uncertainty structure in robust control design.

Related to aspect 3, model uncertainty structures for robust control have been recently refined to enable a nonconservative and efficient synthesis. First, traditional additive and multiplicative structures have been extended toward coprime-factor based uncertainty structures in, e.g., [21]. These coprime-factor based structures have been further extended toward dual-Youla uncertainty structures, see, e.g., [20], [2], [7], [10]. Such dual-Youla structures are particularly suitable for uncertainty modeling in closed-loop situations. However, a common aspect in these uncertainty structures is that the multivariable case typically involves a highly structured and frequency weighted uncertainty model, as is confirmed in [41] for the additive case and in [8] for the dual-Youla case. Recently, a new coprime factorization has been proposed in [26] that extends dual-Youla type of uncertainty structures toward the nonconservative use of unstructured uncertainty models. Thus, it automatically provides a suitable multivariable frequency weighting of the uncertainty block. The key advantage is a subsequent nonconservative robust controller synthesis, see, e.g., [11], since the uncertainty structure is μ-simple [29] for any input–output system dimension. However, the new coprime factorization in [26] cannot be directly identified using existing techniques, including [18]. In the present chapter, it is aimed to develop a frequency domain algorithm for identifying the coprime factorization in [26].

In view of aspect 1 mentioned above, frequency domain system identification is numerically often ill-conditioned, as is evidenced by the development of approaches to enhance numerical conditioning, including [30], [14], [43]. Since the numerical conditioning typically deteriorates for increasing model complexity, the development of numerically reliable algorithms is particularly important for complex systems. A similar reason holds for aspect 2, i.e., the development of efficient algorithms, since the computational time often grows significantly for increasing model complexity.

The main contribution of this chapter is a new algorithm for identifying complex systems for robust control design. The algorithm generalizes well-established results for the open-loop case, including [34] to the closed-loop case, and

1. is numerically optimal by exploiting results in orthonormal basis functions with respect to a data-based discrete inner product, see [3] for details;
2. can be efficiently implemented, both in the required number of iterations and the computational load per iteration that can be rendered $\mathcal{O}(N)$, where N is the data length;
3. is suitable for efficient robust controller synthesis due to the internal coprime-factor based model structure, enabling a fast and nonconservative robust controller synthesis, exploiting the results in [26].

Although the results presented in this chapter aim at closed-loop identification of a certain coprime factorization, the open-loop case is directly recovered as a special case. As a result, the proposed algorithm also enables the numerically reliable frequency domain identification of open-loop systems. In addition, so-called control-relevant weighting functions are used in this chapter. This can directly be replaced by any user-chosen weighting function. Similarly, the approach is presented for discrete time identification. The presented approach can be directly adapted toward continuous time model identification. Finally, the approach is specifically aimed toward handling complex systems. Of course, the approach can also be applied to relatively low-order systems with smaller input–output dimensionality.

This chapter is organized as follows. In Section 5.2, the role of coprime factorizations for modeling uncertainty in complex systems is established, revealing the advantages of recently developed coprime factor-based uncertainty structures. In addition, the considered robust control framework is outlined in Section 5.2. Then, in Section 5.3, new algorithms for identifying coprime factorizations are presented. The presented algorithms generalize commonly used SK-iterations to the closed-loop case. Next, a numerically optimal implementation is presented in Section 5.4, which is based on orthonormal polynomials that are orthonormal with respect to a certain data-dependent inner product. In Section 5.5, the proposed approach is applied to an industrial wafer stage system. Finally, conclusions are provided in Section 5.6.

Notation. The pair $\{N, D\}$ is an RCF of P implies that [40] $P = ND^{-1}$; $N, D \in \mathcal{RH}_\infty$; and $\exists X, Y \in \mathcal{RH}_\infty$ such that $XN + YD = I$. Throughout, P is assumed to have n_y outputs and n_u inputs. To facilitate the notation, the dimensions of matrices are often omitted if no ambiguity can arise.

5.2 Coprime factor identification for refined uncertainty structures in robust control

5.2.1 Robust control framework

The general $\mathcal{H}_\infty$ norm-based criterion

$$\mathcal{J}(P, C) := \|WT(P, C)V\|_\infty \tag{5.1}$$

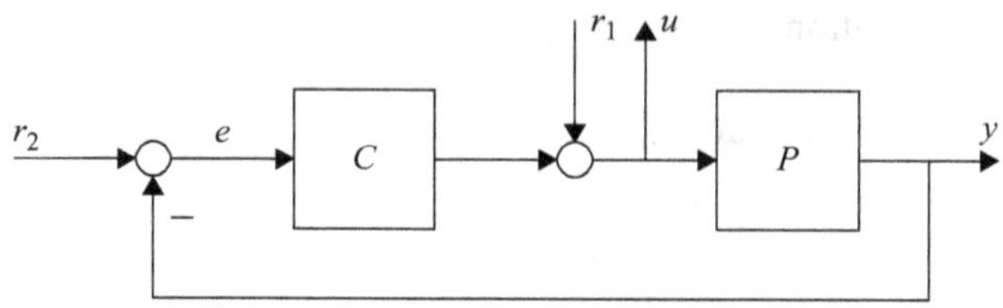

Figure 5.1 Feedback interconnection

is considered, where

$$T(P,C) = \begin{bmatrix} P \\ I \end{bmatrix} (I + CP)^{-1} \begin{bmatrix} C & I \end{bmatrix} \tag{5.2}$$

where $T(P,C) : \begin{bmatrix} r_2 \\ r_1 \end{bmatrix} \mapsto \begin{bmatrix} y \\ u \end{bmatrix}$, see Figure 5.1. The criterion (5.1) in conjunction with the four-block encompasses many relevant $\mathscr{H}_\infty$-design problems, including the loop-shaping approach in [21] and typical mixed-sensitivity problems. Throughout, it is assumed that $W = \text{diag}(W_y, W_u)$, $V = \text{diag}(V_2, V_1)$, and $W, V, W^{-1}, V^{-1} \in \mathscr{RH}_\infty$.

The criterion (5.1) is formulated such that it is to be minimized for P_o, i.e.,

$$C^{\text{o}} = \arg\min_C \mathscr{J}(P_o, C) \tag{5.3}$$

The key idea in robust control is to represent the unknown system P_o by a model set $\mathscr{P}$ such that

$$P_o \in \mathscr{P} \tag{5.4}$$

Consequently, the robust performance controller design

$$C^{\text{RP}} = \arg\min_C \mathscr{J}_{\text{WC}}(\mathscr{P}, C) \tag{5.5}$$

where

$$\mathscr{J}_{\text{WC}}(\mathscr{P}, C) := \sup_{P \in \mathscr{P}} \mathscr{J}(P, C) \tag{5.6}$$

leads to the performance guarantee

$$\mathscr{J}(P_o, C^{\text{RP}}) \leq \mathscr{J}_{\text{WC}}(\mathscr{P}, C^{\text{RP}}) \tag{5.7}$$

A key observation is that the resulting performance guarantee in (5.7) hinges on the shape and size of the model set $\mathscr{P}$. Typically, this model set is structured as a general LFT, i.e., the model set $\mathscr{P}$ is constructed by connecting an $\mathscr{H}_\infty$-norm-bounded perturbation $\Delta_u \in \mathbf{\Delta}_u \subseteq \mathscr{RH}_\infty$ to the nominal model as

$$\mathscr{P} = \{P | P = \mathscr{F}_u(\hat{H}(\hat{P}), \Delta_u), \Delta_u \in \mathbf{\Delta}_u\} \tag{5.8}$$

where $\hat{H}(\hat{P})$ contains the nominal model $\hat{P}$ and the uncertainty structure, as is elaborated on in the next section. Throughout, the model uncertainty set

$$\boldsymbol{\Delta}_u := \{\Delta_u | \|\Delta_u\|_\infty \leq \gamma\} \tag{5.9}$$

is considered, where it is assumed that $\boldsymbol{\Delta}_u$ contains multivariable operators with suitable dimensions.

Two requirements are imposed on the model set in view of high performance robust control:

R1. The model set $\mathscr{P}$ should lead to a small bound (5.7), and
R2. The model set should have low complexity to enable an efficient and non-conservative robust control synthesis.

To anticipate on the results in the next sections, a coprime factor-based approach will be pursued to address both aspects, since

- it enables connecting the size of model uncertainty and the control criterion, addressing Requirement 1, and
- it enables the nonconservative use of unstructured model uncertainty, enabling an efficient and nonconservative robust controller synthesis in view of Requirement 2.

5.2.2 Identification for robust control approach

To achieve high robust performance, i.e., a small bound in (5.7), a model set is identified that is robust control relevant. To obtain a general framework, the robust-control-relevant identification criterion

$$\begin{aligned} &\min_{\mathscr{P}} \mathscr{J}_{\mathrm{WC}}(\mathscr{P}, C^{\mathrm{exp}}) \\ &\quad \text{subject to } (5.4) \end{aligned} \tag{5.10}$$

is considered as in [7], [26]. In (5.10), C^{exp} denotes the controller that is present during the identification experiment. Indeed, many systems are open-loop unstable or need to have suitable feedback controller implemented for safety reasons. As such, the criterion in (5.10) explicitly takes the experimental conditions into account. In addition, note that the controller C^{exp} is often updated iteratively, which is well-established in iterative identification and control [35], [7], [1]. Note that (5.10) also encompasses the open-loop situation, since in case $\hat{P}$ is stable, then $C^{\mathrm{exp}} = 0$ is an internally stabilizing controller.

Next, by evaluating the performance of the feedback interconnection of the controller C^{exp} and all candidate models P in $\mathscr{P}$ for the general uncertainty description (5.8),

$$\begin{aligned} \mathscr{J}_{\mathrm{WC}}(\mathscr{P}, C^{\mathrm{exp}}) &= \sup_{\Delta \in \boldsymbol{\Delta}_u} \|\mathscr{F}_u(\hat{M}, \Delta_u)\|_\infty \\ &= \sup_{\Delta_u \in \boldsymbol{\Delta}_u} \|\hat{M}_{22} + \hat{M}_{21}\Delta_u(I - \hat{M}_{11}\Delta_u)^{-1}\hat{M}_{12}\|_\infty \end{aligned} \tag{5.11}$$

where

$$\hat{M}(\hat{H}, C^{\text{exp}}) = \left[\begin{array}{c|c} \hat{M}_{11} & \hat{M}_{12} \\ \hline \hat{M}_{21} & \hat{M}_{22} \end{array}\right] \tag{5.12}$$

The key point in (5.10) is that in general $\hat{M}_{11} \neq 0$ for many classical uncertainty structures, including additive uncertainty in which case

$$\mathscr{P}^{\text{ADD}} := \{P | P = \hat{P} + \Delta_u, \Delta_u \in \boldsymbol{\Delta}_u\} \tag{5.13}$$

where all considered systems have appropriate dimensions.

To ensure boundedness of (5.11), the dual-Youla-Kučera uncertainty structure has been considered in, e.g., [20], [2], [7], [10], [22]. Specifically,

$$\mathscr{P}^{\text{DY}} := \{P | P = (\hat{N} + D_c\Delta_u)(\hat{D} - N_c\Delta_u)^{-1}, \Delta_u \in \boldsymbol{\Delta}_u\} \tag{5.14}$$

where the pairs $\{\hat{N}, \hat{D}\}$ and $\{N_c, D_c\}$ are *any* RCF of $\hat{P}$ and C^{exp}, respectively. The model set $\mathscr{P}^{\text{DY}}$ leads to

$$\hat{H}^{\text{DY}} = \left[\begin{array}{c|c} \hat{D}^{-1}N_c & \hat{D}^{-1} \\ \hline D_c + \hat{P}N_c & \hat{P} \end{array}\right] \tag{5.15}$$

and

$$\hat{M}^{\text{DY}}(\hat{P}, C^{\text{exp}}) = \left[\begin{array}{c|c} 0 & (\hat{D} + C^{\text{exp}}\hat{N})^{-1}\begin{bmatrix} C^{\text{exp}} & I \end{bmatrix} V \\ \hline W\begin{bmatrix} D_c \\ -N_c \end{bmatrix} & WT(\hat{P}, C^{\text{exp}})V \end{array}\right] \tag{5.16}$$

Interestingly, (5.16) leads to

$$\mathscr{J}_{\text{WC}}(\mathscr{P}^{\text{DY}}, C^{\text{exp}}) = \sup_{\Delta_u \in \boldsymbol{\Delta}_u} \|\hat{M}_{22} + \hat{M}_{21}\Delta_u\hat{M}_{12}\|_\infty \tag{5.17}$$

which is an affine function of Δ_u and hence bounded for all $\Delta_u \in \boldsymbol{\Delta}_u$. Essentially, the dual-Youla-Kučera model uncertainty structure is especially useful from a robust stability perspective, since it excludes candidate models that are not stabilized by C^{exp}. However, it is emphasized that $\hat{M}_{12}$ and $\hat{M}_{21}$ in (5.17) are frequency-dependent and multivariable transfer function matrices. Consequently, the bound in (5.17) is finite but in general arbitrary. This typically necessitates the use of highly structured perturbation models, e.g., as in [8].

Recently, the result (5.17) has been significantly strengtened in [26], leading to

$$\mathscr{J}_{\text{WC}}(\mathscr{P}^{\text{DY}}, C^{\text{exp}}) \leq \mathscr{J}(\hat{P}, C^{\text{exp}}) + \sup_{\Delta_u \in \boldsymbol{\Delta}_u} \|\Delta_u\|_\infty \tag{5.18}$$

The result (5.18) has been established through the introduction of a new coprime factorization, which will be described in detail in the next section. The key advantage is that the result (5.18) holds true for *unstructured* perturbation models and is independent of the input–output dimensions. As a result, the uncertainty structure remains μ-simple [29] for complex systems with high input–output dimensionality. Consequently, this enables an efficient and nonconservative robust controller synthesis, see, e.g., [37], [11].

5.2.3 *Identifying robust-control-relevant coprime factorizations*

The key step in establishing the result (5.18) is a certain robust-control-relevant coprime factorization. In [26], it is shown that

- in the identification of the model set in (5.10), the nominal model $\hat{P}$ can be identified in a separate step, followed by quantifying model uncertainty, i.e., determining γ in (5.9).
- the identification of the nominal model in terms of the relevant coprime factorization has a direct connection to classical control-relevant identification techniques, see [35].

To show the latter aspect, recall that typical control-relevant identification approaches involve the criterion

$$\hat{P} = \arg\min_{P} \| W \left(T(P_o, C^{\exp}) - T(P, C^{\exp}) \right) V \|_\infty \tag{5.19}$$

To proceed, let$\{\tilde{N}_e, \tilde{D}_e\}$ be a left coprime factorization (LCF), see [40] for a definition, with co-inner numerator of $\begin{bmatrix} C^{\exp} V_2 & V_1 \end{bmatrix}$, i.e., $\{\tilde{N}_e, \tilde{D}_e\}$ is an LCF and satisfies the additional condition that $\tilde{N}_e \tilde{N}_e^* = I$. Given $C^{\exp}$, V_2, and V_1, such a coprime factorization can directly be computed, see [44] for details. Next, algebraic manipulations reveal that (5.19) is equivalent to

$$\begin{gathered} \min_{\hat{N},\hat{D}} \left\| W \left(\begin{bmatrix} N_o \\ D_o \end{bmatrix} - \begin{bmatrix} \hat{N} \\ \hat{D} \end{bmatrix} \right) \tilde{N}_e \right\|_\infty \\ \text{subject to } \hat{N}, \hat{D} \in \mathscr{RH}_\infty \end{gathered} \tag{5.20}$$

where

$$\begin{bmatrix} N \\ D \end{bmatrix} = \begin{bmatrix} P \\ I \end{bmatrix} (\tilde{D}_e + \tilde{N}_{e,2} V_2^{-1} P)^{-1} \tag{5.21}$$

and $\tilde{N}_e = \begin{bmatrix} \tilde{N}_{e,2} & \tilde{N}_{e,1} \end{bmatrix}$. In addition, the pairs $\{N_o, D_o\}$ and $\{\hat{N}, \hat{D}\}$ are coprime factorizations of P_o and $\hat{P}$, respectively, as is proved in [26, Theorem 2]. It is emphasized that the pairs $\{N_o, D_o\}$ and $\{\hat{N}, \hat{D}\}$ constitute a new robust-control-relevant factorization, and are not equivalent to normalized coprime factors, e.g., as used in [21] and [40].

The important aspect in (5.20) is that $\tilde{N}_e$ is co-inner and does not influence the $\mathscr{H}_\infty$-norm. Consequently, it can be removed directly, see (5.22), below. As a result, the four-block control-relevant identification problem (5.19) is recast as a two-block coprime factor identification problem.

Solving the identification problem (5.20) is not immediate and several steps are required to arrive at a suitable identification algorithm. First, notice that (5.20) involves an $\mathscr{H}_\infty$ norm. By employing the frequency domain interpretation of the $\mathscr{H}_\infty$-norm, (5.20) is recast as

$$\min_{\hat{N},\hat{D}} \max_{\omega_i \in \Omega^{\text{id}}} \bar{\sigma}\left(W\left(\begin{bmatrix} N_o \\ D_o \end{bmatrix} - \begin{bmatrix} \hat{N} \\ \hat{D} \end{bmatrix}\right)\right) \tag{5.22}$$

$$\text{subject to } \hat{N}, \hat{D} \in \mathscr{RH}_\infty \tag{5.23}$$

Second, $\{N_o, D_o\}$ is unknown. The key idea is that $T(P_o, C^{\text{exp}})$ can be directly identified using frequency response estimation using the results in [31], see also [28, Appendix A], leading to an estimate $\widetilde{T}(P_o, C^{\text{exp}})$ for $\omega_i \in \Omega^{\text{id}}$. A nonparametric estimate of $\{N_o, D_o\}$ is subsequently obtained by

$$\begin{bmatrix} \widetilde{N}_o \\ \widetilde{D}_o \end{bmatrix} = \widetilde{T}(P_o, C^{\text{exp}}) V \widetilde{N}_e^* \text{ for } \quad \omega_i \in \Omega^{\text{id}} \tag{5.24}$$

The key step in the actual identification of a nominal model $\hat{P}$ in terms of the required coprime factorization $\{\hat{N}, \hat{D}\}$ requires solving the optimization problem (5.22). This is investigated next.

5.3 Generalized SK-iterations for closed-loop coprime factor identification

In this section, a key contribution of this chapter is presented, which is a closed-loop generalization of Sanathanan-Koerner (SK) iterations [34]. First, in Section 5.3.1, the model parameterization is introduced. Next, in Section 5.3.2, an auxiliary step is taken to address the ℓ_∞ norm in (5.22). Then, in Section 5.3.3, the generalized SK-iterations for closed-loop coprime factor identification are presented.

5.3.1 Model parameterization

The model needs to be parameterized in terms of coprime factorizations. Before introducing these, the open-loop model is parameterized as

$$\hat{P}(\theta) = B(\theta)A(\theta)^{-1} \tag{5.25}$$

Here, $B \in \mathbb{R}^{n_y \times n_u}[z]$ and $A \in \mathbb{R}^{n_u \times n_u}[z]$, i.e., polynomial matrices in the complex indeterminate z, see [31] for more details. Hence, $\hat{P}(\theta)$ in (5.25) is parameterized as a right matrix fraction description (MFD). An analogous parameterization can be obtained for a left MFD. By using a so-called full polynomial form, see, e.g., [38, Chap. 6], the common dynamics between different input–output channels are taken into account. This leads to models with a low McMillan degree. Due to the one-to-one correspondence between MFDs and state-space models, this directly leads to state-space models with a small state dimension. This is in sharp contrast to the use of common-denominator models, e.g., as in [4], that leads to high-order state-space models.

In case the MFDs are parameterized linearly in the parameters, e.g., if the full polynomial form is employed, then these can be written as

$$\operatorname{vec}\left(\begin{bmatrix} B(\theta) \\ A(\theta) \end{bmatrix}\right) = \sum_q \psi_q \theta_q \tag{5.26}$$

This is a polynomial vector, which is parameterized such that it is linear in $\theta = \begin{bmatrix} \theta_1 & \theta_2 & \dots \end{bmatrix}^T$ and $\psi_q \in \mathbb{R}^{(n_y+n_u)n_u \times (n_y+n_u)n_u}[z]$.

It remains to determine the parametric model $\{\hat{N}, \hat{D}\}$ in (5.20). The tailor-made parameterization

$$\begin{bmatrix} \hat{N}(\theta) \\ \hat{D}(\theta) \end{bmatrix} = \begin{bmatrix} B(\theta) \\ A(\theta) \end{bmatrix} (\tilde{D}_e A(\theta) + \tilde{N}_{e,2} V_2^{-1} B(\theta))^{-1} \tag{5.27}$$

is proposed. This parameterization exploits knowledge of C^{exp} and effectively connects stability of the factors $\{\hat{N}, \hat{D}\}$ and closed-loop stability of the model, see [26, Theorem 4] for a proof. In addition, the dynamics that are introduced by the experimental controller C^{exp} and weighting filters V_2 and V_1 in (5.27) cancel out exactly when constructing $\hat{P}$, since

$$\hat{P}(\theta) = \hat{N}(\theta)\hat{D}(\theta)^{-1} = B(\theta)A(\theta)^{-1} \tag{5.28}$$

which indeed corresponds to (5.25).

5.3.2 Frequency domain identification involving ℓ_∞-norms via Lawson's algorithm

The optimization in (5.22) involves an ℓ_∞-type criterion, resulting in an optimization problem that is not guaranteed to be smooth, hence efficient gradient-based optimization techniques cannot be used directly. In addition, the parametrization (5.27) is nonlinear in the parameters θ, resulting in a generally non-convex optimization problem.

To solve the optimization in (5.22), Lawson's algorithm, see [33] and [5], is employed as follows.

Algorithm 5.1 *Set $\theta^{<o>} = 0$ and $w_i^{<0>} = \frac{1}{n_\omega}$, n_ω denoting the number of frequencies in Ω. Iterate over k until convergence:*

$$\theta^{<k>} = \arg\min_{\theta} \sum_i w_i^{<k>} \|\varepsilon_i(\theta)\|_F^2 \tag{5.29}$$

$$\text{where } w_i^{<k>} = \frac{\bar{\sigma}(\varepsilon_i(\theta^{<k>}))w_i^{<k-1>}}{\sum_i (\bar{\sigma}(\varepsilon_i(\theta^{<k>}))w_i^{<k-1>})} \tag{5.30}$$

Algorithm 5.1 iteratively solves the nonlinear least squares problem (5.29), which is defined in detail in Section 5.3.3. The iteratively adjusted weighting function $w_i^{<k>}$ is employed to minimize (5.22).

5.3.3 A closed-loop generalization of SK iterations

In this section, a new algorithm is presented to solve the actual computational step (5.29) in Algorithm 5.1. Observe that the nonlinear least squares problem in (5.29) is equivalent to

$$\sum_i \left\| W_h^{<k>} \circ \left(W \left(\begin{bmatrix} N_o \\ D_o \end{bmatrix} - \begin{bmatrix} \hat{N}(\theta) \\ \hat{D}(\theta) \end{bmatrix} \right) \right) \right\|_F^2 \tag{5.31}$$

where the elements of $W_h^{<k>}$ are equal to $\sqrt{w_i^{<k>}}$ and $\circ$ denotes the Hadamard product, which is introduced to separate the weighting of Lawson's algorithm and the nonlinear least squares problem. Note that the use of the Hadamard product is by no means necessary, but increases flexibility of the algorithm for incorporating user-chosen weighting functions. Rearranging and using the following facts from Kronecker algebra for matrices of suitable dimensions, see, e.g., [6]:

- $\|A\|_F = \|\text{vec}(A)\|_2$
- $\text{vec}(ABC) = (C^T \otimes A)\text{vec}(B)$
- $\text{vec}(A \circ B) = \text{diag}(\text{vec}(A))\text{vec}(B)$,

reveal that (5.31) can be written as

$$\sum_i \left\| W_{\text{lsq},i}^{<k>}(\theta)\text{vec}\left(\begin{bmatrix} B(\theta) \\ A(\theta) \end{bmatrix} \right) \right\|_2^2 \tag{5.32}$$

where

$$W_{\mathrm{lsq},i}^{<k>}(\theta) = \mathrm{diag}(\mathrm{vec}(W_h^{<k>}))\left(\left[(\tilde{\bar{D}}_e V_1^{-1}A(\theta) + \tilde{\bar{N}}_e V_2^{-1}B(\theta))^{-1}\right]^T \right.$$

$$\left. \otimes \left[W\left(\begin{bmatrix} N_o\tilde{\bar{N}}_e V_2^{-1} & N_o\tilde{\bar{D}}_e V_1^{-1} \\ D_o\tilde{\bar{N}}_e V_2^{-1} & D_o\tilde{\bar{D}}_e V_1^{-1}\end{bmatrix} - I\right)\right]\right) \tag{5.33}$$

The key idea is solving the nonlinear least squares problem (5.32) by introducing another iteration index $<f>$, leading to

$$\sum_i \left\| W_{\mathrm{lsq},i}^{<k>}(\theta^{<f-1>})\mathrm{vec}\left(\begin{bmatrix} B(\theta^{<f>}) \\ A(\theta^{<f>})\end{bmatrix}\right)\right\|_2^2 \tag{5.34}$$

By using the parameterization in (5.26), then (5.32) can be solved by iteratively solving the linear least squares problem

$$\min_{\theta^{<f>}} \| W_{\mathrm{lsq}}^{<k>}(\theta^{<f-1>})\Psi\theta^{<f>} \|_2^2 \tag{5.35}$$

subject to an appropriate degree constraint, i.e., ensuring that the resulting denominator satisfies a monotonicity constraint. In (5.35),

$$W_{\mathrm{lsq}}^{<k>} = \mathrm{diag}\left(W_{\mathrm{lsq},1}^{<k>}, W_{\mathrm{lsq},2}^{<k>}, \ldots\right) \tag{5.36}$$

$$\Psi = \left[\psi(\omega_1)^T \quad \psi(\omega_2)^T \quad \ldots\right]^T \tag{5.37}$$

The iteration (5.35) constitutes a novel generalization of well-known Sanathanan-Koerner iterations, e.g., [42], to the closed-loop case. The fixed point of these iterations is generally not equal to the local minimum of (5.32) but generally very close and achieved after several iterations, as is extensively elaborated on in [42] and confirmed by the experimental results presented below. Therefore, the fixed point of the iteration (5.35) is further refined using a Gauss-Newton optimization step.

Due to non-convexity of the problem, it is not guaranteed that the proposed iterative algorithm converges to the global minimum of (5.22). However, extensive application of the algorithm for various data has delivered good results. In addition, the convergence is generally fast, since for subsequent iterations good initial estimates are available. Similar observations have been obtained in [5], where related algorithms have been used for open-loop system identification.

5.4 Orthogonal polynomials w.r.t. a data-dependent discrete inner product

In the previous section, a new algorithm is presented that enables the identification of a certain coprime factorization from closed-loop data. Although the algorithm in

Section 5.3 can be implemented directly, the accuracy of the estimated model and convergence of the algorithm hinges on a numerically reliable implementation. A key contribution of the present section and Section 5.5 is to experimentally confirm that standard choices in system identification lead to an extremely ill-conditioned identification problem for the considered class of systems. In addition, the underlying mechanism is analyzed and a solution is provided that is shown to lead to a numerically reliable solution for experimental data.

The essential computational step in the identification problem involves the solution of the linear least-squares problem (5.35) and its subsequent Gauss-Newton refinement. The numerical properties are consequently related to the condition number $\kappa(.) = \frac{\overline{\sigma}(.)}{\underline{\sigma}(.)}$ of the matrix $W_{\text{lsq}}^{\langle k-1\rangle}\Psi$.

A common choice regarding the basis ψ is a monomial basis. In this case, the matrix Ψ is a Vandermonde matrix. Even though this leads to a unitary Ψ for discrete time identification problems with equidistantly spaced frequencies, and hence $\kappa(\Psi) = 1$, the matrix $W_{\text{lsq}}^{\langle k-1\rangle}\Psi$ generally is ill-conditioned. Consequently, the resulting problem (5.35) is poorly conditioned and cannot be solved accurately. This ill-conditioning is also evidenced by

1. the use of several procedures to enhance the numerical conditioning, e.g., [30], [14], [43]; and
2. by the experimental results in the next section.

The essential observation is that the conditioning in (5.35) hinges on the choice of the polynomial basis in (5.26). The key idea is to select a basis that is orthonormal with respect to the data-based discrete inner product

$$\langle \psi_m, \psi_l \rangle_{W_{\text{lsq}}} = \sum_i \psi_m(\omega_i)^* W_{\text{lsq},i}^{\langle k-1\rangle *} W_{\text{lsq},i}^{\langle k-1\rangle} \psi_l(\omega_i) \tag{5.38}$$

for a certain m and l. If an orthonormal basis with respect to the inner product (5.38) is selected, then $\kappa(W_{\text{lsq}}^{\langle k-1\rangle}\Psi) = 1$. Hence, this would render the computational step (5.35) optimally conditioned. In addition, subsequent Gauss-Newton iterations are then also typically well-conditioned as the estimate θ^{SK} is often close to optimality.

The orthonormal basis (5.38) can be computed in a numerically stable and efficient manner. This numerical procedure is beyond the scope of the present chapter and is outlined in [16], which extends the results in [3]. Importantly, the actual implementation can be done in $\mathcal{O}(N)$, i.e., linear in the data N. The focus of this chapter is on the experimental comparison of these results with respect to the commonly used monomial basis.

5.5 Experimental application

5.5.1 Experimental system

The considered nanometer-accurate positioning system is an industrial wafer stage system that is depicted in Figure 5.2. Wafer stages are part of wafer scanners, see Figure 5.3 and [41], which are the state-of-the-art machines for the automated

Figure 5.2 *Experimental wafer stage system, where 1: metrology frame, 2: mover, 3: airmount*

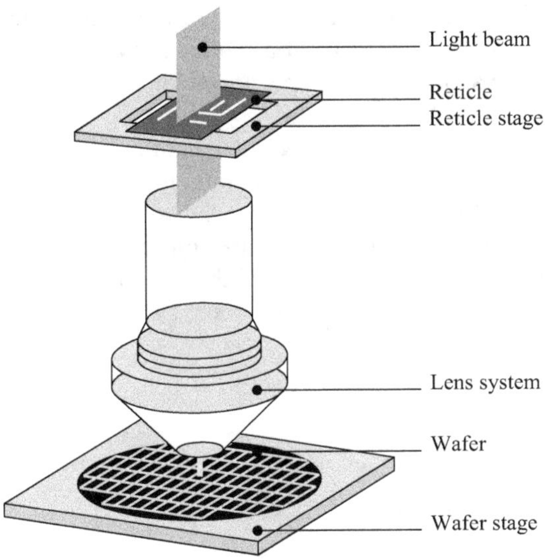

Figure 5.3 *Schematic illustration of a wafer scanner. Light is projected through a sophisticated lens or mirror system. The image of the IC is contained on the reticle stage (top stage), and then projected onto a silicon wafer. This wafer is accurately positioned in six degrees-of-freedom by the wafer stage (bottom stage)*

production of ICs. During the production process, a photoresist is exposed on a silicon disc, called a wafer. During exposure, the image of the desired IC patterns, which is contained on the reticle, is projected through a lens on the photoresist. The exposed photoresist is then removed by means of a solvent. Subsequent chemical reactions enable etching of these patterns, which is repeated for successive layers. Typically, more than 20 layers are required for each wafer. Each wafer contains more than 200 ICs that are sequentially exposed. During this entire process, the wafer must extremely accurately track a predefined reference trajectory in six motion degrees-of-freedom (DOFs). This precision motion task is performed by the wafer stage, which is the motion system considered in this chapter.

The system is equipped with moving-coil permanent magnet planar motors that enable contactless operation, see [9] for the underlying principle. The motion system consists of two parts: a stator, which is a plate consisting of an ordered array of permanent magnets, and a mover, which constitutes the moving part of the wafer stage.

Four actuators are connected to the mover to provide the necessary force, each consisting of three coils, which are powered by a three-phase power source. By means of an appropriate position-dependent commutation of the coils, each actuator delivers a well-defined and reproducible force in two directions. As a result, eight independent forces are available.

Laser interferometers in conjunction with a mirror block, which are connected to the metrology frame and the wafer stage, respectively, enable a high accuracy position measurement in all six motion DOFs, i.e., three translations and three rotations. Specifically, subnanometer measurement accuracy is available for the translational DOFs. Throughout, all signals and systems operate in discrete time with a sampling frequency of 2.5 kHz.

It is emphasized that the approach presented in this chapter is aimed to deal with a large number of actuators and sensors, generally more than the number of motion DOFs. However, to facilitate a clear exposition, a two input two output subsystem is considered in this chapter. The other DOFs are controlled by low performance PID controllers. The translational x and y DOFs in the horizontal plane are considered in this chapter. Hence, the input u and output y to the system are given by

$$u = \begin{bmatrix} u^x \\ u^y \end{bmatrix}, \quad y = \begin{bmatrix} y^x \\ y^y \end{bmatrix} \tag{5.39}$$

whereas the system is partitioned as

$$P = \begin{bmatrix} P^{xx} & P^{xy} \\ P^{yx} & P^{yy} \end{bmatrix} \tag{5.40}$$

In the experimental setup, an initial controller $C^{\exp}$ is implemented for the considered DOFs that achieves a closed-loop bandwidth of 40 Hz. The control goal is to achieve a closed-loop bandwidth of approximately 90 Hz. This is reflected by a suitable choice regarding the weighting filters W and V in (5.1) using the loop-shaping design rules in [21], [39].

5.5.2 Coprime factor identification results

The identification algorithm in Section 5.3 is applied to the experimental wafer stage data. First, the closed-loop frequency response function $\widetilde{T}(P_o, C^{\exp})$ is identified using the approach in [31], see also [28] for details. The resulting $\widetilde{T}(P_o, C^{\exp})$ is depicted in Figure 5.4.

Next, (5.24) is used to compute the coprime factor frequency response function $\{\widetilde{N}_o, \widetilde{D}_o\}$ in (5.24). The resulting frequency response function $\{\widetilde{N}_o, \widetilde{D}_o\}$ is depicted in Figure 5.5.

Subsequently, Algorithm 5.1 is invoked to solve the identification problem (5.22). The model order is selected using the results in [15]. The McMillan degree of the optimal model equals 8. Specifically, the control-relevant identification criterion (5.22) decreases significantly for orders up to 8, whereas increasing the model order beyond 8 does not significantly improve the fit in terms of the criterion (5.22).

The resulting identified coprime factors $\{\hat{N}, \hat{D}\}$ are depicted in Figure 5.5. In addition, the open-loop frequency response function P_o and model $\hat{P} = \hat{N}\hat{D}^{-1} = BA^{-1}$ are compared in Figure 5.6, which facilitates the interpretation of the model in terms of physical system properties. Note that due to the specific parameterization (5.25) in terms of a matrix fraction description, a minimal state-space model of $\hat{P}$ has state dimension 8. From Figure 5.6, the following observations are made. Below 200 Hz, the system is decoupled, revealing a rigid-body behavior in the diagonal elements, corresponding to translations in the x-direction and y-direction, respectively. The first resonance phenomena appear at 208 Hz and 214 Hz in all elements of $\widetilde{P}_o(\omega_i)$. Since these flexible dynamics are not aligned with the motion DOFs, the interaction between the x-direction and y-direction is high, i.e., the four elements of $\widetilde{P}_o(\omega_i)$ have an approximately equal gain beyond 200 Hz.

Analysis of the identified model in Figure 5.6 that has minimal state dimension 8 reveals that it is of low order for two reasons. First, the model only represents a limited number of resonance phenomena of $\widetilde{P}_o$ that are observed in Figure 5.6. This is a direct consequence of the control-relevant identification criterion in (5.19). The coprime factor domain in Figure 5.5 directly connects to control-relevance in terms of (5.19). From visual inspection it is confirmed that the model is accurate in regions where the amplitude of the corresponding coprime factors is large, which are control-relevant in view of (5.19).

Second, the model exploits the inherently multivariable parameterization in (5.25) that has a one-to-one correpondence with state-space realizations. Indeed, observe that four states in the model correspond to the two rigid-body modes in both the x-direction and the y-direction. The other four states correspond to resonance phenomena. Since these resonance phenomena correspond to flexible dynamical behavior, these correspond to complex pole pairs, see, e.g., [12]. Specifically, around 208 Hz and 214 Hz two closely spaced resonance phenomena are present. Interestingly, these

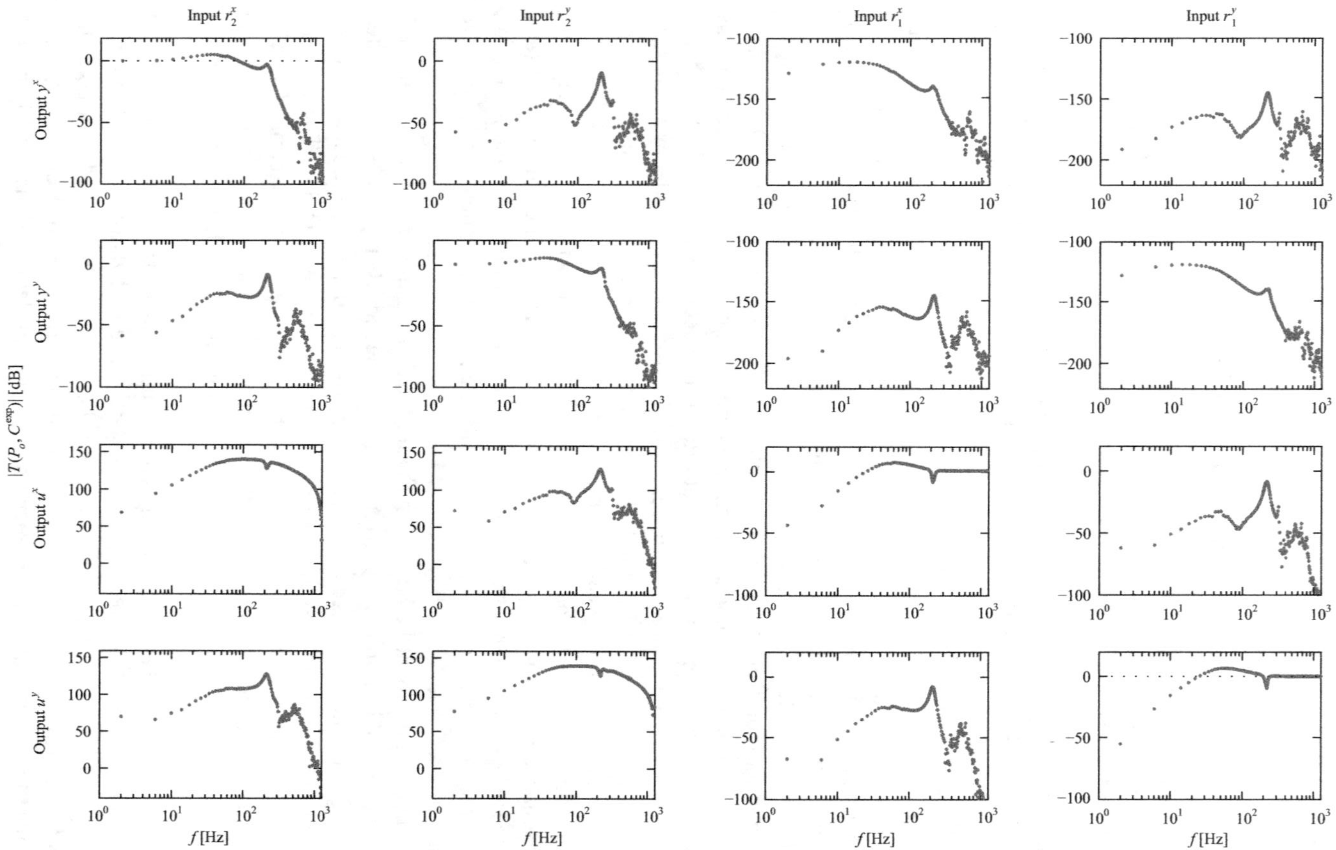

Figure 5.4 *Nonparametric closed-loop frequency response function estimate* $\widetilde{T}(P_o, C^{\text{exp}})$ *for* $\omega_i \in \Omega^{\text{id}}$

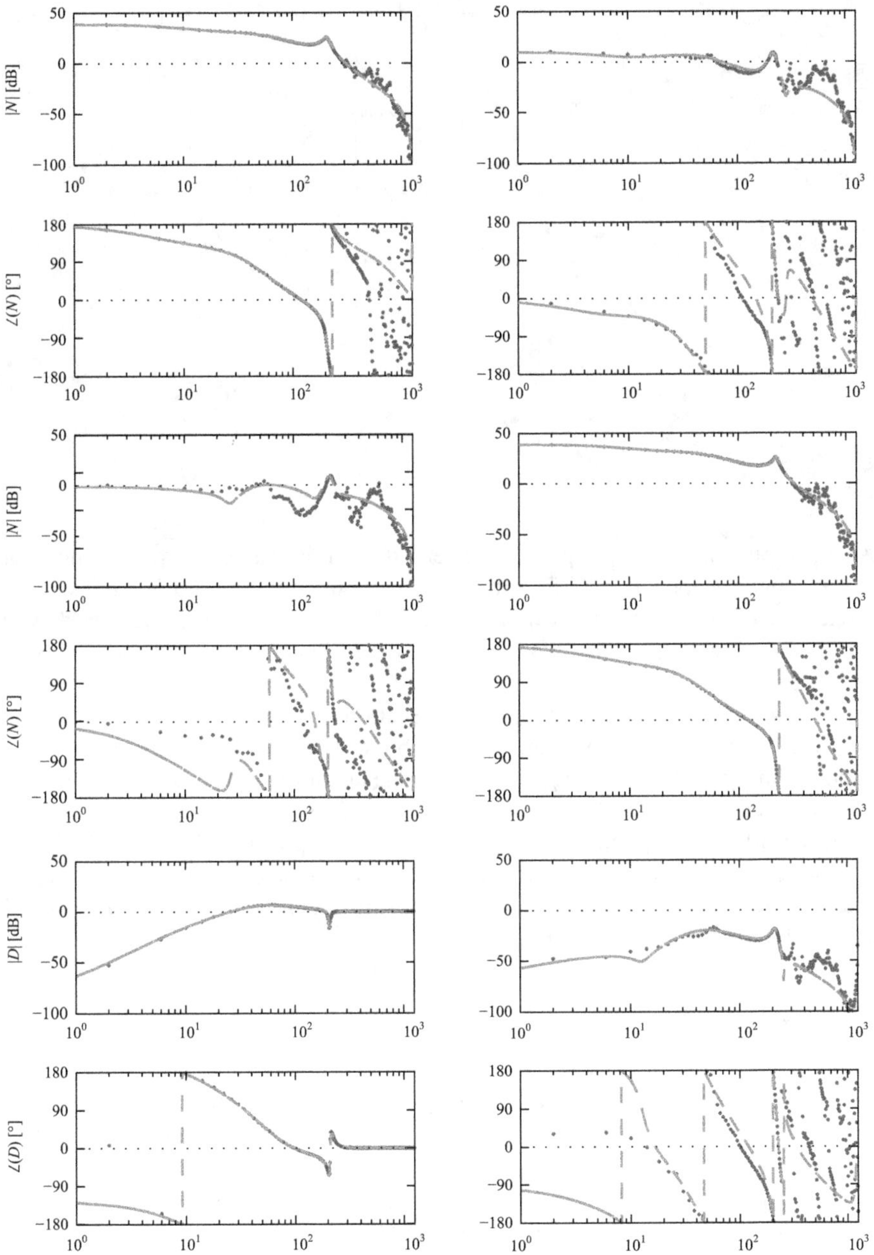

Figure 5.5 Coprime factorization: nonparametric N_o, D_o (dots), eighth order parametric model coprime factorizations $\hat{N}$, $\hat{D}$ (dashed)

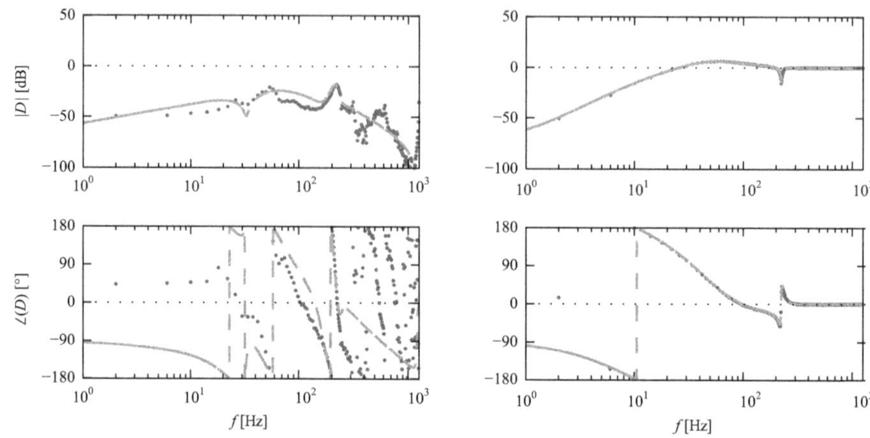

Figure 5.5 (Continued)

resonance phenomena correspond to inherently multivariable behavior, since both these resonances appear in all four transfer functions in Figure 5.6, yet only require two states each. This is a significant improvement over the models as delivered by the procedure in, e.g., [41], that does not take into account common dynamics between the different DOFs.

Finally, the optimization algorithm in Section 5.3 has an additional advantage from an efficiency perspective. In particular, by recasting the control-relevant identification problem (5.19) as the equivalent coprime factor identification problem (5.20), redundant data is effectively removed, leading to 50% reduction of the complexity for the considered two-input two-output wafer stage application. This can directly be understood when comparing Figures 5.4 and 5.5 that involve 4×4 and 4×2 transfer function matrices, respectively. Hence, the coprime factors in Figure 5.5 involve only half the amount of data when compared to Figure 5.4.

5.5.3 Numerical conditioning

The numerically optimal procedure in Section 5.5.3 is essential to obtain the accurate models in Section 5.5.2. To show this, it has first been attempted to solve the least squares problem (5.35) using monomial basis functions. The resulting condition numbers corresponding to (5.35) and subsequent Gauss-Newton iterations are depicted in Figure 5.7 and are in between 10^{20} and 10^{30}, leading to a breakdown of the algorithm even if implemented using a *QR*-factorization. Extensive computations have revealed that the commonly used monomial basis leads to an ill-conditioned least-squares problem (5.35) and subsequent Gauss-Newton iteration for the considered class of motion systems.

To improve the conditioning associated with (5.35), basis functions that are orthonormal with respect to the data-based discrete inner product (5.38) are employed.

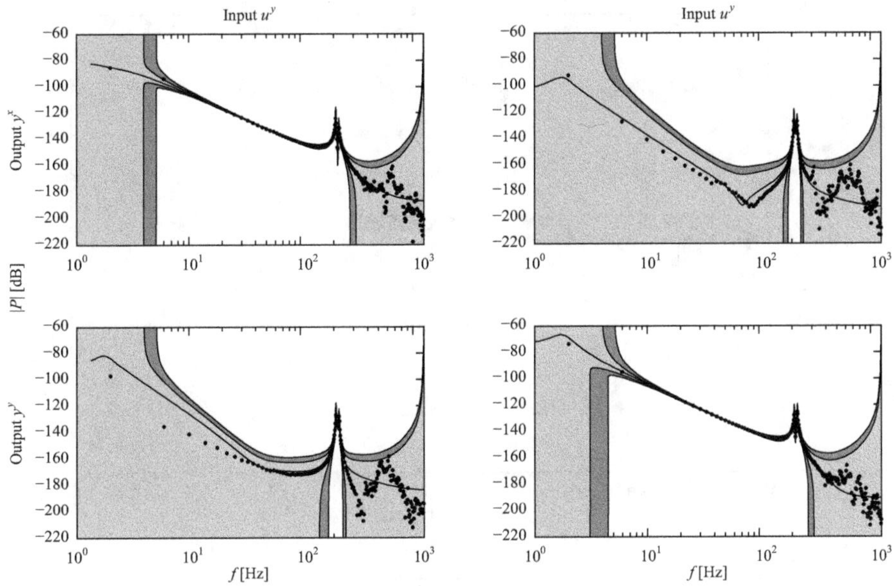

Figure 5.6 Bode magnitude diagram: nonparametric estimate (dot), nominal model $\hat{P}$ (solid), model set $\mathscr{P}$ using γ (dark grey), and $\mathscr{P}$ for only 50% uncertainty, i.e., using $\frac{\gamma}{2}$ (light grey)

As a result of the developed algorithm, the condition number corresponding to the least-squares problem equals the optimal value of one. In addition, the condition number during the Gauss-Newton iterations, where the basis is fixed, remains close to optimal as is observed in Figure 5.7. This leads to a converging algorithm and an accurate computation of the optimal model.

Summarizing, these results underline the importance of numerical aspects in system identification of complex systems such as next-generation motion systems.

5.5.4 Illustration of robust-control-relevance

In Section 5.2, it is claimed that the identified coprime factorization in Figure 5.5 is essential for modeling uncertainty in view of robust control, i.e., for obtaining the result (5.18). To show this from a control design perspective, the model is extended with uncertainty. In particular, the *unstructured* perturbation model (5.9) is considered. Herein, γ is determined using the model validation procedure presented in [24]. The resulting model set $\mathscr{P}$ as defined in (5.8) is depicted in Figure 5.6. In addition, the model set corresponding to $\frac{\gamma}{2}$ is depicted, i.e., only 50% uncertainty. It is observed in Figure 5.6 that the uncertainty is automatically shaped. First, it has a frequency dependent shape, since the uncertainty is the smallest around the target bandwidth of 90 Hz and the resonance phenomena around 200 Hz, which are known to be essential for closed-loop stability. On the other hand, at low frequencies the uncertainty is very

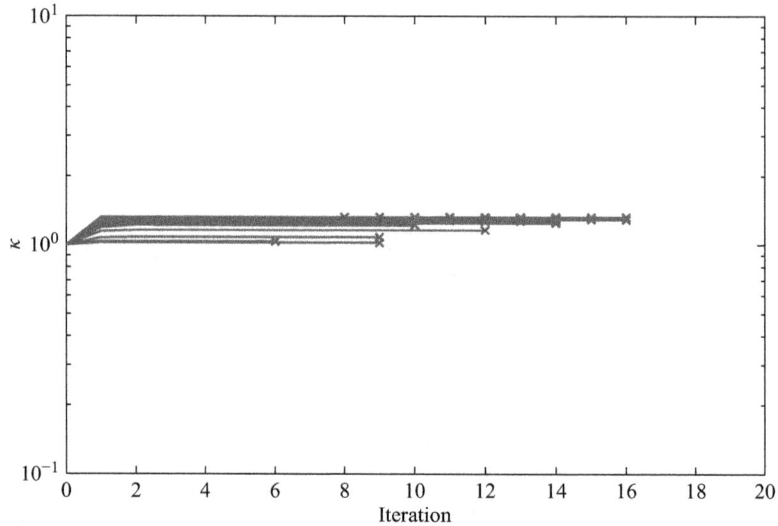

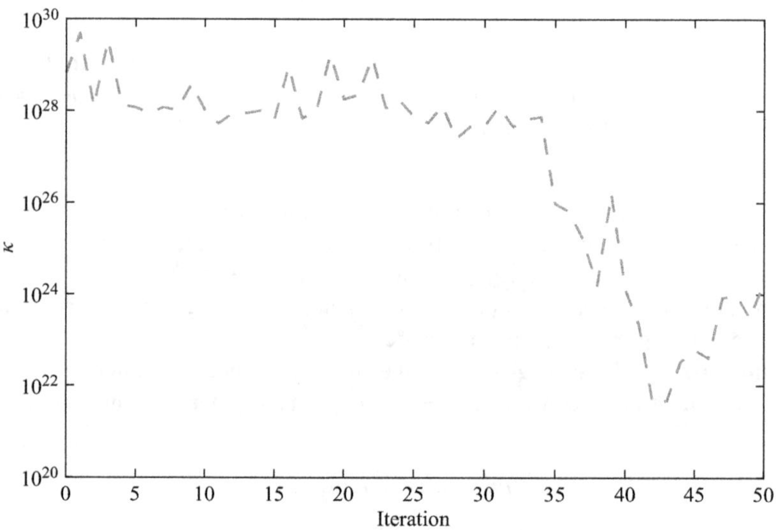

Figure 5.7 Condition number corresponding to linear least-squares and Gauss-Newton iterations. Top: using basis functions that are orthonormal with respect to a data-based discrete inner product (solid). Bottom: using standard monomial basis functions (dashed). In the latter case, the iterations break down after the first Gauss-Newton iteration

high, since the controller will have integral action. In addition, at high frequencies, the control-relevant aspect addresses the fact that the controller has roll-off, hence the high frequent resonance phenomena are not relevant for stability and performance.

Second, the specific coprime factorizations inherently scale the input–output channels of the model uncertainty. Indeed, observe that (5.9) is a multivariable and unstructured perturbation block. Still, from visual inspection the model uncertainty is scaled well with respect to the two inputs and outputs. This aspect is further explained in [25].

Finally, a subsequent robust control design in [28] confirms that the model set $\mathscr{P}$ is adequate for generating a robust controller that significantly enhances the control performance.

5.6 Conclusions

In this chapter, a new algorithm is presented for identifying complex systems. The proposed algorithm has three important advantages. First, a numerically optimal implementation is presented. Second, the new implementation enables an efficient implementation, as is shown in, e.g., [16]. Third, the algorithm enables the identification of a certain coprime factorization. This enables the use of unstructured uncertainty, which is essential for synthesizing robust controllers for complex systems.

The algorithm is demonstrated on an industrial wafer stage system. Experimental results confirm that common algorithm implementations based on monomial basis functions are generally ill-conditioned, prohibiting the identification of accurate models. In addition, it is shown that the presented approach is numerically reliable and provides good identification results.

Present research focuses on extending the algorithm in this section toward rendering Gauss-Newton iterations superfluous, see, e.g., [16] for early results in this direction. In addition, the coprime factor framework is being extended to deal with nonmeasurable performance outputs in [27]. Interestingly, this also leads to an enhanced complexity of the identification problem, as it leads to an additional number of inputs and outputs that have to be addressed.

Acknowledgments

The authors are grateful for the contributions of Okko Bosgra that have led to the results in this chapter. The authors want to thank Robbert van Herpen for his important contributions and suggestions. In addition, Marc van de Wal and Sander Quist are gratefully acknowledged for their help in performing the experiments and sharing their expertise on wafer stage motion control. Finally, the authors would like to thank Philips Innovation Services, Eindhoven, for providing access to the experimental facilities.

This work is supported by the Innovational Research Incentives Scheme under the VENI grant "Precision Motion: Beyond the Nanometer" (no. 13073) awarded by NWO (The Netherlands Organisation for Scientific Research) and STW (Dutch Science Foundation).

References

[1] Albertos, P., Sala, A. (eds.): *Iterative Identification and Control.* Springer, London, UK (2002)

[2] Anderson, B.D.O.: From Youla-Kucera to identification, adaptive and nonlinear control. *Automatica* **34**(12), 1485–1506 (1998)

[3] van Barel, M., Bultheel, A.: Orthonormal polynomial vectors and least squares approximation for a discrete inner product. *Elec. Trans. Numer. Anal.* **3**, 1–23 (1995)

[4] Bayard, D.S.: High-order multivariable transfer function curve fitting: Algorithms, sparse matrix methods and experimental results. *Automatica* **30**(9), 1439–1444 (1994)

[5] Bohn, C., Unbehauen, H.: Minmax and least squares multivariable transfer function curve fitting: Error criteria, algorithms and comparisons. In: *Proc. 1998 Americ. Contr. Conf.*, pp. 3189–3193. Philadelphia, PA (1998)

[6] Brewer, J.W.: Kronecker products and matrix calculus in system theory. *IEEE Trans. Circ. Syst.* **25**(9), 772–781 (1978)

[7] de Callafon, R.A., Van den Hof, P.M.J.: Suboptimal feedback control by a scheme of iterative identification and control design. *Math. Mod. Syst.* **3**(1), 77–101 (1997)

[8] de Callafon, R.A., Van den Hof, P.M.J.: Multivariable feedback relevant system identification of a wafer stepper system. *IEEE Trans. Contr. Syst. Techn.* **9**(2), 381–390 (2001)

[9] Compter, J.C.: Electro-dynamic planar motor. *Prec. Eng.* **28**(2), 171–180 (2004)

[10] Douma, S.G., Van den Hof, P.M.J.: Relations between uncertainty structures in identification for robust control. *Automatica* **41**(3), 439–457 (2005)

[11] Doyle, J.: Analysis of feedback systems with structured uncertainties. *IEE Proc.-Control Theory Appl.* Pt. D **129**(6), 242–250 (1982)

[12] Gawronski, W.K.: *Advanced Structural Dynamics and Active Control of Structures*. Springer, New York, NY (2004)

[13] Gevers, M.: Identification for control: From the early achievements to the revival of experiment design. *Eur. J. Contr.* **11**(4-5), 1–18 (2005)

[14] Hakvoort, R.G., Van den Hof, P.M.J.: Frequency domain curve fitting with maximum amplitude criterion and guaranteed stability. *Int. J. Contr.* **60**(5), 809–825 (1994)

[15] van Herpen, R., Oomen, T., Bosgra, O.: A robust-control-relevant perspective on model order selection. In: *Proc. 2011 Americ. Contr. Conf.*, pp. 1224–1229. San Francisco, CA (2011)

[16] van Herpen, R., Oomen, T., Bosgra, O.: Numerically reliable frequency-domain estimation of transfer functions: A computationally efficient methodology. In: *16th IFAC Symp. Sys. Id.*, pp. 595–600. Brussels, Belgium (2012)

[17] Hjalmarsson, H.: From experiment design to closed-loop control. *Automatica* **41**, 393–438 (2005)

[18] Van den Hof, P.M.J., Schrama, R.J.P., de Callafon, R.A., Bosgra, O.H.: Identification of normalised coprime plant factors from closed-loop experimental data. *Eur. J. Contr.* **1**(1), 62–74 (1995)

[19] Hughes, P.C.: Space structure vibration modes: How many exist? Which ones are important? *IEEE Contr. Syst. Mag.* **7**(1), 22–28 (1987)

[20] Ma, C.C.H.: Comments on "A necessary and sufficient condition for stability of a perturbed system". *IEEE Trans. Automat. Contr.* **33**(8), 796–797 (1988)

[21] McFarlane, D.C., Glover, K.: Robust Controller Design Using Normalized Coprime Factor Plant Descriptions, *Lecture Notes in Control and Information Sciences*, vol. 138. Springer-Verlag, Berlin, Germany (1990)

[22] Niemann, H.: Dual Youla parameterisation. *IEE Proc.-Control Theory Appl.* **150**(5), 493–497 (2003)

[23] Ninness, B., Goodwin, G.C.: Estimation of model quality. *Automatica* **31**(12), 1771–1797 (1995)

[24] Oomen, T., Bosgra, O.: Well-posed model uncertainty estimation by design of validation experiments. In: *15th IFAC Symp. Sys. Id.*, pp. 1199–1204. Saint-Malo, France (2009)

[25] Oomen, T., Bosgra, O.: Selecting uncertainty structures in identification for robust control with an automotive application. In: *16th IFAC Symp. Sys. Id.*, pp. 601–606. Brussels, Belgium (2012)

[26] Oomen, T., Bosgra, O.: System identification for achieving robust performance. *Automatica* **48**(9), 1975–1987 (2012)

[27] Oomen, T., Grassens, E., Hendriks, F., van Herpen, R., Bosgra, O.: Inferential motion control: Identification and robust control with unmeasured performance variables. In: *Proc. 50th Conf. Dec. Contr.*, pp. 964–969. Orlando, FL (2011)

[28] Oomen, T., van Herpen, R., Quist, S., van de Wal, M., Bosgra, O., Steinbuch, M.: Connecting system identification and robust control for next-generation motion control of a wafer stage. *IEEE Trans. Contr. Syst. Techn.* **22**(1), 102–118 (2014)

[29] Packard, A., Doyle, J.: The complex structured singular value. *Automatica* **29**(1), 71–109 (1993)

[30] Pintelon, R., Kollár, I.: On the frequency scaling in continuous-time modeling. *IEEE Trans. Inst. Meas.* **54**(1), 318–321 (2005)

[31] Pintelon, R., Schoukens, J.: *System Identification: A Frequency Domain Approach*, 2nd edn. IEEE Press, New York, NY (2012)

[32] Reinelt, W., Garulli, A., Ljung, L.: Comparing different approaches to model error modeling in robust identification. *Automatica* **38**(5), 787–803 (2002)

[33] Rice, J.R.: *The Approximation of Functions, vol. 2: Nonlinear and Multivariate Theory*. Addison-Wesley Publishing Company, Reading, MA (1964)

[34] Sanathanan, C.K., Koerner, J.: Transfer function synthesis as a ratio of two complex polynomials. *IEEE Trans. Automat. Contr.* **8**(1), 56–58 (1963)

[35] Schrama, R.J.P.: Accurate identification for control: The necessity of an iterative scheme. *IEEE Trans. Automat. Contr.* **37**(7), 991–994 (1992)

[36] Sebastian, A., Salapaka, S.M.: Design methodologies for robust nano-positioning. *IEEE Trans. Contr. Syst. Techn.* **13**, 6 (2005)

[37] Skogestad, S., Postlethwaite, I.: *Multivariable Feedback Control: Analysis and Design*, 2nd edn. John Wiley & Sons, West Sussex, UK (2005)

[38] Söderström, T., Stoica, P.: *System Identification*. Prentice Hall, Hemel Hempstead, UK (1989)

[39] Steinbuch, M., Norg, M.L.: Advanced motion control: An industrial perspective. *Eur. J. Contr.* **4**(4), 278–293 (1998)

[40] Vinnicombe, G.: *Uncertainty and Feedback: $\mathscr{H}_\infty$ Loop-Shaping and the ν-gap Metric*. Imperial College Press, London, UK (2001)

[41] van de Wal, M., van Baars, G., Sperling, F., Bosgra, O.: Multivariable $\mathscr{H}_\infty/\mu$ feedback control design for high-precision wafer stage motion. *Contr. Eng. Prac.* **10**(7), 739–755 (2002)

[42] Whitfield, A.H.: Asymptotic behaviour of transfer function synthesis methods. *Int. J. Contr.* **45**(3), 1083–1092 (1987)

[43] Wills, A., Ninness, B.: On gradient-based search for multivariable system estimates. *IEEE Trans. Automat. Contr.* **53**, 1 (2008)

[44] Zhou, K., Doyle, J.C., Glover, K.: *Robust and Optimal Control*. Prentice Hall, Upper Saddle River, NJ (1996)

Chapter 6
Subspace-based multi-step predictors for predictive control

Marzia Cescon[1] *and Rolf Johansson*[2]

Abstract

In the framework of the subspace-based identification of linear systems, the first step for the construction of a state-space model from observed input–output data involves the estimation of the output predictor. Such construction is based on projection operations of certain structured data matrices onto suitable subspaces spanned by the collected data. To the purpose of predictive control using short-term predictors, this algorithmic step can be elaborated to provide data-based multi-step predictors. Using such an approach, this contribution deals with subspace-based identification methods for the estimation of short-term predictors. One illustrative example is provided: blood glucose prediction in type 1 diabetes mellitus.

6.1 Introduction

Time series filtering and prediction theory based on linear systems have a long history dating back to the 1940s and the works of Kolmogorov [30, 29] and Wiener [43]. From the Wiener–Kolmogorov theory of stochastic processes, the Kalman filter [24] and the ARMAX-based predictor [3] could be formulated, representing solutions suitable for a digital implementation. Thorough descriptions of a variety of predictor structures can be found in [2], [17], [4], [16], and [23]. Usually, the predictors are formulated for one-step-ahead prediction horizon and are based on a known model. Many practical cases, however, face the problem of simultaneously forecasting the time series of interest for multiple look ahead. This prediction scheme is called a multi-step predictor [19, 36]. Efforts in this direction towards adaptive control were made by Mosca et al. [34, 35]. An alternative solution to the prediction problem for multi-input multi-output systems is offered by the so-called subspace identification

[1] Lund University, Lund, Sweden, e-mail: marzia.cescon@control.lth.se
[2] Lund University, Lund, Sweden, e-mail: Rolf.Johansson@control.lth.se

methods (SIMs) [25, 40, 41, 42, 33, 21, 5]. In the SIM framework, indeed, the first step for the construction of a state-space model from observed input–output data involves the estimation of the output predictor. The algorithmic step was exploited in the past by several authors in receding horizon type of control strategies [8, 14, 45, 22, 12, 10, 11].

Based on subspace-based identification methods for linear systems this chapter formulates short-term multi-step-ahead predictors for application in predictive control.

6.1.1 Model description

Consider a discrete-time linear time-invariant system $\mathscr{S}_n(A, B, C, K)$ in innovation form

$$\begin{aligned} x_{k+1} &= Ax_k + Bu_k + Ke_k \\ y_k &= Cx_k + e_k \end{aligned} \tag{6.1}$$

where $u_k \in \mathbb{R}^m$ is the input, $y_k \in \mathbb{R}^l$ the output, $x_k \in \mathbb{R}^n$ the state, $e_k \in \mathbb{R}^l$ the zero-mean white noise innovation process uncorrelated with u_k and $A \in \mathbb{R}^{n\times n}$, $B \in \mathbb{R}^{n\times m}$, $C \in \mathbb{R}^{l\times n}$, $K \in \mathbb{R}^{l\times n}$ are constant matrices. In the following equations, assume that (A, B) is reachable and (C, A) is observable. Further, assume that no linear feedback from the states to the input is present, i.e. input–output data are obtained from an open loop experiment.

Transfer function relationships using z-transform are

$$Y(z) = G(z)U(z) + H(z)E(z) \tag{6.2}$$

$$G(z) = C(zI - A)^{-1}B \tag{6.3}$$

$$H(z) = I + C(zI - A)^{-1}K \tag{6.4}$$

$$H^{-1}(z) = I - C(zI - A + KC)^{-1}K \tag{6.5}$$

The reconstruction of $\{e_k\}$ from input–output data may be accomplished by means of the inverse transfer function relationship

$$E(z) = H^{-1}(z)(Y(z) - G(z)U(z)) \tag{6.6}$$

and the state-space realization of the estimator

$$\hat{x}_{k+1|k} = A\hat{x}_{k|k-1} + Bu_k + K\hat{e}_k \tag{6.7}$$

$$\hat{y}_{k|k-1} = C\hat{x}_{k|k-1} \tag{6.8}$$

$$\hat{e}_k = y_k - \hat{y}_k = y_k - C\hat{x}_k \tag{6.9}$$

Considering the innovation model (6.1) in predictor form, with $\overline{A} = A - KC$, the one-step-ahead predictor can be formulated:

$$\begin{aligned} \hat{x}_{k+1} &= \overline{A}\hat{x}_k + Bu_k + Ky_k \\ \hat{y}_{k|k-1} &= Cx_k + e_k \end{aligned} \tag{6.10}$$

with state error dynamics

$$\tilde{x}_k = x_k - \hat{x}_k \tag{6.11}$$

$$\tilde{x}_{k+1} = A\hat{x}_{k|k-1} + Bu_k + Ke_k\hat{x}_k \tag{6.12}$$

$$= -((A-KC)\hat{x}_k + Ky_k + Bu_k) = (A-KC)\tilde{x} = \overline{A}\tilde{x} \tag{6.13}$$

and covariance

$$P_k = \mathscr{E}\{\tilde{x}_k\tilde{x}_k^{\mathsf{T}}\} \tag{6.14}$$

$$P_{k+1} = \mathscr{E}\{\tilde{x}_{k+1}\tilde{x}_{k+1}^{\mathsf{T}}\} = \mathscr{E}\{\overline{A}\tilde{x}_k\tilde{x}_k^{\mathsf{T}}\overline{A}^{\mathsf{T}}\} = \overline{A}\mathscr{E}\{\tilde{x}_k\tilde{x}_k^{\mathsf{T}}\}\overline{A}^{\mathsf{T}} = \overline{A}P_k\overline{A}^{\mathsf{T}} \tag{6.15}$$

6.1.2 Notation

The available data sequences $\{u_k\}$, $\{y_k\}$, the state $\{x_k\}$ and the innovation process $\{e_k\}$ will be organized in Hankel matrices denoted by uppercase letters. Subscript indices $[\alpha, \beta]$ of a Hankel matrix will be used to indicate the argument of the upper-left and the lower-left element respectively, e.g. $U_{[t_1,t_2]}$ will contain in the first column the input history between instants t_1 and t_2. Accordingly, data records of finite length N will be represented by the block rows of the Hankel data matrices and will be denoted by uppercase letters, the subscript indicating the first time instants of the time series:

$$U_{[t_1,t_2]} = \begin{bmatrix} U_{t_1} \\ U_{t_1+1} \\ \vdots \\ U_{t_2} \end{bmatrix} = \begin{bmatrix} u_{t_1} & u_{t_1+1} & \cdots & u_{t_1+N-1} \\ u_{t_1+1} & u_{t_1+2} & \cdots & u_{t_1+N} \\ \vdots & \vdots & \cdots & \vdots \\ u_{t_2} & u_{t_2+1} & \cdots & u_{t_2+N-1} \end{bmatrix}$$

The orthogonal projection of the rows of a given matrix A onto the row space of a given matrix B will be denoted by $\hat{E}\{A \mid B\}$, whereas the symbol $\hat{E}_{\parallel C}\{A \mid B\}$ will denote the oblique projection of the row space of A onto the row space of B along the row space of C, the projection operator being $\hat{E}\{\cdot\}$. Throughout the chapter k will be the discrete-time index, t shall denote the current time instance in the identification problem, t_0 shall be the initial time from which the data are collected, so that $t - t_0 = p$ is the past horizon in the identification problem, T shall be such that $T - t + 1 = f$ represents the future horizon. The two integers p and f are such that $p \geq \max(f, n)$, where n is the model order. Last, the number of steps in the look ahead that one wishes to investigate will be denoted by τ, where $\tau \leq f$.

6.1.3 Statement of the problem

Let the finite sequences $\{u_k\}_{k=t-p}^{t+f-1}$ and $\{y_k\}_{k=t-p}^{t+f-1}$ be the measured input and the corresponding output, respectively, of the system (6.1). Our aim is to find a multi-step estimator of the output sequence $\{y_k\}_{k=t}^{t+f-1}$ by means of linear combinations of the past joint input–output $\{z_k\}_{k=t-p}^{t-1}$ and future input $\{u_k\}_{k=t}^{t+f-1}$.

6.2 Subspace-based linear multi-step predictors

Suppose ideally that we have observations of the processes $\{u_k\}$, $\{y_k\}$, $\{x_k\}$, $\{e_k\}$, $k \in [t-p, t+f-1]$. In addition, suppose the finite length N of the interval is large. Since the finite-length observed data sequences are realizations of the underlying stochastic processes in (6.1), with the notation previously introduced, the following holds:

$$\begin{aligned} X_{k+1} &= AX_k + BU_k + KE_k \\ Y_k &= CX_k + E_k \end{aligned} \tag{6.16}$$

Furthermore, from the observed samples construct the following block Hankel matrices:

$$U^p = U_{[t-p,t-1]} \in \mathbb{R}^{p\cdot m\times N} \tag{6.17}$$

$$U^f = U_{[t,t+f-1]} \in \mathbb{R}^{f\cdot m\times N} \tag{6.18}$$

called the past and future input data matrices, respectively, and

$$Y^p = Y_{[t-p,t-1]} \in \mathbb{R}^{p\cdot l\times N} \tag{6.19}$$

$$Y^f = Y_{[t,t+f-1]} \in \mathbb{R}^{f\cdot l\times N} \tag{6.20}$$

called the past and future output data matrices, respectively.

Define the extended observability matrix

$$\mathcal{O}_f = \begin{bmatrix} C \\ CA \\ CA^2 \\ \vdots \\ CA^{f-1} \end{bmatrix} \tag{6.21}$$

and the Toeplitz matrices containing the impulse responses of the system due to the input u_k and the innovation process e_k, respectively.

$$\mathcal{H}_f = \begin{bmatrix} 0 & 0 & \cdots & 0 \\ CB & 0 & \cdots & 0 \\ \vdots & \vdots & \ddots & \vdots \\ CA^{f-2}B & CA^{f-3}B & \cdots & 0 \end{bmatrix}, \quad \mathcal{W}_f = \begin{bmatrix} I & 0 & \cdots & 0 \\ CK & I & \cdots & 0 \\ \vdots & \vdots & \ddots & \vdots \\ CA^{f-2}K & CA^{f-3}K & \cdots & 0 \end{bmatrix} \tag{6.22}$$

Then, by iteration of the system equations in (6.1), the following matrix input–output relations may be written to express the future output matrix:

$$Y^f = \mathcal{O}_f X_f + \mathcal{H}_f U^f + \mathcal{W}_f E^f \tag{6.23}$$

In the practical scenario the state sequence X_f is not known, so future outputs cannot be computed. An estimator of future outputs can be found, however, from the available data as a linear combination of the joint input–output past and the future input (see Appendix A.6 in [41] for a proof), provided that u_k and e_k are uncorrelated:

$$\hat{Y}^f = \hat{\Gamma} Z^p + \hat{\Lambda} U^f \tag{6.24}$$

where we have introduced the shorthand notation

$$Z^p = \begin{bmatrix} U^p \\ Y^p \end{bmatrix} = Z_{[t-p,t-1]} \in \mathbb{R}^{p\cdot(m+l)\times N} \tag{6.25}$$

to denote the joint past input–output data matrix. Now, the problem of finding an optimal output predictor can be formulated mathematically as a least-squares problem:

$$\hat{\Gamma}, \hat{\Lambda} = \arg \min_{\substack{\Gamma \in \mathbb{R}^{lp\times(l+m)p} \\ \Lambda \in \mathbb{R}^{lp\times mf}}} \left\| Y^f - \begin{bmatrix} \Gamma & \Lambda \end{bmatrix} \begin{bmatrix} Z^p \\ U^f \end{bmatrix} \right\|_F^2 \tag{6.26}$$

where $\|\cdot\|_F$ stands for the Frobenius norm of a matrix. Note that in this way f prediction problems are solved simultaneously row-wise. Each problem consists in estimating $\hat{Y}_{t+\tau}$, $\tau \in [0, f-1]$:

$$\hat{Y}^f = \begin{bmatrix} \hat{Y}_t \\ \hat{Y}_{t+1} \\ \vdots \\ \hat{Y}_{t+f-1} \end{bmatrix} = \begin{bmatrix} \hat{y}_t & \hat{y}_{t+1} & \cdots & \hat{y}_{t+N-1} \\ \hat{y}_{t+1} & \hat{y}_{t+2} & \cdots & \hat{y}_{t+N} \\ \vdots & \vdots & \vdots & \vdots \\ \hat{y}_{t+f-1} & \hat{y}_{t+f} & \cdots & \hat{y}_{t+f+N-2} \end{bmatrix} \tag{6.27}$$

A geometric interpretation can be given to the least-squares problem as the orthogonal projection of Y^f onto $\begin{bmatrix} Z^p & U^f \end{bmatrix}^T$, i.e.

$$\hat{Y}^f = \hat{E}\left\{ Y^f \mid \begin{bmatrix} Z^p \\ U^f \end{bmatrix} \right\} \tag{6.28}$$

Actually, the orthogonal projection (6.28) corresponds to the sum of two oblique projections [27, lemma 1] under the assumption $span(Z^p) \cap span(U^f) = \{0\}$, with $span(\cdot)$ standing for the space spanned by the row vectors of a matrix:

$$\hat{Y}^f = \hat{E}_{\|U^f}\{Y^f \mid Z^p\} + \hat{E}_{\|Z^p}\{Y^f \mid U^f\} \tag{6.29}$$

$$= \hat{\Gamma} Z^p + \hat{\Lambda} U^f \tag{6.30}$$

Once the operators $\hat{\Gamma}$ and $\hat{\Lambda}$ have been estimated, they can be applied to new data generated by the same underlying mechanisms to forecast the output τ steps ahead. In particular, note that the matrix relation in (6.28) can be expressed vector-by-vector in the following way:

$$\begin{bmatrix} \hat{y}_t \\ \hat{y}_{t+1} \\ \hat{y}_{t+2} \\ \vdots \\ \hat{y}_{t+f-1} \end{bmatrix} = \hat{\Gamma} \begin{bmatrix} z_{t-p} \\ z_{t-p+1} \\ z_{t-p+2} \\ \vdots \\ z_{t-1} \end{bmatrix} + \hat{\Lambda} \begin{bmatrix} u_t \\ u_{t+1} \\ u_{t+2} \\ \vdots \\ u_{t+f-1} \end{bmatrix} \tag{6.31}$$

6.2.1 Computing projections

As just pointed out, linear regression problems have a geometric interpretation that involves appropriate projection operations. Recall (6.28). The orthogonal projection is

$$\hat{Y}^f = \hat{E}\left\{Y^f \mid \begin{bmatrix} Z^p \\ U^f \end{bmatrix}\right\} \tag{6.32}$$

$$= Y^f \begin{bmatrix} Z^p \\ U^f \end{bmatrix}^{\mathsf{T}} \left(\begin{bmatrix} Z^p \\ U^f \end{bmatrix} \begin{bmatrix} Z^p \\ U^f \end{bmatrix}^{\mathsf{T}} \right)^{\dagger} \begin{bmatrix} Z^p \\ U^f \end{bmatrix} \tag{6.33}$$

$$= Y^f \begin{bmatrix} Z^{p\mathsf{T}} & U^{f\mathsf{T}} \end{bmatrix} \begin{bmatrix} Z^p Z^{p\mathsf{T}} & Z^p U^{f\mathsf{T}} \\ U^f Z^{p\mathsf{T}} & U^f U^{f\mathsf{T}} \end{bmatrix}^{\dagger} \begin{bmatrix} Z^p \\ U^f \end{bmatrix} \tag{6.34}$$

Under the no feedback assumption, the orthogonal projection (6.28) can be decomposed as the sum of two oblique projections:

$$\begin{aligned} \hat{Y}^f &= \hat{E}_{\|U^f}\{Y^f \mid Z^p\} + \hat{E}_{\|Z^p}\{Y^f \mid U^f\} \\ &= Y^f \begin{bmatrix} Z^{p\mathsf{T}} & U^{f\mathsf{T}} \end{bmatrix} \begin{bmatrix} Z^p Z^{p\mathsf{T}} & Z^p U^{f\mathsf{T}} \\ U^f Z^{p\mathsf{T}} & U^f U^{f\mathsf{T}} \end{bmatrix}^{\dagger}_{p\cdot(m+l)\text{ columns}} Z^p \\ &\quad + Y^f \begin{bmatrix} Z^{p\mathsf{T}} & U^{f\mathsf{T}} \end{bmatrix} \begin{bmatrix} Z^p Z^{p\mathsf{T}} & Z^p U^{f\mathsf{T}} \\ U^f Z^{p\mathsf{T}} & U^f U^{f\mathsf{T}} \end{bmatrix}^{\dagger}_{f\cdot m\text{ columns}} U^f \end{aligned} \tag{6.35}$$

Numerically, the efficient implementation of such projection operations relies upon LQ decompositions (i.e. transpose of the QR decomposition) [15]. As a matter of fact, LQ decompositions have also a system theoretic interpretation (see, e.g. [26]). Two algorithms are outlined.

Algorithm 6.1 [42, 41]

1. Consider the factorization

$$\begin{bmatrix} U^f \\ Z^p \\ Y^f \end{bmatrix} = \begin{bmatrix} L_{11} & 0 & 0 \\ L_{21} & L_{22} & 0 \\ L_{31} & L_{32} & L_{33} \end{bmatrix} \begin{bmatrix} Q_1{}^T \\ Q_2{}^T \\ Q_3{}^T \end{bmatrix} \tag{6.36}$$

where $L_{11} \in \mathbb{R}^{fm \times fm}$, $L_{22} \in \mathbb{R}^{p(m+l) \times p(m+l)}$, $L_{33} \in \mathbb{R}^{fl \times fl}$ and Q_i, $i = 1, \ldots, 3$ are orthogonal.

2. Define

$$\hat{\Gamma} = L_{32} L_{22}{}^{\dagger} \tag{6.37}$$

$$\hat{\Lambda} = (L_{31} - L_{32} L_{22}{}^{\dagger} L_{21}) L_{11}{}^{-1} \tag{6.38}$$

3. Use (6.24) to calculate the future outputs as $Y^f = \hat{\Gamma} Z^p + \hat{\Lambda} U^f$.

Algorithm 6.2 [14]

1. Set

$$W^p = \begin{bmatrix} Y^p & U^p \end{bmatrix}^{\mathsf{T}} \tag{6.39}$$

2. Let the LQ decomposition be defined by

$$\begin{bmatrix} W^p \\ U^f \\ Y^f \end{bmatrix} = \begin{bmatrix} \mathscr{L}_{11} & 0 & 0 \\ \mathscr{L}_{21} & \mathscr{L}_{22} & 0 \\ \mathscr{L}_{31} & \mathscr{L}_{32} & \mathscr{L}_{33} \end{bmatrix} \begin{bmatrix} \mathscr{Q}_1{}^T \\ \mathscr{Q}_2{}^T \\ \mathscr{Q}_3{}^T \end{bmatrix} \tag{6.40}$$

where $\mathscr{L}_{11} \in \mathbb{R}^{p(m+l) \times p(m+l)}$, $\mathscr{L}_{22} \in \mathbb{R}^{fm \times fm}$, $\mathscr{L}_{33} \in \mathbb{R}^{fl \times fl}$ and $\mathscr{Q}_i$, $i = 1, \ldots, 3$ are orthogonal.

3. Define

$$\hat{\Upsilon} = (\mathscr{L}_{31} - \mathscr{L}_{32} \mathscr{L}_{22}^{-1} \mathscr{L}_{21}) \mathscr{L}_{11}^{\dagger} \tag{6.41}$$

$$\hat{\Lambda} = \mathscr{L}_{32} \mathscr{L}_{22}^{-1} \tag{6.42}$$

4. The output predictor is finally retrieved as $Y^f = \hat{\Upsilon} W^p + \hat{\Lambda} U^f$.

6.3 Example

6.3.1 Diabetes mellitus

Diabetes mellitus is a chronic disease characterized by the inability of the organism to autonomously regulate the blood glucose levels. It is due to defects in either insulin secretion from the pancreatic β-cells or insulin action [44]. In type 1 diabetes (T1DM), focus of this example, there is an absolute deficiency of insulin secretion, which is due to auto-immune β-cells distruction. Without sufficient insulin, the body is not able to metabolize the glucose absorbed after a food intake and use it efficiently as energy source. As a result, diabetic patients present a chronically raised blood glucose concentration (hyperglycemia). This condition produces long-term consequences for the body associated with damage, dysfunction and failure of various organs [44].

Because insulin deficiency defines T1DM, exogenous insulin administration is required to regulate blood glucose concentration. Focusing on tight blood glucose targets, i.e. 70–140 (mg/dL) [37], the philosophy of insulin replacement is to mimic the physiological endogenous insulin secretion pattern of the non-diabetic person. In the non-diabetic subjects, insulin is secreted into the portal circulation at two rates: a slow basal secretion throughout the 24 hours and an augmented rate at meal times. This pattern can be achieved to some extent with the so-called basal-bolus regime: a basal dose of long-acting insulin is sufficient to keep a constant glucose concentration during fasting conditions and a prandial bolus of rapid-acting insulin enhances an increased glucose uptake during and after meals.

Intensive insulin therapy has been strongly promoted during the last decade [38, 39] in order to keep blood glucose levels as close to normal as possible. However, it may induce potentially severe hypoglycemia, resulting from too high levels of insulin, leading to loss of consciousness, coma and, in rare severe cases, death. In current medical practice, the rough calculation of insulin doses required to compensate for the increased glycemia level due to a food intake is based on empirical rules-of-thumb. Many factors have to be considered in this decision process: health status, current blood glucose level, blood glucose target, foreseen activities, insulin sensitivity, expected future glycemia evolution and approximation of the estimated meal carbohydrate content effects as well as insulin's impact on the subject's own blood glucose. The diabetic subject has usually inadequate understanding and overview of the actual physiological state at any time and would therefore benefit from some sort of decision support. Accurate predictions of blood glucose levels could accomplish this task. The early knowledge about the effects of inputs on glycemia would, indeed, provide the patients with invaluable informations for appropriate on-the-spot decision making concerning the management of the disease. Actually, the European FP7-IST research project DIAdvisor™ [9] during the quadriennium 2008–2012 pursued the development of a personalized blood glucose predicting system and an advisory control system, the DIAdvisor™ tool, to be used on the spot by the users in different daily situations, predicting glycemic excursions following meals, insulin intakes and exercise and giving them advices about how to adjust their treatments. Within this background, data-driven predictors may be used to overcome the limitations arising from the lack of the underlying physiological system model. In this example, therefore, we show how the predictors presented in section 6.2 may be exploited to the purpose of predicting blood glucose multiple steps ahead in the future.

6.3.2 Experimental conditions

Data collection was accomplished during a 3-day in-hospital trial within the project DIAdvisor^{TM} [9]. For the duration of the study, the subjects were equipped with an Abbott FreeStyle Navigator^{TM} Continuous Glucose Monitoring Sensor (CGMS) [1], which provided estimates of glycemia levels from interstitial glucose measurements every 10 minutes, and were served standardized meals for breakfast (8:00), lunch (13:00) and dinner (19:00); the amount of administered carbohydrates being 42, 70 and 70 g respectively. During the whole stay at the clinic, the patients adjusted their boluses by themselves on the basis of a HemoCue blood glucose analyser [18] outcomes and were requested to annotate their insulin doses in a logbook. We present results for a population of 14 patients (9 males and 5 females, age 45.8 ± 12.7 (yr), disease duration 18 ± 11.6 (yr), BMI 23.4 ± 2.7 (kg/m^2), 10 MDI and 4 CSII, HbA1_c 7.6 ± 1 (%), total daily insulin 41.9 ± 18.9 (IU)). The selection criteria were the quantity of data collected (>80% of the expected), no sensor failures and logbook correctly filled in.

6.3.3 Prediction strategy

The physiology of glucose metabolism in diabetes was considered having one output, i.e. blood glucose concentration, and two main inputs, i.e. carbohydrate intake and administered insulin [6]. Physiological models from the literature were used to filter the raw information on carbohydrate and insulin intakes collected by the study participants in their diaries. In particular, the glucose intestinal absorption model first presented in [7] was used to retrieve the glucose rate of appearance in plasma after a meal, i.e. $\hat{u}_g$, while the insulin pharmacokinetics model in [31] allowed us to compute the insulin appearing in plasma after subcutaneous injection, i.e. $\hat{u}_i$. Both models exploited mean population values for the parameters appearing therein which were provided us by Dalla Man and co-workers. The Abbott FreeStyle Navigator^{TM} [1] records were taken as assessment of glycemia. A retrospective recalibration against the collected HemoCue [18] samples was performed offline. The algorithm proposed by King et al. [28], later evaluated by Facchinetti et al. in [13], was used. For each patient a recalibration parameter α^r was found by least-squares estimation:

$$\alpha^r = \arg\min_{\alpha} ||y_{BG} - \alpha y_{IG}||^2$$

with y_{BG} and y_{IG} denoting the HemoCue and the CGM samples, respectively. The recalibrated traces $y^r_{IG} = \alpha^r y_{IG}$ were provided by Facchinetti and co-workers within the DIAdvisor^{TM} project and were used as output signals. Input signals were, instead, $\hat{u}_g$ and $\hat{u}_i$.

We were interested in predictions up to 120 (min) ahead. Denoting with $\tau_{\max}$ (min) the maximum prediction horizon we take into consideration, we distinguished two simulation set-ups: case A where we wanted to optimize predictions up to 30 (min) ahead, i.e. $\tau_{\max} = 30$, and case B with focus on predictions up to 120 (min), i.e. $\tau_{\max} = 120$. Below the scheme used for calculation of multi-step predictions is reported.

Algorithm 6.3 (multi-step prediction)

---

1. Choose the maximum prediction horizon $\tau_{\max}$.
2. Set $p = f = \tau_{\max}$.
3. Estimate $\hat{\Gamma}$ and $\hat{\Lambda}$ from (6.37) and (6.38) applying Algorithm 6.1 to the first half of the data.
4. Let t denote the current time step. Form the predictions $\hat{y}$ of the second half of the data (validation) using the relation

$$\begin{bmatrix} \hat{y}_t \\ \hat{y}_{t+1} \\ \hat{y}_{t+2} \\ \vdots \\ \hat{y}_{t+f-1} \end{bmatrix} = \hat{\Gamma} \begin{bmatrix} z_{t-p} \\ z_{t-p+1} \\ z_{t-p+2} \\ \vdots \\ z_{t-1} \end{bmatrix} + \hat{\Lambda} \begin{bmatrix} \hat{u}_t \\ \hat{u}_{t+1} \\ \hat{u}_{t+2} \\ \vdots \\ \hat{u}_{t+f-1} \end{bmatrix} \tag{6.43}$$

where $\begin{bmatrix} z_{t-p} & z_{t-p+1} & \cdots & z_{t-1} \end{bmatrix}^{\mathsf{T}}$ are the joint input–output data up to time $t-1$ and $\begin{bmatrix} \hat{u}_t & \hat{u}_{t+1} & \cdots & \hat{u}_{t+f-1} \end{bmatrix}^{\mathsf{T}}$ are the future inputs. Notation: $\hat{u} = \begin{bmatrix} \hat{u}_i & \hat{u}_g \end{bmatrix}$ and $y = y^r_{IG}$.

6.3.4 Results

The predictors per se were evaluated with respect to prediction performance on validation data on the basis of the prediction error variance, i.e. $\mathscr{E}[(y^r_{IG} - \hat{y}^r_{IG})(y_{IG} - \hat{y}^r_{IG})^{\mathsf{T}}]$. In addition, given the importance of ultimately predicting glucose levels in blood, the accuracy of the overall predictions $\hat{y}^r_{IG}$ was evaluated according to diabetes-relevant metrics. Numerical accuracy was assessed pointwise by

- absolute difference (AD) [mg/dL]: $\mathrm{AD}(k) = |y_{BG}(k) - \hat{y}^r_{IG}(k)|$
- relative difference (RD): $\mathrm{RD}(k) = \dfrac{(y_{BG}(k) - \hat{y}^r_{IG}(k))}{y_{BG}(k)}$
- International Organization for Standardization (ISO) [20] criteria.

Clinical accuracy was evaluated with Clarke error grid analysis [32]. Performance was assessed for the prediction horizons.

- Case A: $\tau = 10, 20, 30$ (min)
- Case B: $\tau = 30, 60, 90, 120$ (min)

Table 6.1 Multi-step short-term predictors performance evaluation. Case A: Prediction error variance (mg/dL)^2 vs. prediction horizon (min) on validation data

Patient ID	10 (min)	20 (min)	30 (min)
102	7.41	47.31	119.34
103	20.08	107.65	247.32
104	18.31	157.28	495.43
105	13.17	91.54	243.34
106	20.49	129.38	341.73
107	11.99	69.48	174.09
108	7.28	54.56	158.25
115	6.77	38.83	96.32
119	19.92	212.34	714.18
120	5.50	44.79	147.20
121	12.06	72.03	180.02
128	9.39	49.71	114.12
130	13.28	93.56	276.08

Table 6.2 Multi-step short-term predictors performance evaluation. Case B: Prediction error variance (mg/dL)^2 vs. prediction horizon (min) on validation data

Patient ID	30 (min)	60 (min)	90 (min)	120 (min)
102	136.8	482.9	772.6	952.7
103	531.0	1448.3	2336.1	2644.8
104	606.0	2914.4	6277.8	9574.6
105	308.0	946.1	1316.5	1488.3
106	549.7	2492.6	5032.0	7642.2
107	252.2	964.7	1781.7	2181.5
108	364.4	1453.2	1772.7	1202.3
115	131.4	467.6	966.3	1544.0
119	1498.1	6013.8	9100.3	10287.0
120	244.8	1472.8	3364.6	5043.8
121	210.8	805.5	1500.7	2338.7
128	154.8	442.9	814.9	1383.4
130	449.5	1864.3	3548.0	4916.2

Tables 6.1 and 6.2 present assessment of prediction per se relative to cases A and B, respectively, while Tables 6.3–6.6 present numerical and clinical accuracy of prediction with respect to blood glucose relative to Case B. Figure 6.1 shows predictions for one representative subject.

Table 6.3 Numerical and clinical accuracy of predictions. Case B: 30 (min) prediction

Patient	AD (mg/dL)		RD		ISO (%)	CG-pEGA (%)		CG-rEGA (%)	
	Mean	Median	Mean	Median	*	A	B	A	B
CHU102	18.56	13.40	0.00	0.01	77.27	78.26	21.73	60.00	28.88
CHU103	27.61	23.39	−0.12	−0.16	43.58	45.65	52.17	60.00	24.44
CHU104	35.41	29.67	−0.05	0.00	50.00	48.93	42.55	52.17	26.08
CHU105	33.42	22.56	−0.04	−0.05	56.09	55.55	37.77	54.54	11.36
CHU106	29.49	24.45	−0.09	−0.12	58.13	56.52	39.13	64.44	15.55
CHU107	31.83	29.26	−0.11	−0.03	52.17	51.06	46.80	54.34	26.08
CHU108	26.27	16.86	−0.01	0.02	55.55	54.34	41.30	57.77	20.00
CHU115	20.94	14.56	−0.01	−0.01	78.26	78.26	21.73	68.88	15.55
CHU118	21.61	16.43	0.01	0.00	65.85	63.82	31.91	58.69	19.56
CHU119	58.30	36.35	−0.02	0.05	46.80	46.80	38.29	41.30	34.78
CHU120	37.96	28.12	−0.10	−0.01	58.69	58.69	41.30	48.88	26.66
CHU121	30.30	25.27	−0.05	−0.02	52.38	55.31	42.55	58.69	17.39
CHU128	25.30	20.41	0.04	0.00	68.75	68.75	31.25	61.29	19.35
CHU130	39.61	32.76	−0.01	−0.01	60.00	62.50	31.25	54.83	19.35

*Prediction within 20% from reference y_{BG} when $y_{BG} \geq 75$ (mg/dL).

Table 6.4 Numerical and clinical accuracy of predictions. Case B: 60 (min) prediction

Patient	AD (mg/dL)		RD		ISO (%)	CG-pEGA (%)		CG-rEGA (%)	
	Mean	Median	Mean	Median	*	A	B	A	B
CHU102	28.62	24.29	−0.01	0.00	65.11	66.66	33.33	45.45	34.09
CHU103	42.34	39.67	−0.19	−0.28	31.57	31.11	62.22	52.27	36.36
CHU104	64.03	46.76	−0.06	−0.01	66.66	43.47	28.26	46.66	24.44
CHU105	51.54	33.36	−0.06	−0.12	37.50	36.36	47.72	39.53	13.95
CHU106	50.85	45.56	−0.16	−0.12	35.71	35.55	60.00	56.81	27.27
CHU107	44.05	32.67	−0.22	−0.08	53.33	52.17	39.13	46.66	13.33
CHU108	47.62	29.48	0.01	0.04	43.18	42.22	46.66	47.72	20.45
CHU115	27.11	21.27	−0.04	−0.01	60.00	60.00	37.77	50.00	18.18
CHU118	49.74	44.15	0.08	0.04	37.50	36.95	54.34	55.55	24.44
CHU119	76.71	46.17	0.01	0.08	43.47	43.47	36.95	33.33	33.33
CHU120	68.03	43.18	−0.22	0.02	37.77	37.77	51.11	43.18	27.27
CHU121	44.69	35.22	−0.04	0.09	43.90	47.82	43.47	44.44	28.88
CHU128	30.44	27.46	0.08	0.08	59.37	59.37	37.50	64.51	19.35
CHU130	58.96	54.99	−0.00	0.07	33.33	37.50	56.25	48.38	22.58

*Prediction within 20% from reference y_{BG} when $y_{BG} \geq 75$ (mg/dL).

Table 6.5 Numerical and clinical accuracy of predictions. Case B: 90 (min) prediction

Patient	AD (mg/dL)		RD		ISO (%)	CG-pEGA (%)		CG-rEGA (%)	
	Mean	Median	Mean	Median	*	A	B	A	B
CHU102	37.77	33.69	−0.01	−0.02	48.83	51.11	48.88	63.63	18.18
CHU103	48.87	37.49	−0.19	−0.11	40.54	36.36	52.27	55.81	25.58
CHU104	86.46	65.39	−0.09	0.03	33.33	32.60	36.95	35.55	35.55
CHU105	51.04	38.16	−0.14	−0.16	30.00	29.54	61.36	53.48	20.93
CHU106	69.20	51.80	−0.13	−0.14	29.26	29.54	52.27	37.20	30.23
CHU107	52.60	42.39	−0.19	−0.11	40.90	40.00	51.11	45.45	22.72
CHU108	47.19	33.75	0.01	0.09	34.09	33.33	55.55	47.72	29.54
CHU115	29.16	21.43	−0.04	−0.05	65.90	65.90	34.09	48.83	32.55
CHU118	64.22	58.67	0.19	0.22	22.50	26.08	58.69	55.55	26.66
CHU119	86.04	52.44	0.07	0.11	37.77	37.77	37.77	34.09	31.81
CHU120	88.84	62.76	−0.28	−0.03	26.66	26.66	55.55	47.72	20.45
CHU121	49.05	37.15	0.00	0.06	37.50	40.00	51.11	54.54	15.90
CHU128	47.59	39.35	0.13	0.18	45.16	45.16	48.38	53.33	26.66
CHU130	61.08	63.03	−0.01	0.01	33.33	35.48	58.06	43.33	16.66

*Prediction within 20% from reference y_{BG} when $y_{BG} \geq 75$ (mg/dL).

Table 6.6 Numerical and clinical accuracy of predictions. Case B: 120 (min) prediction

Patient	AD (mg/dL)		RD		ISO (%)	CG-pEGA (%)		CG-rEGA (%)	
	Mean	Median	Mean	Median	*	A	B	A	B
CHU102	49.33	42.67	0.05	0.04	39.02	41.86	58.13	59.52	19.04
CHU103	51.09	43.87	−0.25	−0.12	32.43	27.27	61.36	53.48	20.93
CHU104	98.97	63.47	−0.14	0.09	26.82	27.27	36.36	23.25	30.23
CHU105	54.57	49.65	−0.27	−0.28	31.57	28.57	66.66	58.53	24.39
CHU106	78.62	60.58	−0.15	−0.09	36.58	34.09	36.36	37.20	27.90
CHU107	62.77	69.62	−0.17	−0.08	25.00	24.44	68.88	43.18	34.09
CHU108	38.05	30.37	0.01	0.07	40.47	39.53	55.81	42.80	30.95
CHU115	37.77	30.40	−0.06	−0.01	51.16	51.16	46.51	52.38	23.80
CHU118	73.38	56.88	0.23	0.25	26.31	27.27	52.27	37.20	32.55
CHU119	91.51	58.52	0.09	0.11	40.00	40.00	35.55	43.18	27.27
CHU120	113.98	98.49	−0.30	0.13	9.30	9.30	55.81	47.61	19.04
CHU121	60.08	35.0	−0.18	0.07	42.50	42.22	37.77	47.72	20.45
CHU128	59.04	44.16	0.18	0.16	43.33	43.33	40.00	31.03	41.37
CHU130	75.17	58.84	0.05	0.11	26.66	26.66	60.00	48.27	17.24

*Prediction within 20% from reference y_{BG} when $y_{BG} \geq 75$ (mg/dL).

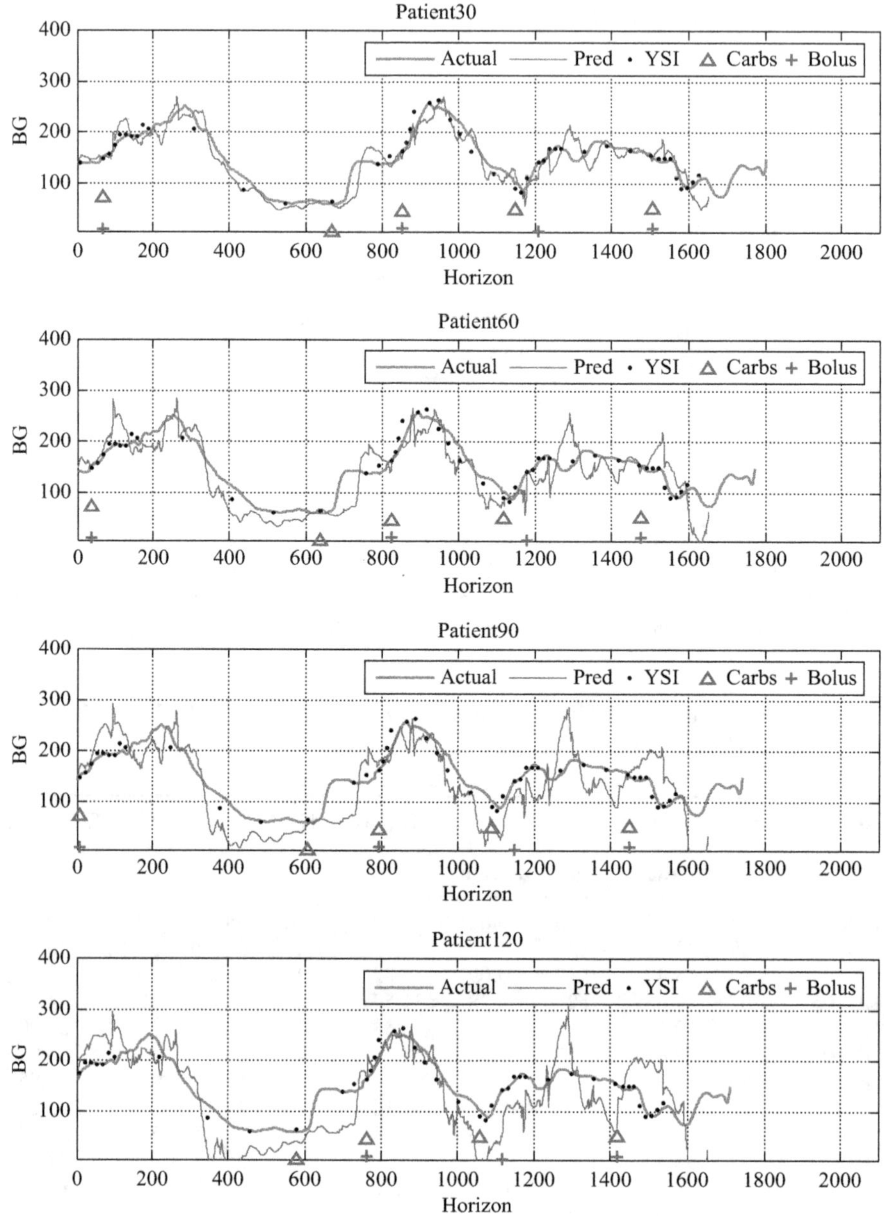

Figure 6.1 Patient CHU102. Multi-step predictions. Case B, $p = f = 120$. Evaluation on validation data. Predictor (thin) and recalibrated interstitial glucose $\hat{y}^r_{IG}$ (thick) (mg/dL) vs. time (min). Top: 30-min ahead; top center: 60-min ahead; bottom center: 90-min ahead; bottom: 120-min-ahead prediction. Meals and injections are indicated with triangles and pluses respectively

6.4 Discussion and conclusions

In this chapter subspace-based data-driven linear multi-step predictors are presented. Resorting to geometric operations on appropriate subspaces spanned by the measured input–output data sequences, predictor coefficients have been estimated directly from the collected data without any prior knowledge of the system generating the data. Neither model structure selection nor model order determination was, therefore, required at this stage.

Some conditions need to be fulfilled in order to have unbiased predictors. Specifically:

- Input data should be 'rich' enough, that is, the persistency of excitation of the data should be of sufficiently high order.
- The space spanned by the row vectors of matrix Z_p and the space spanned by the row vectors of matrix U_f should have zero intersection: $\text{span}(Z_p) \cap \text{span}(U_f) = \{0\}$. This corresponds to no feedback interconnection.

The first condition is related to the LQ decompositions required to solve the problem. Matrix inversions in (6.37) and (6.42) is possible if and only if L_{11} and $\mathscr{L}_{22}$ are full-rank matrices. This is guaranteed when the input signal u is PE of order at least $f \cdot m$.

The second condition is necessary for the orthogonal projection to split uniquely into the sum of two oblique projections (6.29). When data are generated in closed loop this condition is no longer satisfied.

Throughout the chapter, p is chosen so to satisfy the following condition:

$$p \geq \max\{n, \tau\}$$

in order for the system (6.1) to be observable (condition $p \geq n$) and to guarantee predictions up to the largest future horizon we wish to investigate (condition $p \geq \tau$). The parameter has strong connections with model order, and to optimize results it should be given by the user from knowledge or intuitions on the system dynamics. Indeed, requesting to predict with a look-ahead $\tau \gg n$ implies to choose also $p \gg n$, which leads to overparametrization of the model and suboptimal performances.

The example examined the application of the proposed subspace-based multi-step data-driven predictors to the problem of short-term T1DM glycemia prediction. The user was required to choose one parameter: p. Actually the parameter p should be given by the user so as to optimize performances. In other words, if the objective is to predict $\tau = 30$ (min) ahead p should not be given $\gg 30$. Inspection of Tables 6.1 and 6.2 confirmed this statement. The highlighted columns refer to the same prediction horizon, i.e. $\tau = 30$ (min), and show better performances in the case where $p = \tau$ (Case A). Tables 6.1 and 6.2 evaluate the predictors per se, while Tables 6.3–6.6 considered diabetes-relevant metrics.

References

[1] Abbott: FreeStyle Navigator™ (2013). http://www.abbottdiabetescare.co.uk/your-products/freestyle-navigator. Page retrieved September 2013

[2] Anderson, B., Moore, J.: *Optimal Filtering*. Prentice-Hall, Englewood Cliffs, NJ (1979)

[3] Åström, K.J.: *Introduction to Stochastic Control Theory*. Academic Press, New York (1970)

[4] Caines, P.: *Linear Stochastic Systems*. John Wiley, New York (1988)

[5] Chiuso, A., Picci, G.: Prediction error vs. subspace methods in closed-loop identification. In: *Proc. 16th IFAC World Congress (IFAC2005)*. Prague, Czech Republic (2005)

[6] Cobelli, C., Dalla Man, C., Sparacino, G., Magni, L., De Nicolao, G., Kovatchev, B.: Diabetes: Models, signals and control (2009). *IEEE Reviews in Biomedical Engineering* **2**, 54–96

[7] Dalla Man, C., Camilleri, M., Cobelli, C.: A system model of oral glucose absorption: validation on gold standard data. *IEEE Transactions on Biomedical Engineering* **53**(12), 2472–2477 (2006)

[8] Di Ruscio, D.: Model based predictive control: an extended state space approach. In: *Proc. of the 36th Conference on Decision and Control (CDC1997)*, pp. 3210–3217. San Diego, CA (1997)

[9] DIAdvisor: The DIAdvisor™ (2012). http://www.diadvisor.eu. Page retrieved May 2013

[10] Dong, J., Verhaegen, M.: Model free probabilistic design with uncertain Markov parameters identified in closed loop. In: *Proc. 47th Conference on Decision and Control (CDC2008)*, pp. 2637–2643. Cancun, Mexico (2008)

[11] Dong, J., Verhaegen, M.: On the equivalence of closed-loop subspace predictive control with LQG. In: *Proc. 47th Conference on Decision and Control (CDC2008)*, pp. 4085–4090. Cancun, Mexico (2008)

[12] Dong, J., Verhaegen., M., Holweg, E.: Closed-loop subspace predictive control for fault tolerant MPC design. In: *Proc. 17th IFAC World Congress (IFAC2008)*, pp. 3216–3221. Seoul, South Korea (2008)

[13] Facchinetti, A., Sparacino, G., Cobelli, C.: Reconstruction of glucose in plasma from interstitial fluid continuous glucose monitoring data: role of sensor calibration. *Journal of Diabetes Science and Technology* **1**(5), 617–623 (2007)

[14] Favoreel, W., De Moor, B., Van Overschee, P., Gevers, M.: Model-free subspace-based LQG-design. In: *Proc. of the American Control Conference (ACC1999)*, pp. 3372–3376. San Diego, CA (1999)

[15] Golub, G., Van Loan, C.: *Matrix Computations*. Johns Hopkins University Press, Baltimore, MD (1996)

[16] Hamilton, J.: *Time Series Analysis*. Princeton University Press, Princeton, NJ (1994)

[17] Hannan, E., Deistler, M.: *The Statistical Theory of Linear Systems*. John Wiley, New York (1988)

[18] Hemocue: HemoCue® (2013). http://www.hemocue.com/international/Products/Glucose-1160.html. Page retrieved September 2013

[19] Holst, J.: Adaptive prediction and recursive estimation. Ph.D. thesis, Department of Automatic Control, Lund Institute of Technology, Sweden (1977)

[20] International Organization for Standardization (ISO), Publication 15197: In vitro diagnostic test systems: requirements for blood glucose monitoring systems for self-testing in managing diabetes mellitus, Geneva (2003)

[21] Jansson, M.: Subspace identification and ARX modeling. In: *Proc. 13th IFAC Symposium on System Identification (SYSID2003)*. Rotterdam, the Netherlands (2003)

[22] Kadali, R., Huang, B., Rossiter, A.: A data driven subspace approach to predictive controller design. *Control Engineering Practice* **11**, 261–278 (2003)

[23] Kailath, T., Hassibi, B.: *Linear Estimation*. Prentice Hall, Upper Saddle River, NJ (2000)

[24] Kalman, R.: A new approach to linear filtering and prediction problems. *Transactions of the ASME–Journal of Basic Engineering* **82**, 35–45 (1960)

[25] Katayama, T.: *Subspace Methods for System Identification*. Springer-Verlag, London (2005)

[26] Katayama, T.: A system theoretic interpretation of LQ decomposition in subspace identification methods. In: *Proc. of the 17th Int. Symposium on Mathematical Theory of Networks and Systems (MTNS2006)*, pp. 1089–1095. Kyoto, Japan (2006)

[27] Katayama, T., Picci, G.: Realization of stochastic systems with exogenous inputs and subspace identification methods. *Automatica* **35**, 1635–1652 (1999)

[28] King, C., Anderson, S., Breton, M., Clarke, W., Kovatchev, B.: Modeling and calibration effectiveness and blood-to-interstitial glucose dynamics as potential confounders of the accuracy of continuous glucose sensors during hyperinsulinemic clamp. *Journal of Diabetes Science and Technology* **1**(3), 317–322 (2007)

[29] Kolmogorov, A.: Interpolation und extrapolation von stationaren zufalligen folgen. *Izvestiya Akademii Nauk SSSR. Seriya Matematicheskaya* **5**, 3–14 (1939)

[30] Kolmogorov, A.: Sur l'interpolation et extrapolation des suites stationnaires. *Comptes Rendus de l'Académie des Sciences*, Paris **208**, 2043–2045 (1939)

[31] Kovatchev, B., Breton, M., Dalla Man, C., Cobelli, C.: *In Silico Model and Computer Simulation Environment Approximating the Human Glucose/Insulin Utilization*. Master File MAF-1521, Food and Drug Administration (FDA), Silver Spring, MA (2008)

[32] Kovatchev, B., Gonder-Frederik, L., Cox, D., Clarke, W.: Evaluating the accuracy of continuous glucose-monitoring sensors. *Diabetes Care* **27**(8), 1922–1928 (2004)

[33] Ljung, L., McKelvey, T.: Subspace identification from closed-loop data. *Signal Processing* **52**(2), 209–215 (1996)

[34] Menga, G., Mosca, E.: MUSMAR: Multivariable Adaptive Regulators Based on Multistep Cost Functionals. D. Reidel Publishing Company, Dordrecht (1980)

[35] Mosca, E., Zappa, G., Lemos, J.: Robustness of multipredictor adaptive regulator: MUSMAR. *Automatica* **25**(4), 521–529 (1989)
[36] Stoica, P., Nehorai, A.: On multistep prediction error methods for time series models. *Journal of Forecasting* **8**, 357–368 (1989)
[37] The American Diabetes Association: Standards of medical care in diabetes 2013. *Diabetes Care* **36**(Suppl 1), S11–S66 (2013)
[38] The Diabetes Control and Complications Trial Research Group: The effect of intensive treatment of diabetes on the development and progression of long-term complications in insulin-dependent diabetes mellitus. *New England Journal of Medicine* **329**(14), 977–986 (1993)
[39] The Diabetes Control and Complications Trial/Epidemiology of Diabetes Interventions and Complications Study Research Group: Intensive diabetes treatment and cardiovascular disease in patients with type 1 diabetes. *New England Journal of Medicine* **353**(25), 2643–2653 (2005)
[40] Van Overschee, P., De Moor, B.: N4SID: Subspace algorithms for the identification of combined deterministic-stochastic systems. *Automatica* **30**, 75–93 (1994)
[41] Van Overschee, P., De Moor, B.: *Subspace Identification for Linear Systems: Theory-Implementation-Application*. Kluwer Academic Publishers, Boston, MA (1996)
[42] Verhaegen, M.: Identification of the deterministic part of MIMO state space models given in innovations form from input–output data. *Automatica* **30**(1), 61–74 (1994)
[43] Wiener, N.: *The Extrapolation, Interpolation and Smoothing of Stationary Time Series with Engineering Applications*. Wiley, New York (1949)
[44] Williams, G., Pickup, J.: *Handbook of Diabetes* (2nd edn.). Blackwell Science, Oxford (1992)
[45] Woodley, B., How, J., Kosut, R.: Subspace based direct adaptive H_∞ control. *International Journal of Adaptive Control and Signal Processing* **15**, 535–561 (2001)

Chapter 7
Closed-loop subspace predictive control

Gijs van der Veen[1], Jan-Willem van Wingerden[2] and Michel Verhaegen[3]

Abstract

This chapter considers subspace predictive control of systems whose dynamics can be described locally by LTI models. The control algorithm is based on the predictor-based subspace identification framework. In a linear least-squares problem, the observer Markov parameters of the system are recursively estimated. Those parameters are used to construct an output predictor which is in turn used to solve a predictive control problem subject to constraints.

7.1 Introduction

In this chapter we consider subspace predictive control (SPC) as an option of unifying the fields of identification and control design for slowly time-varying systems. The SPC framework offers a way to combine the advantages of a model predictive control (MPC) law with the capability to deal with unknown and slowly time-varying dynamics. Obvious advantages of MPC are the ability to deal with constraints and the intuitive tuning which is not dissimilar to tuning of a linear quadratic regulator. The adaptivity is useful for systems which can be considered locally linear, but have dynamics that vary mildly with the operating condition. SPC has been presented primarily as a framework for fault-tolerant control design [9, 16], because the recursive estimation scheme will automatically detect events such as actuator failures. This is the main strength of the SPC framework.

SPC was first presented in [11, 12] and later in [36] for data-driven H_∞ control. The identification framework used in these versions was the framework of open-loop subspace identification. Since in SPC the identification will by definition take place under closed-loop conditions, modified versions were presented in [9, 16] using the predictor-based subspace identification (PBSID) framework of Chiuso [6]. Compared

[1] Delft University of Technology, Delft, The Netherlands, e-mail: G.J.vanderVeen@tudelft.nl
[2] Delft University of Technology, Delft, The Netherlands, e-mail: J.W.vanWingerden@tudelft.nl
[3] Delft University of Technology, Delft, The Netherlands, e-mail: M.Verhaegen@tudelft.nl

to the original algorithms, the predictor-based version is asymptotically unbiased in closed-loop operation. The PBSID framework is the main result of the developments in closed-loop subspace identification in the last two decades, and essentially combines the high-order ARX modelling proposed by Ljung [24] with the state predictor introduced by Jansson [17]. Several authors contributed to this result (see [33] for a recent review with a brief historical perspective).

SPC comes with a number of challenges which are still largely open problems. First, estimation is performed recursively and to be able to identify the dynamics consistently, all modes must be excited continuously. The challenge is to add the minimal amount of perturbations to maintain a sufficient signal-to-noise ratio and persistence of excitation, since those perturbations will usually adversely affect the process. This problem has been considered in [7]. A partial remedy to this requirement of persistent excitation is to update parameters only when new information is available using a directional forgetting estimation scheme. Such a strategy is recommended later in this chapter. A further challenge is the stability of the control scheme. Efforts have been made to take into account the uncertainty on the estimated parameters to result in cautious H_2 control [8, 21], or robust H_∞ control [36] (using an open-loop identification framework). Similar techniques could be applied to take the uncertainty on the Markov parameters into account in a stochastic robust MPC scheme [3], possibly using the recent framework presented in [10].

Despite these challenges, SPC is an attractive framework due to its conceptual simplicity. In the literature it has been applied primarily to numerical examples, exceptions being [36, 19, 9], where only the latter considered the closed-loop identification framework. Important aspects of ordinary MPC, such as offset-free control and feedforward control are readily implemented in SPC as shown in [19, 29]. While this chapter deals mainly with the theoretical aspects of the SPC framework, the main focus of the next chapter is to show the successful application of SPC to a real example.

7.2 Discrete-time identification framework

In this section we present the framework for closed-loop identification of discrete-time systems that forms the basis for the methods described in this chapter. It is assumed throughout that the system to be identified is a finite-dimensional, linear time-invariant (LTI) system, subject to measurement and/or process noise. It is assumed that the system operates in the vicinity of a steady operating point during some time and that any variations in the operating point or the system dynamics occur slowly. To make this more precise, it is assumed that there is a clear separation between the time constants of the system dynamics and the much slower time constants of parameter variations. A state-space representation of such a system is then given by

$$\mathscr{S}:\begin{cases}\mathbf{x}_{k+1} = A\mathbf{x}_k + B\mathbf{u}_k + \mathbf{w}_k \\ \mathbf{y}_k = C\mathbf{x}_k + D\mathbf{u}_k + \mathbf{v}_k\end{cases} \tag{7.1}$$

with $A \in \mathbb{R}^{n\times n}$, $B \in \mathbb{R}^{n\times n_u}$, $C \in \mathbb{R}^{n_y\times n}$ and $D \in \mathbb{R}^{n_y\times n_u}$. The vectors $\mathbf{x}_k \in \mathbb{R}^n$, $\mathbf{u}_k \in \mathbb{R}^{n_u}$, $\mathbf{y}_k \in \mathbb{R}^{n_y}$ are the state vector, input signal and output signal, respectively. The signals $\mathbf{w}_k \in \mathbb{R}^n$ and $\mathbf{v}_k \in \mathbb{R}^{n_y}$ describe the process noise and measurement noise, respectively, with joint covariance matrix:

$$\mathbb{E}\left\{\begin{pmatrix}\mathbf{w}_j\\ \mathbf{v}_j\end{pmatrix}\begin{pmatrix}\mathbf{w}_k\\ \mathbf{v}_k\end{pmatrix}^T\right\} = \begin{bmatrix} Q & S\\ S^T & R\end{bmatrix}\delta_{jk} \tag{7.2}$$

Furthermore, it is assumed that $R \succ 0$.

It is well-known that such systems admit a so-called innovation state-space representation given by

$$\mathscr{P}: \begin{cases} \mathbf{x}_{k+1} = A\mathbf{x}_k + B\mathbf{u}_k + K\mathbf{e}_k & (7.3a)\\ \mathbf{y}_k = C\mathbf{x}_k + D\mathbf{u}_k + \mathbf{e}_k & (7.3b)\end{cases}$$

with a Kalman gain $K \in \mathbb{R}^{n\times n_y}$ and the (*unique*) ergodic, zero-mean, white noise innovation sequence $\mathbf{e}_k \in \mathbb{R}^{n_y}$ with covariance matrix $\mathbb{E}\{\mathbf{e}_j\mathbf{e}_k^T\} = W\delta_{jk}$, with $W \succ 0$.

It is important to note that the model (7.3) incorporates both the dynamics of the system to be identified as well as the dynamics of the process and measurement noise. This is possible if it is assumed that all noise sources can be modelled as being filtered white noise processes. This, in turn, holds for arbitrarily coloured process and measurement noise sequences when those noise sequences have nonsingular and rational spectra. Then, it can also be assumed that the eigenvalues of $A - KC$ are strictly inside the unit circle, which is equivalent to the natural assumption that the model to be identified is observable.

In representation (7.3), $\mathbf{e}_k$ may be eliminated from the first equation to yield a system description in one-step-ahead predictor form:

$$\mathbf{x}_{k+1} = \tilde{A}\mathbf{x}_k + \tilde{B}\mathbf{u}_k + K\mathbf{y}_k \tag{7.4a}$$

$$\mathbf{y}_k = C\mathbf{x}_k + D\mathbf{u}_k + \mathbf{e}_k \tag{7.4b}$$

where $\tilde{A} \equiv A - KC$ and $\tilde{B} \equiv B - KD$ have been introduced for brevity. We will use the notation $\tilde{\cdot}$ whenever a parameter refers to the predictor model (7.4). This representation forms the basis for the PBSID framework [6].

It is assumed that the plant $\mathscr{P}$ operates in closed-loop as shown in Figure 7.1. In this figure, it is already assumed that the noise effects are modelled as if originating from a filtered innovation sequence. We assume that the feedback system is well-posed, implying that the output is uniquely determined by the states. The feedback system is well-posed if the controller or the plant or both have no direct feedthrough component. If the system and the controller are both LTI, the condition for well-posedness is satisfied if $I_{n_y} + DD_c$ is nonsingular, where D_c is the feedthrough matrix of the controller [31, 20]. Hence, without any means of constraining the structure of D, we can either choose to include it or not in the identification procedure, depending on the feedback system having direct feedthrough or not. Note that a correct

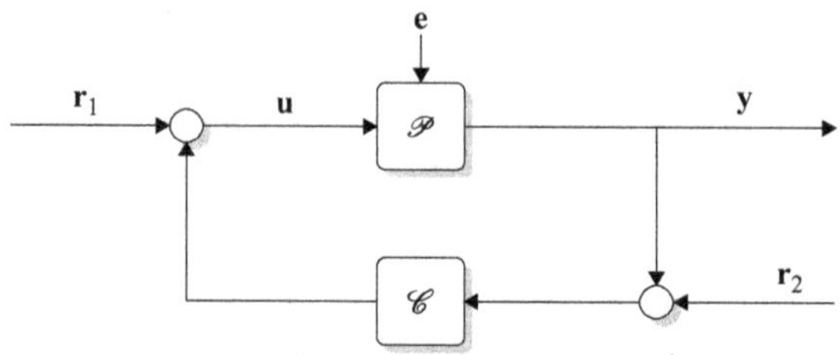

Figure 7.1 The closed-loop configuration considered in this chapter

choice is necessary to obtain consistent estimates of the Markov parameters, see Section 7.2.5.

It is further assumed that the reference signals r_1 and r_2 are chosen such that $\mathbf{u}_k$ and $\mathbf{y}_k$ are both persistently exciting (see Section 7.2.5 for more details).

7.2.1 Preliminaries and notation

Before deriving the data equations for the predictor-based identification framework, we introduce the notation. We introduce a stacked sample of input and output data $\mathbf{z}_k$ according to

$$\mathbf{z}_k = \begin{bmatrix} \mathbf{u}_k \\ \mathbf{y}_k \end{bmatrix}$$

The stacked vector $\mathbf{z}_k^{(p)}$ is then defined as

$$\mathbf{z}_k^{(p)} = \left[\mathbf{z}_{k-p}^\top, \mathbf{z}_{k-p+1}^\top, \ldots, \mathbf{z}_{k-1}^\top\right]^\top$$

where p denotes the *past window* size. We also define a reversed extended controllability matrix $\tilde{\mathcal{K}}^{(p)}$:

$$\tilde{\mathcal{K}}^{(p)} = \left[\tilde{A}^{p-1}\bar{B}, \tilde{A}^{p-2}\bar{B}, \ldots, \bar{B}\right] \tag{7.5}$$

where we have defined $\bar{B} = [\tilde{B}, K]$ for brevity. We emphasise that this matrix contains parameters related to the *predictor* representation (7.4). We will further denote block matrices constructed from data sequences according to

$$Y_{i,N} = \left[\mathbf{y}_i \; \mathbf{y}_{i+1} \; \cdots \; \mathbf{y}_{i+N-1}\right]$$

such that $Y_{i,N}$ has $\mathbf{y}_i$ as its first element and possesses N columns.

7.2.2 Data equations

Starting from some initial state $\mathbf{x}_k$, the state equation (7.4a) can be propagated forward in time, resulting in the expression:

$$\mathbf{x}_{k+p} = \tilde{A}^p \mathbf{x}_k + \tilde{\mathcal{K}}^{(p)} \mathbf{z}_{k+p}^{(p)} \tag{7.6}$$

Based on (7.6) and the output equation (7.4b), the output at time $k + p$ can then be written as

$$\mathbf{y}_{k+p} = C\tilde{A}^p \mathbf{x}_k + C\tilde{\mathscr{K}}^{(p)}\mathbf{z}_{k+p}^{(p)} + D\mathbf{u}_{k+p} + \mathbf{e}_{k+p} \tag{7.7}$$

By the assumption that $\tilde{A}$ has all its eigenvalues inside the open unit disc, the term $\tilde{A}^p$ can be made arbitrarily small, i.e., $\|\tilde{A}^p\|_2 \approx 0$, by choosing p sufficiently large. For that reason, the first term on the right-hand sides of (7.6) and (7.7) will be neglected.

Based on this assumption and depending on the number of samples available, (7.7) can be collected for all available samples k to obtain expressions for $\mathbf{y}_p$ up to $\mathbf{y}_{N-1}$, resulting in the data equation:

$$Y_{p,N_p} = C\tilde{\mathscr{K}}^{(p)} Z_{0,p,N_p} + DU_{p,N_p} + E_{p,N_p} \tag{7.8}$$

Here we have defined $N_p = N - p$ for brevity. In the remainder of this chapter the equality in (7.8) is understood to hold under the aforementioned assumption that $\|\tilde{A}^p\|_2 \approx 0$ in (7.7).

As noted before, the feedthrough term D should only be included when it is present in the actual feedback system to retain consistency of the identification problem. From (7.8) it is clear that if the controller and the system both have direct feedthrough, U_{p,N_p} is correlated with E_{p,N_p} and the Markov parameters can no longer be estimated consistently.

7.2.3 Relation to the ARX model structure

Taking a closer look at the data equation (7.7), neglecting the first term on the right-hand side, it is seen to have a vector-ARX (VARX) structure. Usually, an ARX model structure prescribes a severely restrictive noise model because it forces the system and noise model to have a common set of poles as seen from the following equation:

$$A(z)\mathbf{y}_k = B(z)\mathbf{u}_k + \mathbf{e}_k \tag{7.9}$$

with z^{-1} the unit backshift operator and

$$\begin{aligned} A(z) &= I - a_1 z^{-1} - \cdots - a_p z^{-p} \\ B(z) &= b_0 + b_1 z^{-1} + \cdots + b_p z^{-p} \end{aligned}$$

In this context, based on the assumption that p is chosen sufficiently large and working with the predictor form (7.4), it follows that the high-order VARX model is equivalent to the predictor model. The fact that a high-order ARX model can approximate a predictor model with arbitrary accuracy is well-known in prediction error identification, cf. [35, Example 10.11].

Regarding the ARX model structure defined in (7.9), the parameters a_i and b_i can explicitly be given as the Markov parameters of the predictor form (7.4):

$$a_i = C\tilde{A}^{i-1}K, \quad \text{for } i = 1 \ldots p \tag{7.10a}$$

$$b_i = C\tilde{A}^{i-1}\tilde{B}, \quad \text{for } i = 1 \ldots p \tag{7.10b}$$

$$b_0 = D \tag{7.10c}$$

This follows by a direct comparison of (7.9) with (7.7) after neglecting the contribution of the initial state.

These similarities with ARX model estimation imply that the SPC framework derived in this chapter is in fact very similar to the ARX control suggested, e.g., in [37]. The main difference is that initiating our derivations from a state-space formulations naturally leads to a multi-input-multi-output formulation.

7.2.4 Closed-loop identification issues

Traditional formulations of identification methods often require the plant to operate under open-loop conditions. If such methods are applied to data obtained under closed-loop conditions, the fact that the input signal to the system is correlated with the noise processes (through the feedback mechanism) is disregarded or neglected. The implicit assumption in such methods is that the input signal $\mathbf{u}_k$ is uncorrelated with the past noise process $\mathbf{e}_k$. In a closed-loop situation, however, it is clearly seen that this condition is violated:

$$\mathbb{E}\{\mathbf{u}_k \mathbf{e}_j^{\top}\} \neq 0 \quad \text{for } j < k$$

In the PBSID framework, resulting in (VARX) data equations of the form (7.8), the estimation is not affected by such correlation issues by separating the data into collections of "past" and "future" samples. Thus, asymptotically in the number of samples N and the past window size p, the parameters can be consistently estimated. This is a well-known fact for ARX-type models [23].

7.2.5 Estimating the predictor Markov parameters

Based on the assumption that e_k is the zero-mean white noise innovation sequence, the predictor Markov parameters in (7.8) can be consistently estimated in a least-squares sense:

$$\arg \min_{[C\tilde{\mathcal{K}}^{(p)}\ D]} \left\| Y_{p,N_p} - [C\tilde{\mathcal{K}}^{(p)}\ D] \begin{bmatrix} Z_{0,p,N_p} \\ U_{p,N_p} \end{bmatrix} \right\|_F^2 \tag{7.11}$$

For a full-rank data matrix $\left[Z_{0,p,N_p}^{\top}, U_{p,N_p}^{\top}\right]^{\top}$, the least-squares solution can be found uniquely, for instance, from an RQ decomposition [15] of the data.

It is obvious that uniqueness of the parameter estimate requires $\bar{Z} = \left[Z_{0,p,N_p}^{\top}, U_{p,N_p}^{\top}\right]^{\top}$ to be of full rank. This requirement depends on the experimental data, and thus on the true system, reference excitation and the nature of the feedback mechanism [2]. For interesting accounts regarding model identifiability and experiment requirements see, e.g., [13, 4, 14, 2].

7.2.6 *Recursive solution of the parameter estimation problem*

In the context of SPC we do not consider a batch formulation of the Markov parameter estimation problem as in (7.11), but instead a recursive formulation where the estimates are updated as new samples of input-output data arrive in each sampling instant. In this section we discuss the procedure of recursively estimating the parameters in an online setting using recursive least squares (RLS). The starting point for our derivations is the least-squares estimation of the predictor Markov parameters (7.11).

To maintain a compact notation, the least-squares problem (7.11) will be concisely written in the standard form:

$$\widehat{\Theta}_k = \arg\min_{\Theta} \| Y_k - \Phi_k \Theta \|_F^2 \tag{7.12}$$

where the subscript k signifies that data up to time instant k is available and used. Hence, the involved matrices are defined as (see (7.11)):

$$Y_k \triangleq \begin{bmatrix} \mathbf{y}_{p+1}^{\top} \\ \mathbf{y}_{p+2}^{\top} \\ \cdots \\ \mathbf{y}_k^{\top} \end{bmatrix}, \quad \Phi_k \triangleq \begin{bmatrix} \varphi_{p+1}^{\top} \\ \varphi_{p+2}^{\top} \\ \cdots \\ \varphi_k^{\top} \end{bmatrix}, \quad \Theta \triangleq \left[C\tilde{\mathcal{K}}^{(p)}\ D\right]^{\top}$$

with

$$\varphi_k \triangleq \begin{bmatrix} \mathbf{z}_k^{(p)} \\ \hline \mathbf{u}_k \end{bmatrix}$$

Furthermore, let $n_{\Theta} = \dim_{\mathbb{R}} vec(\Theta)$ denote the number of parameters to be estimated.

Obviously, the data matrices one time instant later, i.e., Y_{k+1} and Φ_{k+1} contain the new samples $\mathbf{u}_{k+1}$ and $\mathbf{y}_{k+1}$ and these matrices therefore grow in time. Explicitly, these matrices become:

$$Y_{k+1} \triangleq \begin{bmatrix} \mathbf{y}_{p+1}^{\top} \\ \mathbf{y}_{p+2}^{\top} \\ \cdots \\ \mathbf{y}_k^{\top} \\ \hline \mathbf{y}_{k+1}^{\top} \end{bmatrix}, \quad \Phi_{k+1} \triangleq \begin{bmatrix} \varphi_{p+1}^{\top} \\ \varphi_{p+2}^{\top} \\ \cdots \\ \varphi_k^{\top} \\ \hline \varphi_{k+1}^{\top} \end{bmatrix}$$

The full-rank least-squares solution of (7.12) can now be written as

$$\widehat{\Theta}_k = (\Phi_k^\top \Phi_k)^{-1} \Phi_k^\top Y_k \tag{7.13}$$

Suppose a new sample, described by $(\mathbf{y}_{k+1}, \varphi_{k+1})$ becomes available. It is straightforward to show that the new least-squares solution satisfies:

$$\widehat{\Theta}_{k+1} = \underbrace{(\Phi_k^\top \Phi_k + \varphi_{k+1}\varphi_{k+1}^\top)^{-1}}_{=P_{k+1}=\mathscr{I}_{k+1}^{-1}} \left(\Phi_k^\top Y_k + \varphi_{k+1}\mathbf{y}_{k+1}\right)$$

We now define the *covariance matrix* P_k as

$$P_k = (\Phi_k^\top \Phi_k)^{-1}$$

It is well-known that the new least-squares solution can be obtained using the RLS algorithm, which is given by the following update relations [27]:

$$P_k = \frac{1}{\lambda}\left(P_{k-1} - P_{k-1}\varphi_k\left(\lambda I_{n_y} + \varphi_k^\top P_{k-1}\varphi_k\right)^{-1}\varphi_k^\top P_{k-1}\right) \tag{7.14a}$$

$$\hat{\Theta}_k = \hat{\Theta}_{k-1} - P_k\varphi_k(\mathbf{y}_k - \varphi_k^\top \hat{\Theta}_{k-1}) \tag{7.14b}$$

Note that since n_y may be greater than 1 we are in fact solving a multiple linear regression problem. For $n_y = 1$ these equations simplify to their more well-known forms. It is finally noted that in propagating the covariance matrix $P_k = (\Phi_k \Phi_k^\top)^{-1}$ in finite precision arithmetic the RLS algorithm is not guaranteed to retain a positive definite covariance matrix due to accumulation of round-off errors. This effect becomes more pronounced as the condition number of P_k grows. Therefore in practice square root implementations of the RLS update are highly preferred [34, 28]. Such algorithms propagate the Cholesky factor of the covariance matrix (or the information matrix) by executing a sequence of orthogonal transformations in each time step. Square root algorithms are numerically superior to their counterpart described above. The computational complexity of the standard RLS as well as the square root algorithms is $\mathscr{O}(n_\Theta^2)$ per iteration ($n_\Theta = p(n_u + n_y) + n_u n_y$ is the number of parameters in the LS problem for the Markov parameters (7.11)). That is, quadratic in the number of parameters. See [32, Chap. 3] for a derivation of this algorithm for SPC.

7.2.7 Using directional forgetting

A drawback of the RLS scheme with exponential forgetting (7.14) is that all information is forgotten uniformly. As a consequence, the matrix $\mathscr{I}$ may become singular over the course of time if the observations φ_k do not excite all directions. Suppose the information matrix is an n-by-n square matrix $\mathscr{I} \in \mathbb{R}^n$, then over time $\varphi_k \in \mathbb{R}^n$ should uniformly 'visit' all directions in the space $\mathbb{R}^n$. This effect is coupled to the notion of persistent excitation [18].

In [5] a directional forgetting algorithm was proposed that uses a decomposition of the information matrix. A significant advantage of this scheme is that old information is only discounted if it can be replaced with new information, thus relaxing persistency of excitation requirements. In [5] it is proven that the condition number of the information matrix remains bounded and that the singular values remain bounded from below and above. Since the added computational complexity is minor we have used the directional forgetting scheme in the case study presented in Chapter 12. The reader is referred to [32, Chap. 3] for a computationally reliable and efficient square-root implementation of this algorithm and for numerical estimation experiments demonstrating the boundedness properties.

7.3 Deriving the subspace predictor

Having estimated the predictor Markov parameters from least-squares problem (7.11), which were explicitly given in (7.10), the next step is to construct an output predictor for a sequence of future outputs. As a starting point (7.7) is revisited:

$$\mathbf{y}_{k+p} = \widehat{C\mathscr{K}}\mathbf{z}_k^{(p)} + \widehat{D}\mathbf{u}_{k+p}$$

where the term $\mathbf{e}_{k+p}$ has been omitted since for purposes of prediction we have $\mathbb{E}\{\mathbf{e}_{k+p}\} = 0$.

At time instant k, we consider the outputs at time instant $k+1$ up to $k+N$, where the arbitrary length of the prediction interval is denoted by $N \in [1, \infty)$.[1] Define by $\mathbf{y}_k^{(N)}$ a stacked sequence of outputs according to

$$\mathbf{y}_k^{(N)} = \left[\mathbf{y}_{k+1}^{\top}, \ldots, \mathbf{y}_{k+N}^{\top}\right]^{\top}$$

A stacked sequence $\mathbf{u}_k^{(N)}$ is defined analogously. Then it is readily derived that

$$\mathbf{y}_k^{(N)} = \tilde{\Gamma}^{(N)}\tilde{\mathscr{K}}^{(p)}\mathbf{z}_k^{(p)} + \tilde{H}^{(N)}(B, D)\mathbf{u}_k^{(N)} + \tilde{H}^{(N)}(K, 0)\mathbf{y}_k^{(N)} \tag{7.15}$$

where $\tilde{H}^{(N)}(B, D)$ is defined as

$$\tilde{H}^{(N)}(B, D) = \begin{bmatrix} D & 0 & 0 & \cdots & 0 \\ CB & D & 0 & \cdots & 0 \\ \vdots & \ddots & \ddots & \ddots & \\ C\tilde{A}^{N-2}B & C\tilde{A}^{N-3}B & \cdots & CB & D \end{bmatrix}$$

[1] It is remarked here that the prediction horizon N is not limited by the past window dimension p.

and $\tilde{H}^{(N)}(K,0)$ likewise. The matrix $\tilde{\Gamma}^{(N)}\tilde{\mathscr{K}}^{(p)}$ is approximated as

$$\tilde{\Gamma}^{(N)}\tilde{\mathscr{K}}^{(p)} = \begin{bmatrix} C\tilde{A}^{p-1}\bar{B} & C\tilde{A}^{p-2}\bar{B} & \cdots & \cdots & C\bar{B} \\ 0 & C\tilde{A}^{p-1}\bar{B} & \cdots & \cdots & C\tilde{A}\bar{B} \\ \vdots & & \ddots & \ddots & \vdots \\ 0 & 0 & C\tilde{A}^{p-1}\bar{B} & \cdots & C\tilde{A}^{N-1}\bar{B} \end{bmatrix}$$

These matrices can be constructed directly from the estimated predictor Markov parameters $\widehat{C\tilde{\mathscr{K}}^{(p)}}$, $\widehat{D}$. Note that since the predictor parameters are used, the output prediction $\mathbf{y}_k^{(N)}$ appears on both sides of the equation. There are at least two efficient ways to obtain the open-loop output predictor. First, one can simply pre-multiply the data equation with $(I - \tilde{H}^{(N)}(K,0))^{-1}$ to obtain

$$\mathbf{y}_k^{(N)} = \Gamma^{(N)}\tilde{\mathscr{K}}^{(p)}\mathbf{z}_k^{(p)} + H^{(N)}(B,D)\mathbf{u}_k^{(N)} \tag{7.16}$$

Note that matrix $(I - \tilde{H}^{(N)}(K,0))$ is lower unit-triangular, allowing an efficient forward-substitution to be used. An alternative method, derived in [7], uses a recursive formulation of the forward-substitution procedure to immediately build the predictor matrices, without explicitly forming $(I - \tilde{H}^{(N)}(K,0))$. This procedure thus requires less storage, but a higher amount of looped operations, so the preferred method is implementation-dependent. In both cases it is possible to show, by carrying out the pre-multiplication, that the open-loop predictor contains the Markov parameters of the *innovation model* (7.3) as opposed to those of the *predictor model* (7.4).

7.4 Setting up the predictive control problem

In this section the derivation of the predictive control problem is considered, based on the predictor (7.16) derived in the previous section. A typical MPC performance index with output weighting is considered. In this context, the goal is to, at each time instant, minimise the value of the following objective function:

$$J_k = \sum_{i=1}^{N} (\mathbf{y}_{k+i} - \mathbf{r}_{k+i})^{\top} Q_y (\mathbf{y}_{k+i} - \mathbf{r}_{k+i}) + \mathbf{u}_{k+i}^{\top} R_u \mathbf{u}_{k+i} + \Delta\mathbf{u}_{k+i}^{\top} R_{\Delta u} \Delta\mathbf{u}_{k+i}$$

subject to

$$\begin{aligned} \mathbf{u}_{\min} \leq \mathbf{u}_{k+i} \leq \mathbf{u}_{\max}, \qquad & i = 1 \ldots N, \\ \Delta\mathbf{u}_{\min} \leq \Delta\mathbf{u}_{k+i} \leq \Delta\mathbf{u}_{\max}, \qquad & i = 1 \ldots N \end{aligned}$$

This objective expresses that the output should track a reference signal $\mathbf{r}_{k+i}$, while minimising the control effort $\mathbf{u}_{k+i}$ and control rate $\Delta\mathbf{u}_{k+i}$.[2,3] Using the notation introduced before, the objective can be reduced to

$$J_k = \|\mathbf{y}_k^{(N)} - \mathbf{r}_k^{(N)})\|_{\tilde{Q}_y} + \|\mathbf{u}_k^{(N)}\|_{\tilde{R}_u} + \|\Delta\mathbf{u}_k^{(N)}\|_{\tilde{R}_{\Delta u}} \tag{7.17}$$

after defining block-diagonal matrices $\tilde{Q}_y$, $\tilde{R}_u$ and $\tilde{R}_{\Delta u}$ appropriately. In this equation, $\|\mathbf{x}\|_Q \triangleq \mathbf{x}^\top Q\mathbf{x}$ denotes the weighted Euclidean norm.

To obtain the control rate $\Delta\mathbf{u}$, the following matrices are defined:

$$S_\Delta = \begin{bmatrix} I_m & & & \\ -I_m & I_m & & \\ & \ddots & \ddots & \\ & & -I_m & I_m \end{bmatrix}, \quad S_0 = \begin{bmatrix} 0 & \cdots & I_m & 0 \\ \vdots & & & \\ 0 & \cdots & 0 & 0 \end{bmatrix}$$

so that

$$\Delta\mathbf{u}_k^{(N)} = S_0\mathbf{z}_k^{(p)} + S_\Delta\mathbf{u}_k^{(N)} \tag{7.18}$$

Furthermore, a control horizon N_c is introduced so that $\mathbf{u}_{k+j} = \mathbf{u}_{k+N_c}$ for $j = [N_c+1, N]$:

$$\mathbf{u}_k^{(N)} = S_{N_c}\mathbf{u}_k^{(N_c)} \tag{7.19}$$

with

$$S_{N_c} = \begin{bmatrix} I_m & & \\ & \ddots & \\ & & I_m \\ & & I_m \\ & & \vdots \\ & & I_m \end{bmatrix} \tag{7.20}$$

Evaluating the objective function results in an inequality constrained quadratic programming (QP) problem in the variables $\mathbf{u}_{k+i}$, $i = 1\ldots N_c$. Here, we have only shown an elementary formulation, but obviously more sophisticated (in)equality constraints or objective terms can be incorporated while retaining convexity of the optimisation problem.

[2] $\Delta\mathbf{u}_k$ can be related to physical rate limits using the approximation $(\Delta\mathbf{u}_k)_{\max} \approx \left(\frac{\mathrm{du}}{\mathrm{d}t}\right)_{\max} \cdot T_s$, with T_s the sample time and $\left(\frac{\mathrm{du}}{\mathrm{d}t}\right)_{\max}$ the actual rate limit.
[3] Note that due to this formulation of the objective, perfect steady-state nonzero reference tracking can never be achieved when $\mathbf{u}_{ss} \neq 0$ if the second cost term is included, except when $R_u = 0$ is chosen.

7.4.1 Real time solution of the QP

For the real time solution of quadratic programming problems very efficient and freely available solvers are available. See, for instance, [1, 26, 30] to mention a few.

7.4.2 Parameter selection

The presented algorithm requires a few parameters to be tuned and these parameters deserve some attention. The parameters are

- The value of past window dimension p determines the number of system parameters that are estimated and the order of the underlying ARX model (7.9). In off-line identification, this parameter is usually taken quite large, e.g., several times the maximum expected model order. In the recursive formulation (7.14), however, this choice directly affects the computational complexity of the update. Furthermore, estimating a larger number of parameters reduces the convergence rate of the RLS scheme. In an off-line analysis, a suitable value of p can be obtained, for instance, by validating the hypothesis that the least-squares residual resembles a zero-mean white noise sequence and whether the predictor Markov parameters indeed tend to zero at p.
- The value of forgetting factor λ should always be strictly less than 1 to avoid overflow issues when updating the covariance matrix with new observations (that is, for $\lambda = 1$ the information matrix will grow unbounded). This is also necessary to retain adaptivity of the parameters by forgetting, in an exponential sense, older data. The effective window length is usually expressed as $N \approx 1/(1 - \lambda)$. Depending on the characteristics of new observations, if the effective window is too short, information will be discarded too rapidly and the covariance matrix will become ill-conditioned. This choice of λ is essentially a trade-off between the ability to track parameter changes and the mean square errors of the estimated parameters. Again, in off-line analysis a value for λ can be chosen that suits the process at hand. Sometimes, λ itself may be an adaptive value, which is chosen as a function of the regressors and excitation levels [22].
- The prediction horizon N_p is typically chosen such that the prediction interval contains the crucial dynamics of the process; its value can for instance be based on the system's step response [25].
- The control horizon N_c determines the number of future inputs that is free. A small N_c reduces the complexity of the QP and further has a smoothing effect on the control signal, which is important for stability.
- The GPC weights, Q_y, R_u and $R_{\Delta u}$, determine the trade-off between control authority (disturbance rejection) and actuator use. In the finite-horizon predictive control problem, these parameters also to a large extent affect the stability of the closed-loop system. These parameters are usually tuned so as to ensure that the control signals stay within their saturation and rate limits during regular operation.

7.5 Concluding remarks

In this chapter we derived the basic SPC framework. It was shown that the subspace predictive control framework can be derived as a direct extension of the predictor-based identification framework for closed-loop systems. In that sense it forms a first step towards the fusion of identification and control into data-driven control. The ultimate goal of such an approach is that overall controller design and tuning becomes more straightforward by directly exploiting system information embedded in measured data. In the subspace predictive control framework this consists mainly in replacing the usual output predictor used in MPC law with a predictor derived from data.

7.5.1 Algorithm summary

Finally, the complete SPC algorithm is briefly summarised below.

Algorithm 7.1 Subspace predictive control

Given: $k = 0, p > 0, P_0, \Theta_0, \varphi_0, \tilde{Q}_y, \tilde{R}_u, \tilde{R}_{\Delta u}, N > 0, 0 < N_c \leq N$

loop

1. Update the covariance matrix P_k using (7.14a) or a square-root equivalent and update the parameters Θ_k using (7.14b) or a square-root equivalent;
2. Construct the parameter matrices in (7.15) using Θ_k;
3. Solve for the open-loop predictor (7.16);
4. Construct the QP based on (7.17);
5. Solve the QP;
6. Implement the first element $\mathbf{u}_{k+1}$ of the optimal input sequence;
7. $k \leftarrow k + 1$.

end loop

References

[1] Ferreau, H.J., Kirches, C., Potschka, A., Bock, H.G., Diehl, M.: qpOASES: a parametric active-set algorithm for quadratic programming. *Mathematical Programming Computation* 1–37 (2014). URL http://link.springer.com/article/10.1007/s12532-014-0071-1

[2] Bazanella, A., Gevers, M., Mišković, L.: Closed-loop identification of MIMO systems: a new look at identifiability and experiment design. *European Journal of Control* **16**(3), 228–239 (2009)

[3] Bemporad, A., Morari, M.: Robust model predictive control: a survey. *Robustness in Identification and Control* **245**, 207–226 (1999)

[4] Bombois, X.J.A., Scorletti, G., Gevers, M., Van den Hof, P.M.J., Hildebrand, R.: Least costly identification experiment for control. *Automatica* **42**(10), 1651–1662 (2006). DOI 10.1016/j.automatica.2006.05.016

[5] Cao, L., Schwartz, H.: A directional forgetting algorithm based on the decomposition of the information matrix. *Automatica* **36**(11), 1725–1731 (2000). DOI 10.1016/S0005-1098(00)00093-5

[6] Chiuso, A.: On the relation between CCA and predictor based subspace identification. *IEEE Transactions on Automatic Control* **52**(10), 1795–1812 (2007)

[7] Dong, J.: Data driven fault tolerant control: a subspace approach. Ph.D. thesis, Delft University of Technology (2009)

[8] Dong, J., Verhaegen, M.: Cautious H_2 optimal control using uncertain Markov parameters identified in closed loop. *Systems & Control Letters* **58**(5), 378–388 (2009)

[9] Dong, J., Verhaegen, M., Holweg, E.: Closed-loop subspace predictive control for fault tolerant MPC design. In: *Proceedings of the 17th IFAC World Congress*, vol. 17, Seoul, South Korea (2008)

[10] Evans, M., Cannon, M., Kouvaritakis, B.: Robust and stochastic linear MPC for systems subject to multiplicative uncertainty. In: *Preprints of 4th IFAC Nonlinear Model Predictive Control Conference*, Leeuwenhorst, The Netherlands (2012)

[11] Favoreel, W., De Moor, B.: SPC: Subspace Predictive Control. In: *Proceedings IFAC World Congress*, Beijing, pp. 235–240 (1998)

[12] Favoreel, W., De Moor, B., Van Overschee, P., Gevers, M.: Model-free subspace-based LQG-design. In: *Proceedings of the American Control Conference*, vol. 5, pp. 3372–3376 (1999). DOI 10.1109/ACC.1999.782390

[13] Gevers, M.: Identification for control: from the early achievements to the revival of experiment design. *European Journal of Control* **11**(4–5), 335–352 (2005). DOI 10.3166/ejc.11.335-352

[14] Gevers, M., Bazanella, A.S., Bombois, X.J.A., Mišković, L.: Identification and the information matrix: how to get just sufficiently rich? *IEEE Transactions on Automatic Control* **54**(12), 2828–2840 (2009). DOI 10.1109/TAC.2009.2034199

[15] Golub, G.H., Van Loan, C.F.: Matrix Computations, 3rd edn. The Johns Hopkins University Press, Baltimore, Maryland (1996)

[16] Hallouzi, R.: Multiple-model based diagnosis for adaptive fault-tolerant control. Ph.D. thesis, Delft University of Technology (2008)

[17] Jansson, M.: Subspace identification and ARX modeling. In: *Proceedings of the 13th IFAC Symposium on System Identification*, pp. 1625–1630. Rotterdam, Netherlands (2003)

[18] Johnstone, R.M., Johnson Jr., C.R., Bitmead, R.R., Anderson, B.D.O.: Exponential convergence of recursive least squares with exponential forgetting factor. *Systems & Control Letters* **2**(2), 77–82 (1982). DOI 10.1016/S0167-6911(82)80014-5

[19] Kadali, R., Huang, B., Rossiter, A.: A data driven subspace approach to predictive controller design. *Control Engineering Practice* **11**(3), 261–278 (2003). DOI 10.1016/S0967-0661(02)00112-0

[20] Katayama, T.: *Subspace Methods for System Identification*. Springer (2005)

[21] Kulcsár, B., Verhaegen, M.: Robust cautious data driven control with guaranteed mean square stability. In: *Proceedings of the 49th IEEE Control Conference on Decision and Control*, pp. 2276–2281. Atlanta (2010). DOI 10.1109/CDC.2010.5717081

[22] Kulhavý, R.: Restricted exponential forgetting in real-time identification. *Automatica* **23**(5), 589–600 (1987). DOI 10.1016/0005-1098(87)90054-9

[23] Ljung, L.: *System Identification: Theory for the User*, 2nd edn. Prentice Hall PTR, Upper Saddle River, New Jersey (1999)

[24] Ljung, L., McKelvey, T.: Subspace identification from closed loop data. *Signal Processing* **52**(2), 209–215 (1996)

[25] Maciejowksi, J.M.: *Predictive Control with Constraints*. Pearson Education Limited, Upper Saddle River, New Jersey (2002)

[26] Mattingley, J., Boyd, S.: Cvxgen: a code generator for embedded convex optimization. *Optimization and Engineering* **13**(1), 1–27 (2012). DOI 10.1007/s11081-011-9176-9. URL http://dx.doi.org/10.1007/s11081-011-9176-9

[27] Åström, K.J., Wittenmark, B.: *Adaptive Control*, 2nd edn. Prentice Hall, Upper Saddle River, New Jersey (1994)

[28] Sayed, A.: *Fundamentals of Adaptive Filtering*. Wiley, New York (2003)

[29] Soliman, M., Malik, O.P., Westwick, D.: Fault tolerant control of variable-speed variable-pitch wind turbines: A subspace predictive control approach. In: *Proceedings of the 16th IFAC Symposium on System Identification*. Brussels (2012)

[30] Ullmann, F.: FiOrdOs: *A MATLAB Toolbox for C-Code Generation for First Order Methods*. Master's thesis, ETH Zürich (2011)

[31] Van Overschee, P., De Moor, B.: Closed loop subspace system identification. In: *Proceedings of the 36th IEEE Conference on Decision and Control*, vol. 2, pp. 1848–1853 (1997). DOI 10.1109/CDC.1997.657851

[32] van der Veen, G.J.: Identification of wind energy systems. Ph.D. thesis, Delft University of Technology (2013)

[33] van der Veen, G.J., van Wingerden, J.W., Bergamasco, M., Lovera, M., Verhaegen, M.: Closed-loop subspace identification methods: an overview. *IET Control Theory and Applications* **7**(10), (2013)

[34] Verhaegen, M.: Round-off error propagation in four generally-applicable, recursive, least-squares estimation schemes. *Automatica* **25**(3), 437–444 (1989). DOI 10.1016/0005-1098(89)90013-7

[35] Verhaegen, M., Verdult, V.: *Filtering and System Identification: A Least Squares Approach*, 1st edn. Cambridge University Press, Cambridge, UK (2007)

[36] Woodley, B.R.: Model free subspace based H_∞ control. Ph.D. thesis, Stanford University (2001)

[37] Zhu, Y.: Multivariable process identification for MPC: the asymptotic method and its applications. *Journal of Process Control* **8**(2), 101–115 (1998). DOI http://dx.doi.org/10.1016/S0959-1524(97)00035-8

Chapter 8
Structured nonlinear system identification

Tyrone Vincent[1], *Kameshwar Poolla*[2] *and Carlo Novara*[3]

Abstract

To obtain an identified model from data, the system identification practitioner has to make an important choice: to specify the set of candidate models, or model structure. This choice can play an outsized role on the success or failure of the identification process. If the model structure is specified too restrictively, so that the true system is not represented, then the identified model will be biased. On the other hand, if the model structure is specified too generally, then the identified model can have a high variance, and a significant amount of data may be needed to reduce the sensitivity to measurement noise. Ideally, the practitioner should choose the model structure so that it encodes all the information that is known with high confidence. The structured nonlinear system identification approach is designed to give the practitioner a very flexible model structure that can easily be configured to be as restrictive or permissive as the a-priori information about the system warrants. In this chapter, a complete introduction to structured identification is developed, with examples relevant to many different real-world applications integrated throughout.

8.1 Introduction

The identification of dynamical models for nonlinear systems remains a challenging problem with a wide variety of approaches [28, 29, 16]. Often, the distinguishing feature between competing methods is the *model class*, that is, the set of candidate models among which the identified model is to be selected. A key property of this model class is its complexity, which intuitively is the number of input–output relationships that can be captured by this model class, but can also be given a more precise definition, such as the Vapnik–Chervonenkis (VC) dimension [33]. As the model complexity increases, it is easier for the true system to be contained in the model class, thus decreasing a potential bias error. On the other hand, increasing model complexity

[1]Colorado School of Mines, Golden, USA, e-mail: tvincent@mines.edu
[2]University of California, Berkeley, USA, e-mail: poolla@berkeley.edu
[3]Politecnico di Torino, Torino, Italy, e-mail: carlo.novara@polito.it

is also associated with a larger sensitivity to measurement noise and other unmeasured signals, increasing the variance of the identified model. Thus, one wishes to find the model class of *least* complexity that also includes (or well approximates) the true system while at the same time allows for an efficient algorithm to search among the model class for the best match to measured data. To create this reduced complexity model structure, the user will wish to apply as much a-priori information about the system to be identified as possible. In this chapter, we discuss a methodology that we term *structured system identification* [9, 22, 10]. The significant feature of this approach is the ease by which a user can specify model structures that respect knowledge concerning signal interconnection behavior, and by doing so, significantly reduce the complexity of the model structure.

This methodology uses the linear fractional representation (LFR) as a data structure for specifying the interconnection patterns of signals within the system [13]. (This has also been called the linear fractional transformation (LFT) model structure.) Because we are emphasizing the interconnection structure, there is some similarity with the block-oriented model structures, such as Hammerstein, Wiener, and linear-nonlinear-linear (LNL) [20, 2, 3, 7]. However, there are some important differences. As will be discussed in detail, the LFR model structure is very flexible and allows for the specification of a wide variety of system interconnections. But, since we are allowing a more general model structure, there will be some identification methods that are possible for Wiener or Hammerstein models (such as specialized inputs [3] or correlation analysis [2]) that we do not take advantage of.

In the remainder of this chapter, we outline the structured system identification methodology. The first section introduces the LFR model structure. This is followed by some specific examples of how system structure can be exploited, and how the structure is represented using the LFR. Since the LFR is so flexible, different realizations of the model structure will have radically different properties. In the third section, we discuss some of the properties that are relevant for system identification, especially those that give some insight into when the LFR system structure implies that the identification problem is well posed and straightforward, or when it is more difficult. The fourth section gives both parametric and nonparametric methods for performing the system identification task from sampled data, and the chapter concludes with an example.

8.2 Specification of model structures using the LFR

We suppose that the system identification experiment consists of L samples of input and output signals at equal intervals, so that have available samples $u_1, \ldots, u_L$ for the input and $y_1, \ldots, y_L$ for the output. The signals u_k and y_k may be multidimensional, with dimension n_u and n_y, respectively. For convenience, we use $u = (u_1, \ldots, u_L)$ and similarly for y. Since we have discrete data, we assume that we are seeking to identify a discrete-time dynamical system.

The data structure that encodes the model set in structured identification is the linear fractional transformation (LFR) shown in Figure 8.1. The term q^{-1} represents

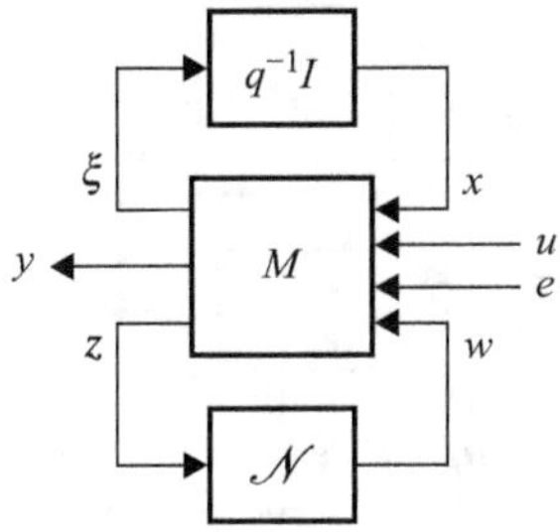

Figure 8.1 Linear fractional representation (LFR) model structure. The signals are u-measured input, e-unmeasured inputs, y-measured output, x-internal state, ξ-prior internal state; z, w satisfy the mapping $w = \theta z$. M is a known matrix that defines the interconnection between signals, while $\mathcal{N}$ is a static nonlinear mapping to be identified

the unit delay operator. All signals and mappings are assumed to be of compatible dimension. This model structure has the following features, in addition to input u and output y:

- Signals
  - x, ξ: signals capturing the system state, such that $x_{k+1} = \xi_k$
  - e: unmeasured input (measurement noise and disturbances)
  - w, z: signals that define a specific model (see $\mathcal{N}$, below)
- Mappings
  - M: known static matrix
  - $\mathcal{N}$: unknown static mapping to be identified

The matrix M determines the signal interconnection structure of the model, and plays an important role in determining the model class. In terms of the familiar notation of a state-space realization, the matrix M may be decomposed as

$$M = \left[\begin{array}{c:c:c:c} A & B_u & B_e & B_w \\ \hdashline C_y & D_{yu} & D_{ye} & D_{yw} \\ \hdashline C_z & D_{zu} & D_{ze} & D_{zw} \end{array}\right], \tag{8.1}$$

where the block columns have the dimensions of x, u, e and w respectively, and the block rows have the dimensions of ξ, y and z.

We may view the top two blocks as defining a dynamical system with inputs u, e and w, and outputs y and z. Thus, an alternate view of the model structure is as shown in Figure 8.2, with two blocks. $\mathcal{L}$ is a linear time invariant (LTI) dynamical system with state-space realization (8.1), and since M is known, $\mathcal{L}$ is known. In order to aid in notation when referring to particular input–output channels, $\mathcal{L}$ will also be decomposed with subscripts denoting the relevant input and output pair

$$\mathcal{L} = \left[\begin{array}{c:c:c} \mathcal{L}_{yu} & \mathcal{L}_{ye} & \mathcal{L}_{yw} \\ \hdashline \mathcal{L}_{zu} & \mathcal{L}_{ze} & \mathcal{L}_{zw} \end{array}\right]$$

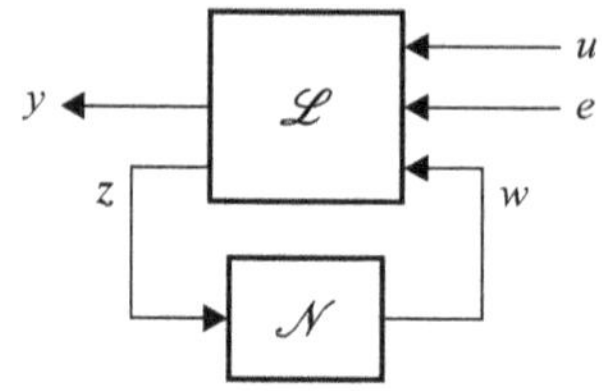

Figure 8.2 Alternate linear fractional representation (LFR) model structure. In this case, the upper two blocks have been represented by the fixed linear dynamical system $\mathscr{L}$

However, one element is still missing; the model class specification is only completed when the allowable set of functions $\mathscr{N}$ is defined. That is, as part of the specification of the model class of dynamical systems represented by the LFR we specify the model class of static functions $\mathscr{N}$. To differentiate these model classes, we will use the term static model class to refer to the latter, which will be denoted by $\mathbb{N}$. Thus, we specify that $\mathscr{N} \in \mathbb{N}$. For later use, we define Ω to be the domain over which $\mathscr{N}$ is to be identified. There are two other important aspects to complete the specification of $\mathbb{N}$: structure and complexity.

- **Structure:** Although unknown, the mapping $\mathscr{N}$ can have a known structure, by which we mean that each output of the mapping $\mathscr{N}$ can have a known subset of the signal z as input. When $\mathscr{N}$ is a linear mapping, this would imply that $\mathscr{N}$ could be represented as a matrix that has a specific structure with some elements known to be zero, such as the block diagonal matrix

$$\mathscr{N} = \begin{bmatrix} \mathscr{N}_1 & & \\ & \ddots & \\ & & \mathscr{N}_{nw} \end{bmatrix}$$

- **Complexity:** There are two approaches to specifying complexity of the static model class: parametric and nonparametric.
  - Parametric: $\mathscr{N}$ is assumed to be represented by a basis expansion $\phi^{[\ell]}$, with coefficients θ_ℓ. That is

$$\mathbb{N} = \left\{ \mathscr{N} : \mathscr{N}(z) = \sum_{\ell=1}^{N} \theta_\ell \phi^{[\ell]}(z) \right\}$$

  where θ_ℓ and $\phi^{[\ell]}$ satisfy the structure discussed above. Note that this basis expansion would also easily allow the user to specify some elements of $\mathscr{N}$ to be linear.
  - Nonparametric: $\mathscr{N}$ satisfies a dispersion function $\mathscr{D}$. This will be discussed in detail in section 8.5.2, but informally a dispersion function returns a

positive real number that rates the complexity or "wigglyness" of a point cloud (w, z). By satisfying a dispersion function, it is meant that $\mathcal{N} \in \mathbb{N}$ satisfies

$$\mathcal{D}(\mathcal{N}(z), z) < \delta$$

for a specified δ.

8.2.1 Simple examples with linear $\mathcal{N}$

Although our main purpose is not identification of linear systems, to help illustrate the flexibility of the LFR model structure we show here how it can be used to capture some classical linear model structures. In addition, these may be useful when used as elements in a larger network of systems, as discussed in section 8.2.3. In this section, we look at some SISO examples with $\mathcal{N}$ restricted to a matrix with a single row, although MIMO systems follow easily with a more general structure for $\mathcal{N}$.

ARX

One of the most common and useful linear model structures is the autoregressive with exogenous input (ARX) structure, which assumes an input–output relationship of the form

$$y_k = a_1 y_{k-1} + a_2 y_{k-2} + \cdots + a_n y_{k-n} + b_0 u_k + b_1 u_{k-1} + \cdots + b_m u_{k-m} + e_k$$

This can be written in transfer function form as

$$y_k = \frac{B(q)}{A(q)} u_k + \frac{1}{A(q)} e_k$$

where $A(q) = 1 - a_1 q^{-1} - \cdots - a_n q^{-n}$ and $B(q) = b_0 + b_1 q^{-1} + \cdots + b_m q^{-m}$. The parameters to be identified are the coefficients of these polynomials in q^{-1}. For this illustration we assume the polynomial orders are the same, although this is not required in general.

We can use the LFR model structure to define the ARX model class by setting the system state to be the past input and output

$$x_k = [u_{k-1}\ u_{k-2} \cdots u_{k-m} \quad y_{k-1}\ y_{k-2} \cdots y_{k-m}]'$$

The signal z appends the current input to the system state so that

$$z_k = \begin{bmatrix} u_k \\ x_k \end{bmatrix}$$

With z as defined above, we can take $\mathcal{N}$ as a $1 \times 2m$ linear mapping, with elements corresponding to the coefficients of $A(q)$ and $B(q)$, so that it has the form

$$\mathcal{N} = [b_0 \ldots b_m \quad a_1 \cdots a_n]$$

The output y is then given by

$$y_k = \mathcal{N} z_k + e_k$$

The matrix M needs to be created so that $\xi_k = x_{k-1}$, and y and z are appropriately defined. With the structure of x, the shift matrix $\Phi_m \in \mathscr{R}^{m\times m}$ is useful, where

$$\Phi_m = \begin{bmatrix} 0 & 0 \\ I_{m-1} & 0 \end{bmatrix} \in \mathbb{R}^{m\times m}$$

I_{m-1} is the identity matrix of dimension $m-1$, and 0 indicates a vector or matrix of zeros of compatible dimension. Also let $\Gamma_m \in \mathscr{R}^{m\times 1}$ be given by

$$\Gamma_m = \begin{bmatrix} 1 \\ 0 \\ \vdots \\ 0 \end{bmatrix}$$

Then the ARX model class is defined by the matrix

$$M = \left[\begin{array}{cc:c:c:c} \Phi_m & 0 & \Gamma_m & 0 & 0 \\ 0 & \Phi_m & 0 & \Gamma_m & \Gamma_m \\ \hdashline 0 & 0 & 0 & 1 & 1 \\ \hdashline 0 & 0 & 1 & 0 & 0 \\ I_m & 0 & 0 & 0 & 0 \\ 0 & I_m & 0 & 0 & 0 \end{array}\right] \tag{8.2}$$

along with $\mathbb{N} = \mathscr{R}^{1\times 2m}$.

ARMAX

Other parameterized linear models can also be represented using the LFR. For example, by adding a moving average term to the ARX model structure, we get the ARMAX model structure

$$y_k = \frac{B(q)}{A(q)} u_k + \frac{C(q)}{A(q)} e_k$$

where $A(q)$ and $B(q)$ are as before, and $C(q) = 1 + c_1 q^{-1} + \cdots + c_m q^{-m}$. In this case, we define

$$x_k = [u_{k-1} \cdots u_{k-m} \quad e_{k-1} \cdots e_{k-m} \quad y_{k-1} \cdots y_{k-m}]'$$

$$z_k = \begin{bmatrix} u_k \\ x_k \end{bmatrix}$$

$\mathscr{N}$ is a linear mapping with matrix representation

$$\mathscr{N} = [b_0 \cdots b_m \quad c_1 \cdots c_m \quad a_1 \cdots a_n]$$

Then we can take

$$M = \left[\begin{array}{ccc:c:c:c} \Phi_m & 0 & 0 & \Gamma_m & 0 & 0 \\ 0 & \Phi_m & 0 & 0 & \Gamma_m & 0 \\ 0 & 0 & \Phi_m & 0 & \Gamma_m & \Gamma_m \\ \hdashline 0 & 0 & 0 & 0 & 1 & 1 \\ \hdashline 0 & 0 & 0 & 1 & 0 & 0 \\ I_m & 0 & 0 & 0 & 0 & 0 \\ 0 & I_m & 0 & 0 & 0 & 0 \\ 0 & 0 & I_m & 0 & 0 & 0 \end{array}\right]$$

8.2.2 *Simple examples with nonlinear $\mathscr{N}$*

In a similar vein, here we define a few simple examples when $\mathscr{N}$ is nonlinear.

Static Mapping

The simplest nonlinear system is simply a static mapping with measurement noise, such as

$$y_k = \mathscr{N}(u_k) + e_k$$

The LFR simply routes the signal u to z, and w summed with e to the output y, as follows:

$$M = \left[\begin{array}{c:c:c:c} 0 & 0 & 0 & 0 \\ \hdashline 0 & 0 & 1 & 1 \\ \hdashline 0 & 1 & 0 & 0 \end{array}\right]$$

Strictly speaking, the signals ξ and x could be eliminated, but we have left them here for convenience.

NARX

A common extension of the ARX model structure to nonlinear systems, termed nonlinear autoregressive with exogenous input (NARX) has the form

$$y_k = \mathscr{N}(y_{k-1}, \ldots, y_{k-n}, u_k, \ldots, u_{k-n}) + e_k$$

Since the only difference between ARX and NARX is that $\mathscr{N}$ is now allowed to be nonlinear, it is clear that the LFR of a NARX system will utilize the same M as in the ARX case, as given by (8.2). However, in this case $\mathbb{N}$ is expanded, from linear functions to the appropriate set of nonlinear functions.

8.2.3 LFRs of block-oriented models

A particular strength of the LFR model structure is when it is used to represent block-oriented systems, as it allows for a single data structure that can capture a wide variety of system interconnections. As an example consider the interconnection of three systems shown in Figure 8.3. Each of these individual systems is represented as an LFR model structure, and thus may be one of the simple examples given above, such as a linear dynamic system or a static nonlinearity. These systems are augmented by multiplexers and de-multiplexers represented by the black bars, and interconnected so that the input of one system may be the output of another.

This interconnection may be represented using a single LFR as shown in Figure 8.4 by simply collecting signals together, such as

$$\xi = \begin{bmatrix} \xi_1 \\ \xi_2 \\ \xi_3 \end{bmatrix}, \quad z = \begin{bmatrix} z_1 \\ z_2 \\ z_3 \end{bmatrix}$$

and similarly for x, e and w. The signals y and u are as defined in Figure 8.3. The matrix M contains elements from M^1, M^2 and M^3, along with additional elements to account for the multiplexers. In this particular case (purely for convenience, under

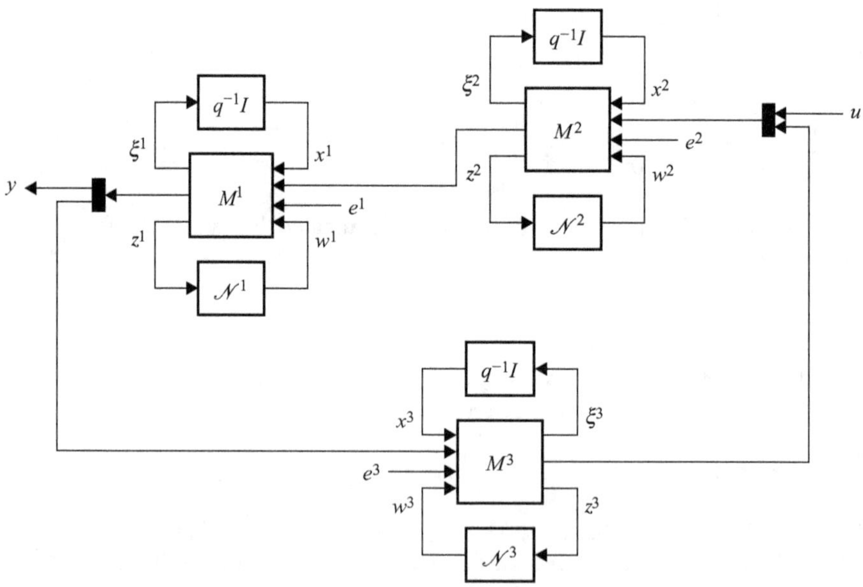

Figure 8.3 Block-oriented system description

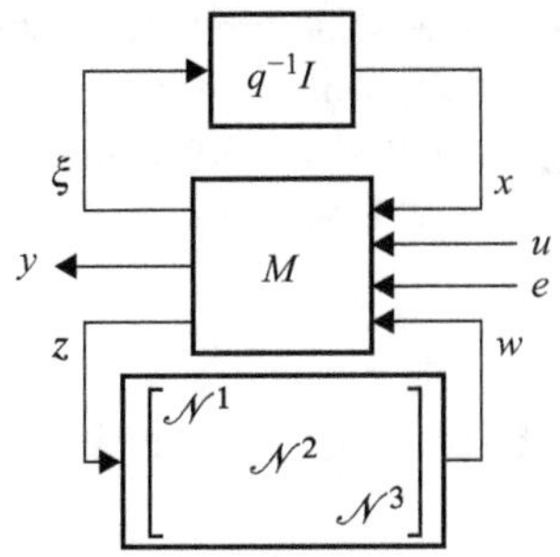

Figure 8.4 Block-oriented system description represented as a single LFR

the assumption of no direct feed-through, so that $D_{yu}=0$, $D_{ye}=0$ and $D_{yw}=0$ for all systems), we would have M of the form

$$M = \left[\begin{array}{ccc|c|ccc|ccc}
A^1 & B_u^1 C_y^2 & 0 & 0 & B_e^1 & 0 & 0 & B_w^1 & 0 & 0 \\
0 & A^2 & B_{u2}^2 C_y^3 & B_{u1}^2 & 0 & B_e^2 & 0 & 0 & B_w^2 & 0 \\
B_u^3 C_{y2}^1 & 0 & A^3 & 0 & 0 & 0 & B_e^3 & 0 & 0 & B_w^3 \\
\hline
C_{y1}^1 & 0 & 0 & 0 & 0 & 0 & 0 & 0 & 0 & 0 \\
\hline
C_z^1 & 0 & 0 & 0 & D_{ze}^1 & 0 & 0 & D_{zw}^1 & 0 & 0 \\
0 & C_z^2 & 0 & 0 & 0 & D_{ze}^2 & 0 & 0 & D_{zw}^2 & 0 \\
0 & 0 & C_z^3 & 0 & 0 & 0 & D_{ze}^3 & 0 & 0 & D_{zw}^3
\end{array}\right]$$

where the matrices

$$B_u^2 = \begin{bmatrix} B_{u1}^2 & B_{u2}^2 \end{bmatrix} \quad C_y^1 = \begin{bmatrix} C_{y1}^1 \\ C_{y2}^1 \end{bmatrix}$$

have been partitioned compatibly with the multiplexer and de-multiplexer.

8.2.4 Discussion: $\mathscr{L}$ known or unknown?

At first viewing the model structure as shown in Figure 8.2, with an fixed linear system connected in feedback with an unknown static function, may seem quite restrictive. It may be thought that this model structure would only be useful if the linear part of the system were completely known. Indeed, there has been some notable work that utilizes the LFR model structure with the aim of identifying both $\mathscr{L}$ and $\mathscr{N}$ [32, 23, 31, 30]. The philosophy here is somewhat different, in that $\mathscr{L}$ is merely a tool to help define the model structure. As should be clear from the previous section, if desired, an unknown linear dynamical system in feedback with a static nonlinearity can be accommodated by restricting a portion of $\mathscr{N}$ to be linear. However, as discussed in the introduction, one should always choose the model structure that most tightly expresses the known a-priori information about the system while at the same time allowing one to efficiently search for the optimal realization, and it certainly may occur that a more specialized model structure than the one we are proposing is more appropriate for a particular application.

8.3 Examples of model structure specification

In this section we give some examples of systems where the structured models enabled by the LFR model can be useful. We start with a general modeling philosophy that aims to reduce the dimensionality of the nonlinear functions to be identified, discuss two specific examples, and then conclude the section with a discussion on the use of the LFR model structure when identifying linear parameter varying (LPV) systems.

8.3.1 High-dimensional model representation

Suppose we have a system with a single output but multiple inputs, so that

$$u = \begin{bmatrix} u^1 \\ \vdots \\ u^m \end{bmatrix}$$

As a model structure, one could take a NARX structure of the form

$$y_k = \mathcal{N}(y_{k-1}, \ldots, y_{k-n}, u_k^1, \ldots, u_{k-n}^1, \ldots, u_k^m, \ldots, u_{k-n}^m)$$

However, the dimension of the input of this nonlinear function is $n \times m + 1$. By allowing for a nonlinear relationship between every possible combination of input and output, the input dimension is quite large, and would require a very large data set in order to identify successfully, unless there was a specific known low-order basis that could be used, or if it were assumed that $\mathcal{N}$ were extremely smooth. The former is unlikely in application, while the latter is somewhat contradictory; why use nonlinear system identification techniques if the system is essentially linear?

An approach is suggested in [24], which is based on the observation that in many physical systems, even if the number of inputs is large, the output depends on input interactions that are relatively low order. Thus, instead of identifying a single high-dimensional nonlinear function, the model structure is a sum of multiple nonlinear functions that include the combinations of inputs for which interactions are thought to occur. For example, if we wished to allow nonlinear behavior between all pairs of inputs, we could assume a model structure of the form

$$\begin{aligned} y_k = {} & \mathcal{N}^1(y_{k-1}, \ldots, y_{k-n}) + \mathcal{N}^2(u_k^1, \ldots, u_{k-n}^1, u_k^2, \ldots, u_{k-n}^2) + \ldots \\ & + \mathcal{N}^{\binom{m}{2}}(u_k^{m-1}, \ldots, u_{k-n}^{m-1}, u_k^m, \ldots, u_{k-n}^m) \end{aligned}$$

While there are more nonlinear functions to be identified, the maximum input dimension of these functions has reduced to $2n$. Naturally, this dimension could be reduced even further with more restrictions in the nonlinear interactions, for example, by considering interactions only in inputs with the same delay. In [24] this is termed a high-dimensional model representation (HDMR). Because the input to the nonlinear functions are delayed values of the input and output, such a representation can be easily described by the LFR model representation in a manner similar to the NARX model structure.

8.3.2 Automobile suspension

Models of vehicle vertical dynamics are important tools in the automotive field, especially in view of the increased deployment of controlled suspension systems [12, 18]. Indeed, accurate models may allow for efficient design and tuning of control algorithms in a computer simulation environment, thus significantly reducing the expensive in-vehicle tuning process.

An automobile with semi-active suspensions is considered in this example. The vertical dynamics of such a vehicle can be described by the half-car model shown in Figure 8.5. The main variables describing this model are

- p_{rf} and p_{rr}: front and rear road profiles;
- i_f and i_r: control currents of front and rear suspensions;
- a_f and a_r: front and rear chassis vertical accelerations;
- Δp_f and Δp_r: front and rear suspension displacements;
- $\Delta \dot{p}_f$ and $\Delta \dot{p}_r$: front and rear suspension displacement velocities;
- F_f and F_r: forces applied to chassis by front and rear suspensions.

The half-car model of Figure 8.5 can be decomposed into the block diagram of Figure 8.6. The block $\mathscr{L}_{\mathsf{CE}}$ represents the behavior of the chassis and engine. The inputs to $\mathscr{L}_{\mathsf{CE}}$ are F_f and F_r, the outputs are a_f and a_r. Since the chassis pitch angles are generally small (<5~6 degrees), this block can be considered LTI. The blocks $\mathscr{L}_{\mathsf{WT}_f}$ and $\mathscr{L}_{\mathsf{WT}_r}$ represent the overall behavior of front and rear wheels and tires, and

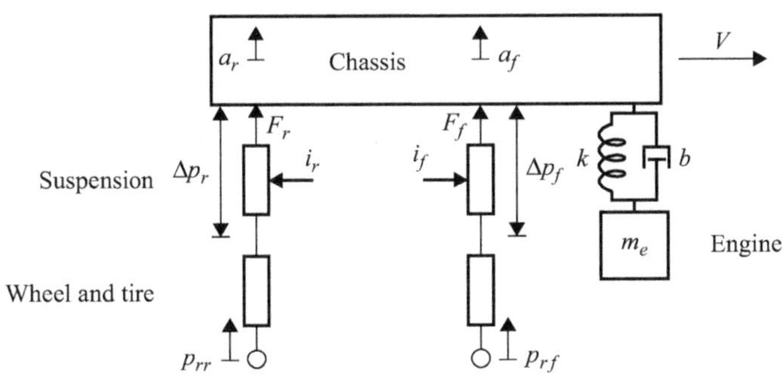

Figure 8.5 Half-car model

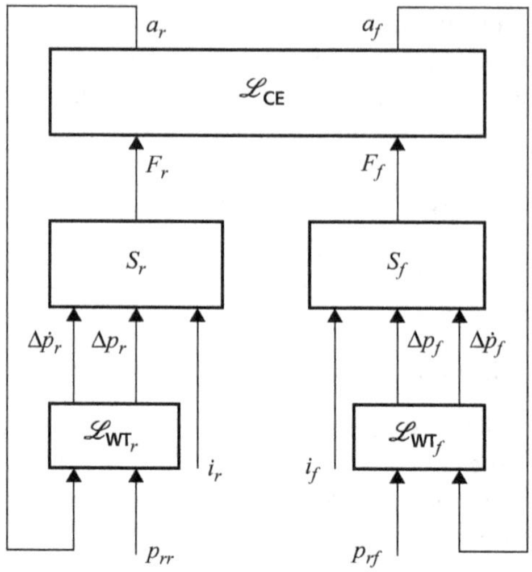

Figure 8.6 Block decomposition of the half-car model

can be accurately described by LTI models. The inputs of $\mathscr{L}_{WT_*}$, where $*$ stands for f or r, are p_{r*} and a_*; the outputs are Δp_* and $\Delta\dot{p}_*$. The blocks S_f and S_r represent the front and rear suspensions. Considering that a real suspension system is characterized by a nonlinear dynamical behavior, mainly due to the fluid motion in the damper, the following dynamic nonlinear model structure can be used to describe S_*:

$$F_{*,k} = \mathscr{N}_*\left(z_{*,k}\right)$$

where

$$z_{*,k} = [\Delta p_{*,k-1} \cdots \Delta p_{*,k-n_1} \quad \Delta\dot{p}_{*,k-1} \cdots \Delta\dot{p}_{*,k-n_1} \quad \Delta i_{*,k-1} \cdots i_{*,k-n-2}]' \tag{8.3}$$

Note that $\mathscr{N}_*$ is a static nonlinear function.

Under this block diagram representation, the half-car model can be recast in the LFR form using the following:

$$\begin{aligned} &\mathscr{L} = \{\mathscr{L}_{CE}, \mathscr{L}_{WT_f}, \mathscr{L}_{WT_r}, D_f, D_r\}, \quad \mathscr{N} = \begin{bmatrix} \mathscr{N}_f & 0 \\ 0 & \mathscr{N}_r \end{bmatrix}, \\ &u_t = \begin{bmatrix} p_{ft} & p_{rt} & i_{ft} & i_{rt} \end{bmatrix}', \qquad y_t = \begin{bmatrix} a_{ft} & a_{rt} \end{bmatrix}' \\ &z_t = \begin{bmatrix} z_{ft} & z_{rt} \end{bmatrix}', \qquad w_t = \begin{bmatrix} w_{ft} & w_{rt} \end{bmatrix}' \end{aligned} \tag{8.4}$$

where D_f and D_r are the LTI systems that generate the lag values in (8.3).

Note that in this model structure it is assumed that the dynamical systems $\mathscr{L}_{CE}$, $\mathscr{L}_{WT_f}$ and $\mathscr{L}_{WT_r}$ are known, and only the behavior of the suspension systems is to be identified. These could be obtained from first principles, or separate identification experiments. An example of a complete identification experiment for this system can be found in [22].

8.3.3 Nonlinear friction: drill-string

In order to develop oil and gas reservoirs, a well is created using a drilling rig that rotates a drill-string that connects the drill bit at the bottom of the hole with the drilling rig at the top. Analysis and control of drill-string systems are fundamental to improve the efficiency and to reduce the wear, the probability of damage and the operation costs of these systems [21]. This motivates the need of accurate drill-string models, which could be suitably used for the required analyses and for the design of effective control systems.

In this example, the model presented in [21] is considered. The drill-string can be modeled as a series of interconnected rotational mass-spring-damper systems, as illustrated in Figure 8.7. As shown in the figure, there is a nonlinear term that describes the friction between the drill bit and the rock, which has the form

$$\tau_d\left(\dot{\theta}_b\right) = WR\left[\mu_{CB} + (\mu_{SB} - \mu_{CB})\, e^{-\frac{\gamma_B}{\nu_F}|\dot{\theta}_b|}\right]\operatorname{sign}\left(\dot{\theta}_b\right) \tag{8.5}$$

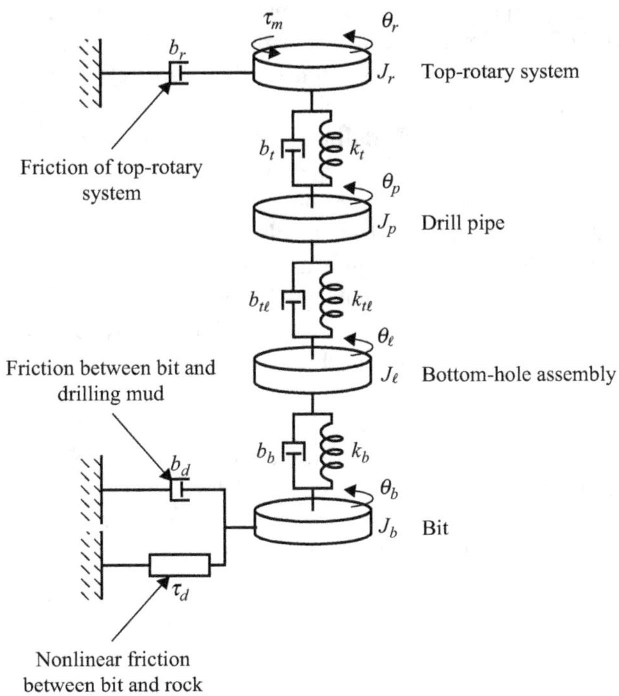

Figure 8.7 Drill-string model

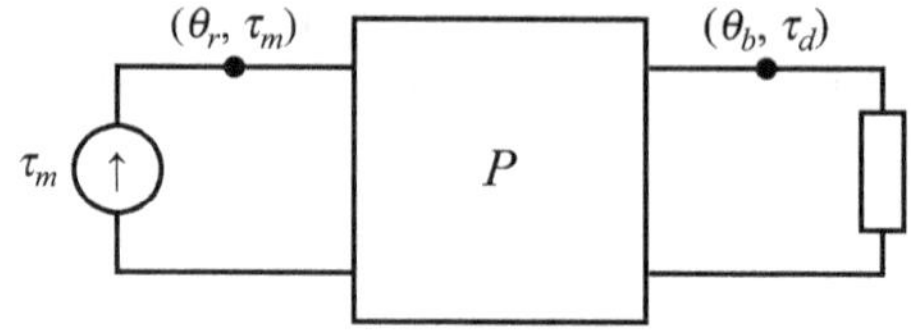

Figure 8.8 Two-port representation of drill-string

where W is the weight on the bit, R is the bit radius, μ_{CB} and μ_{SB} are the static and Coulomb friction coefficients, $0 < \gamma_B < 1$, and $\nu_F > 0$.

Since the remainder of the system is linear, we can use a two-port representation as shown in Figure 8.8, where P represents the two-port system with the applied torque τ_m on the right, and the nonlinear friction element on the left.

In the case when the linear dynamics of the two port system P are known, the system of Figure 8.8 can be represented as a nonlinear LFR system, composed of a continuous-time LTI block with outputs $y = z = \dot{\theta}_b$ and input $u = \tau_m$ and $w = \tau_d$. The static nonlinearity defined in (8.5) is then connected in feedback.

8.3.4 Linear parameter varying systems

Linear parameter varying (LPV) models represent an important means to achieve advanced control of complex dynamical systems that can be effectively implemented in practice. Indeed, one of the main advantages of LPV models is that they permit the application of linear control design techniques to nonlinear or time-varying systems. These techniques include gain-scheduling, optimal control and robust control (see, e.g., [25], [8], and [35]).

LPV models are commonly represented in the form of linear state-space or regression equations, where some of the involved parameters are time-varying. The time-varying parameters, collected into a parameter vector p, called scheduling parameter, are considered as an external signal and are assumed to be measured [25].

The increasing importance of LPV models in control system design has motivated the rise of a new identification problem. Several methods have been proposed to solve this problem, including regression approaches, subspace techniques, nonlinear programming, and interpolation, (see [17] and [1] and the references therein). Among these techniques, the most used in practical applications are probably those based on interpolation, in which several LTI models are estimated performing small signal perturbation experiments, where p is a maintained constant in each of these experiments. The overall LPV model is then obtained blending the LTI models by means of a suitable interpolation rule, written as a function of the scheduling parameter p. However, this rule is in general not easy to design and has often to be tuned using experimental data. This relevant issue can be addressed using the LFR.

Suppose the system is operated at several distinct values of p, with small signal perturbation experiments performed that result in linear models $\mathscr{L}_i$ that are valid at the various operating points. A block-oriented model structure that can be used to identify

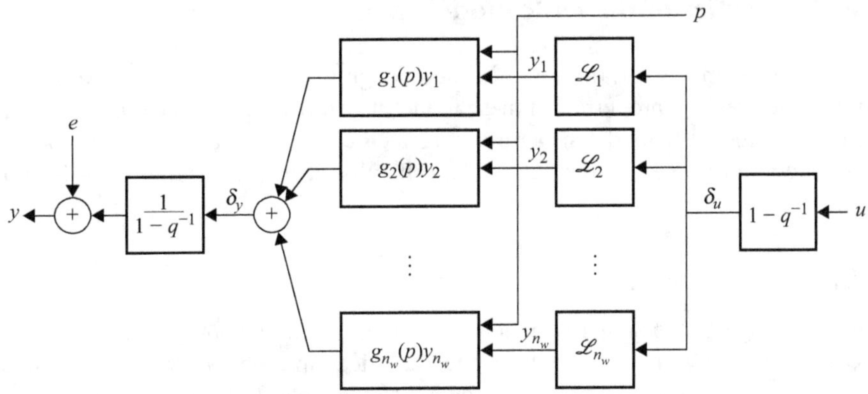

Figure 8.9 Gain scheduled LPV model structure

blending functions g_i is illustrated in Figure 8.9. Note that a velocity implementation has been chosen [11], so that $\delta u = u_k - u_{k-1}$ is applied to each known linear system, and the outputs of the blending functions are summed together to create $\delta y = y_k - y_{k-1}$, which is integrated to produce y. The result is a block-oriented model with known linear systems and unknown static functions that have the specific structure of a nonlinear function g_i multiplied by the signal y_i. As with all such block-oriented models, this can be easily represented within the LFR framework, treating p as an additional input. With this model structure in hand, an additional experiment that explores the entire operating space can be used to identify the functions g_i. In this case, additional a priori information about the static functions is known, in that the function g_i should be between 0 and 1, and equal to 1 (and all other blending functions equal to zero) when p is equal to the operating point at which system i was identified. These requirements can be easily added as convex constraints in the nonparametric identification method given in section 8.5.2. An example of this method applied to identification of the dynamics of a solid oxide fuel cell can be found in [26].

In some cases, the user does not have control over the scheduling parameter p, so small signal experiments are not possible and a gain scheduled approach cannot be taken. As an alternate approach, one could extend the ARX model structure by allowing the parameters a_i, b_i to become functions of the scheduling parameters. That is, the model could be described by the following LPV–ARX model structure:

$$y_k = a_1(p_{k-1})y_{k-1} + \cdots + a_n(p_{k-n})y_{k-n} + b_0(p_k)u_k + \cdots + b_m(p_{k-m})u_{k-m} + e_k$$

Similar to the HDMR case, we have a sum of nonlinear functions with inputs that are delayed values of the input (considering p an additional input) and output, and the LFR can again be utilized.

8.4 Properties of the LFR model structure

Since the LFR model structure is flexible enough to encompass a wide variety of model structures, the properties of these model structures may also be quite different. Fortunately, the LFR model structure gives us a consistent framework to establish some important properties that relate the difficulty of the underlying identification problem.

8.4.1 Measurability

Our knowledge of the signal z, which is the input to the mapping $\mathscr{N}$, plays a key role in determining the complexity of the identification problem. Specifically, since $\mathscr{N}(z)$ is linear in the output y, if z is known then the identification problem is greatly simplified.

Definition 8.1. The signal z is *measurable* if there exists an LTI system Φ_M such that

$$\begin{bmatrix} \mathscr{L}_{ze} & \mathscr{L}_{zw} \end{bmatrix} = \Phi_m \begin{bmatrix} \mathscr{L}_{ye} & \mathscr{L}_{yw} \end{bmatrix}$$

This definition has an important implication for identification: when z is measurable, it can be determined from knowledge of y and u alone, and thus can be specified before identification begins. Specifically, when z is measurable

$$z = \mathscr{L}_{zu}u + \Phi_m(y - \mathscr{L}_{yu}u)$$

In this case, the identification algorithms given in the next section involve a convex optimization problem, and thus are easily solved. When z is not measurable, then the identification algorithm requires the solution of a (generally) nonconvex optimization problem.

As particular examples, note that for the LFR model structure for the ARX system (as well as the NARX system) z is measurable, while the LFR model structure for the ARMAX system does not have z measurable. This demarcation is compatible with what we already know about ARX and ARMAX systems when using prediction error methods [14]. That is, the parameters of the ARX model structure appear linearly in the prediction output, and can be easily found using least squares. On the other hand, the parameters of the ARMAX model structure appear nonlinearly in the output, and in this case an iterative method must be used to solve for the parameters, and in most cases the resulting optimization problem contains local minima.

Measurability is thus a desirable property, and this gives a strong guide when setting up our identification problem. Namely, if we have some choice as to which internal signals to measure, then we should try to do so with the aim of ensuring that z is measurable. For example, in the drill-string example of section 8.3.3, z is obviously measurable if the bottom rotational motion θ_b is directly measured. In this case, we have a strong impetus for implementing a sensor that can take this measurement and transmit it to the surface.

In the case that z is not measurable, and additional measurements cannot be applied, successful identification may still be possible using an iterative method. This is especially true if $\mathscr{L}_{zw} = \Phi_m \mathscr{L}_{yw}$, that is, if z is measurable when e is considered known. In this case, we may proceed by initially calculating a provisional z assuming $e = 0$, performing the system identification method in the next section to obtain an estimate of $\mathscr{N}$, and e, and then iterating this procedure. However, just as with the ARMAX model structure in the linear case, it is not guaranteed to converge to a global minimum.

8.4.2 *Identifiability*

A central property of any model structure is the question of identifiability. For the LFR model structure, it is the function $\mathscr{N}$ that determines a particular input–output realization. Identifiability relates to the question of whether the function $\mathscr{N}$ can be determined uniquely based on input–output information. Let $\psi(\mathscr{L}, \mathscr{N})$ denote the input–output operator from u to y, with $e = 0$.

Definition 8.2. The LFR model structure is *identifiable on* $\mathbb{N}$ if for all $\mathscr{N}_1, \mathscr{N}_2 \in \mathbb{N}$:

$$\mathscr{N}_1 \neq \mathscr{N}_2 \implies \psi(\mathscr{L}, \mathscr{N}_1) \neq \psi(\mathscr{L}, \mathscr{N}_1)$$

When the LFR model structure is identifiable, there exists an experiment that would allow one to distinguish all functions within the set $\mathbb{N}$. Unfortunately, checking identifiability in general for nonlinear model structures is very difficult. However, in [9] a simple test is given under some additional assumptions on $\mathscr{L}$.

Definition 8.3. The signal z is *co-measurable* if there exists an LTI system Φ_c such that

$$\begin{bmatrix} \mathscr{L}_{ze} & \mathscr{L}_{zw} \end{bmatrix} = \mathscr{L}_{zu} \Phi_c$$

Note that co-measurability implies that the range space of $\mathscr{L}_{zu}$ is the same as $\begin{bmatrix} \mathscr{L}_{ze} & \mathscr{L}_{zw} \end{bmatrix}$. Roughly speaking, this means that z can be completely explored using u, and in fact in this case the set of achievable z is a subspace.

Theorem 8.1 [9]. *Let $\mathscr{L}$ be such that z is measurable and co-measurable, and let $\mathscr{N} \in \mathbb{N}$ be differentiable everywhere. Let $\mathbb{X}$ be the set of matrices that have the same structure as $\mathscr{N}$. Then the LFR model structure is identifiable on $\mathbb{N}$ if and only if for any $X \in \mathbb{X}$*

$$\mathscr{L}_{yw} X \mathscr{L}_{zu} = 0 \implies X = 0$$

Note that in the case of z measurable and co-measurable, identifiability depends only on the linear system $\mathscr{L}$ and the input–output structure of $\mathbb{N}$, so that the system is identifiable over nonlinear functions with a particular structure if and only if it is identifiable over linear functions with that structure.

8.4.3 Persistence of excitation

While identifiability gives the possibility of uniquely determining $\mathscr{N}$ using an appropriate experiment, we still need to ensure this can occur for a *specific* experiment. When the experiment is sufficient to uniquely determine the unknown function $\mathscr{N}$, it is usually termed persistently exciting, and in our context it will be called PEID (since identifiability is also a necessary condition). Given a sequence x, let $\pi_L x$ be the first L elements of the sequence.

Definition 8.4. The experimental inputs u and e are PEID on Ω if for every $\mathscr{N} \in \mathbb{N}$

$$\liminf_{L\to\infty} \frac{1}{L} \|\pi_L \mathscr{L}_{ye}^{-1} \mathscr{L}_{yw} \mathscr{N}(z)\|^2 = 0 \implies \mathscr{N} = 0$$

The PEID condition requires that for all feasible nonlinear functions, the signal z cannot be filtered to an essentially zero power signal unless the nonlinear function is identically zero. This ensures that the input sufficiently probes the nonlinear function so that the algorithms presented in the next section will converge to the true nonlinearity under some mild technical conditions. We note that for a fixed input signal u, the resulting loop signal z is random (because of the noise e). Note that the PEID property states in terms of the the signal z, and therefore only indirectly relates to u and e through the LFT model structure.

8.5 Identification algorithms

In this section, we introduce some methods for performing the system identification process by which estimates of $\mathscr{N}$ can be obtained. For the implementation described here, we assume that $\mathscr{L}$ is given as a stabilizable and detectable realization and that z is measurable. As discussed earlier, when z is not measurable, but would be measurable if e were known, a straightforward extension is possible via iteration between estimates of z and e. It is also possible that in some other cases extensions using overparameterization [5] or convex relaxations [6] can be used, but this area is currently open to further research.

8.5.1 Parametric estimates

A large number of real-world systems can be described by a finite set of basis functions. When possible, a suitable basis function representation for a given system can be directly inferred from the first principle laws. In other situations, due to the fact that these laws are not sufficiently known or too complex, this is not possible or not convenient and a black-box approach is taken, where the basis functions are chosen considering given families of functions, known to have strong approximation properties (e.g., Gaussian, sigmoidal, wavelet, polynomial, and trigonometric [28, 29, 16]).

In both cases, the nonlinearity $\mathscr{N}$ can be represented as a finite linear combination of known basis functions:

$$\mathscr{N}(z) = \sum_{\ell=1}^{N} \theta_\ell \phi^{[\ell]}(z) \tag{8.6}$$

where $\phi^{[\ell]} : \mathbb{R}^{n_z} \to \mathbb{R}^{n_w}$, $\ell = 1, \ldots, N$ are known vector-valued functions, $\theta = [\theta_1 \ \cdots \ \theta_N]' \in \Theta$ is a parameter vector to be estimated, and Θ is a convex compact subset of $\mathbb{R}^N$. In [22] an algorithm has been developed for the estimation of the parameter vector θ. The algorithm can be described as follows.

Consider the relation between the signals of the LFR system, described by the equation

$$y = \mathscr{L}_{yu}u + \mathscr{L}_{ye}e + \mathscr{L}_{yw}\Phi(z)\theta \tag{8.7}$$

where $\Phi(z) = [\phi^{[1]}(z), \ldots, \phi^{[N]}(z)]$. Perform a stable left co-prime factorization:

$$[\mathscr{L}_{yu} \quad \mathscr{L}_{ye} \quad \mathscr{L}_{yw}] = G^{-1}[H_{yu} \quad H_{ye} \quad H_{yw}]$$

where G and $H = [H_{yu} \quad H_{ye} \quad H_{yw}]$ are asymptotically stable LTI systems. Then, we can rewrite (8.7) as

$$Gy = H_{yu}u + H_{ye}e + H_{yw}\Phi(z)\theta \tag{8.8}$$

In this way, all systems appearing in (8.8) are stable and all the signals can be actually computed. This operation allows us to deal with cases where $\mathscr{L}_y$ is unstable. Clearly, if $\mathscr{L}_y = [\mathscr{L}_{yu} \quad \mathscr{L}_{ye} \quad \mathscr{L}_{yw}]$ is stable, we can choose

$$G = I, \quad H = \mathscr{L}_y$$

See [22] for more details about the co-prime factorization.

The noise sequence e and the parameter vector θ in (8.8) are not known and have to be estimated. We wish to choose estimates for which the norm of e is minimum. This is accomplished with the following optimization problem:

$$\begin{array}{ll} \min_{\varepsilon,\theta} & \|e\|^2 \\ \text{subject to} & Gy = H_{yu}u + H_{ye}e + H_{yw}\Phi(z)\theta \end{array} \tag{8.9}$$

where the constraint restricts the signal e to be consistent with input–output data.

If z is measurable, the optimization problem (8.9) is convex. Moreover, if the experimental inputs u and e are PEID, the optimization problem (8.9) has a unique solution (see [22]), and this solution can be computed in closed form as follows.

Let $\mathscr{K}$ be the Kalman filter for the LTI system H_{ye}. By application of this filter to (8.8), we obtain

$$\mathscr{K}H_{ye}e = \mathscr{K}(Gy - H_{yu}u - H_{yw}\Phi(z)\theta) \tag{8.10}$$

Since e is assumed to be a white i.i.d. sequence and

$$\varepsilon = \mathscr{K} H_{ye} e \tag{8.11}$$

is the innovation sequence provided by the Kalman filter, minimizing $\|e\|^2$ is equivalent to minimizing $\|\varepsilon\|^2$. See [22] for more details about the Kalman filter. The estimation of the parameter vector θ can thus be performed solving the following optimization problem:

$$\begin{array}{ll} \min_{\varepsilon,\theta} & \|\varepsilon\|^2 \\ \text{subject to} & \mathscr{K} Gy = \mathscr{K} H_{yu} u + \varepsilon + \mathscr{K} H_{yw} \Phi(z)\theta \end{array}$$

which can be posed as the following convex least-squares problem:

$$\min_{\theta} \ \|Y - \mathscr{Q}(z)\theta\|^2$$

where $Y = \mathscr{K} Gy - \mathscr{K} H_{yu} u$ and $\mathscr{Q}(z) = H_{yw}\Phi(z)$. The solution of this problem can be computed in closed form as

$$\hat{\theta} = \left(\mathscr{Q}(z)'\mathscr{Q}(z)\right)^{-1} \mathscr{Q}(z)'Y$$

where, thanks to the PEID assumption, $\mathscr{Q}(z)'\mathscr{Q}(z)$ is an invertible matrix [22]. Note that, by using the Kalman filter, we do not require the invertibility of H_{ye}.

Under the measurability assumption and other technical assumptions, the estimate obtained solving the optimization problem (8.9) converges as the number of data L tends to infinity. If, additionally, the experimental inputs u and e are PEID, this estimate converges to the true parameter vector. See [22] for the formal proofs of these results.

8.5.2 Nonparametric estimates

In some cases, an appropriate (small) set of basis elements for $\mathscr{N}$ will not be known. An alternative is to use a nonparametric approach. In this section, we consider an approach that estimates $\mathscr{N}(z_k)$ for all values of z_k that occur during the identification experiment. In other words, we attempt to estimate the signal w, since $w_k = \mathscr{N}(z_k)$. This can be thought of as "identification by estimation," and it is instructive to connect this problem with a related Kalman filtering problem, shown in Figure 8.10. A fixed, known linear system $\mathscr{L}$ has a known input u, and the output y is measured. Additional unmeasured inputs are the measurement noise e and disturbances w, which are modeled as i.i.d. random sequences with zero mean and unit variances. While the Kalman filter is often stated as estimating the internal states of the system, this is equivalent to estimating the unknown input sequences e and w, and it is in this sense

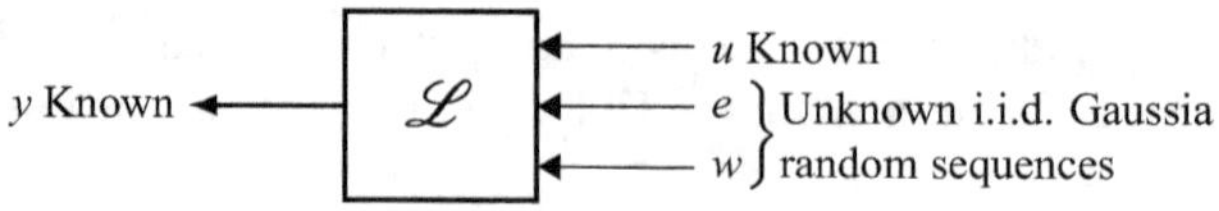

Figure 8.10 An estimation problem

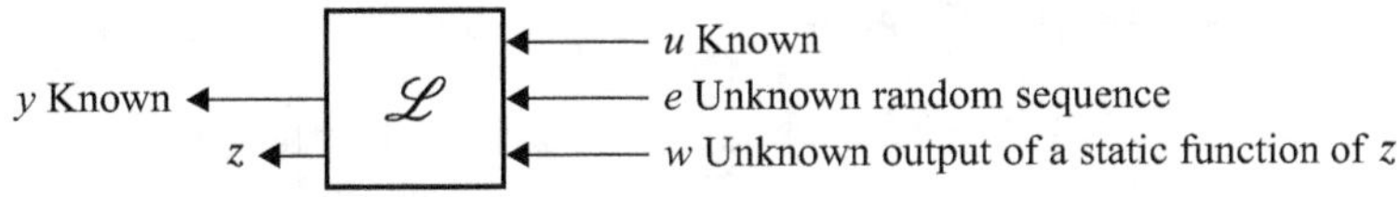

Figure 8.11 The nonparametric identification problem

that we will consider the problem. In any case, if (e, w) are unknown, how can they be estimated? The Kalman filter can be viewed as imposing the following requirements on (e, w) [4]:

- (e, w) should be consistent with the measurements, in that (u, e, w) will re-create y.
- Among all (e, w) consistent with the measurements, select the pair that is *most likely*, as measured the likelihood function.

If e and w are assumed to be independent with unit variance, then maximizing the likelihood for e and w is equivalent to minimizing the 2-norm, and these requirements can be fulfilled by solving the following optimization problem:

$$\begin{array}{ll} \min_{e,w} & \|e\|^2 + \|w\|^2 \\ \text{subject to} & y = \mathscr{L}_{yu}u + \mathscr{L}_{ye}e + \mathscr{L}_{yw}w \end{array} \tag{8.12}$$

This optimization problem is convex, and the Kalman filter provides an efficient, recursive solution.

Now let's return to the identification problem, which is illustrated in Figure 8.11. Again, we have two unknown inputs, but our assumptions on these inputs will now be different. Since e is meant to model measurement noise or disturbances, it is reasonable to utilize the assumption of an i.i.d. Gaussian random sequence as before. However, for w, instead of a random sequence, it is expected to be the output of a static function of z. What is needed is a metric that rates the sequence w that is analogous to the likelihood used for random sequences. This metric, denoted $\mathscr{D}(w, z)$ and called here a dispersion function, takes z and w as inputs, and returns a real number that rates the likelihood that w is the output of a static function of z.

There are several different paths to creating a usable dispersion function; we will describe one such path that focuses on evaluating the interpolant of the point cloud created by plotting w versus z. To give some intuition, consider three systems: a hysteresis, an LTI system, and $\tan^{-1}(z)$. White noise is applied as an input to each system and the output is measured. A scatter plot for each system is shown in Figure 8.12. It is intuitive that the third plot is most representative of a static function. This intuition may be quantified by fitting a piecewise linear interpolant to the data. In Figure 8.13, it is clear that the interpolant has high variation for the cases where the points did not result from a static map. One can measure the complexity of the interpolant via a metric that increases with variation. Thus, our metric of staticness will measure the complexity of interpolating the points (z, w).

Using Delaunay triangulation, partition the convex hull of $\{z_k\}_{k=1}^{L}$ into a disjoint set of n_T simplicies whose vertices are points in $\{z_k\}_{k=1}^{L}$. These simplices will be referred to as "triangles." When each point z_k is paired with a scalar w_k, associated with each triangle is a facet, as illustrated in Figure 8.14.

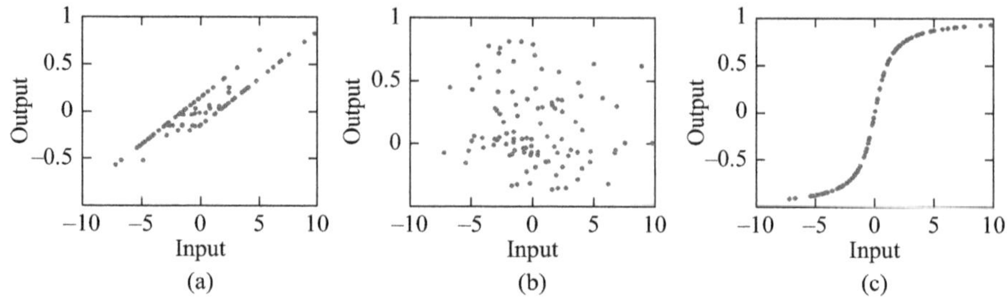

Figure 8.12 Data sampled using white noise input for (a) hysteresis, (b) LTI system, and (c) $\tan^{-1}$

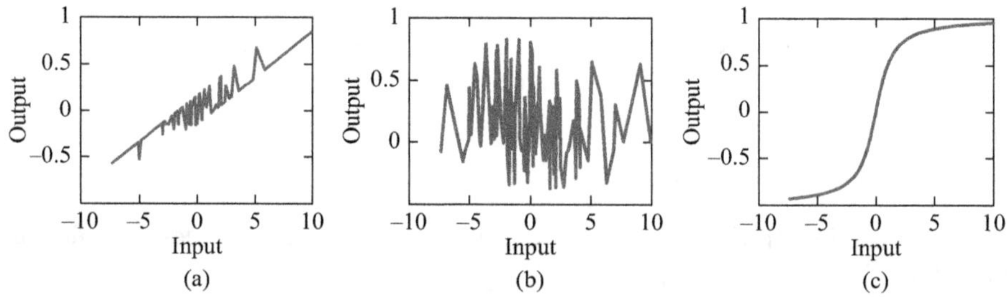

Figure 8.13 Linear interpolants for data from (a) hysteresis, (b) LTI system, and (c) $\tan^{-1}$

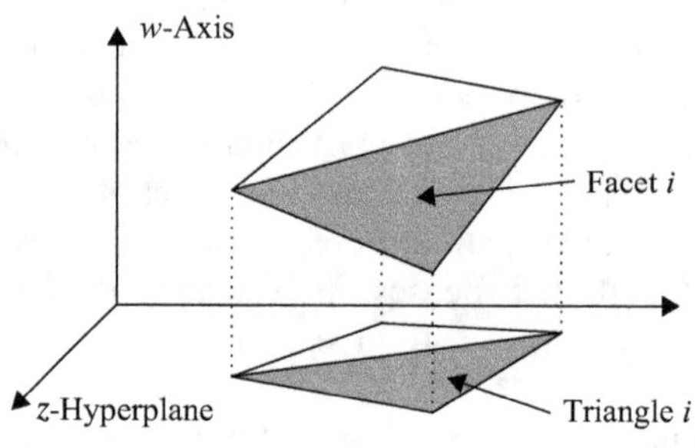

Figure 8.14 Triangulation and corresponding facets

Definition 8.5. Let $\mathscr{I}(z)$ be the linear interpolation of w defined by the n_T facets of the triangulation given above. Define

$$\mathscr{D}(w, z) = \left\| \left[\nabla \mathscr{I}|'_{z \in T_1} \quad \nabla \mathscr{I}|'_{z \in T_2} \cdots \nabla \mathscr{I}|'_{z \in T_{n_T}} \right] \right\|^2$$

where $\nabla \mathscr{I}|_{z \in T_i}$ is the gradient of triangle i.

Note that for fixed z, $\nabla\mathscr{I}|_{z\in T_i}$ is a *linear* function of w. This implies $\mathscr{D}(z,w)$ is a quadratic function in w, so that there exists a matrix Q (that is a function of z) such that

$$\mathscr{D}(z,w) = (w)^* Q w$$

and the matrix Q can be calculated explicitly.

For example, if z and w are both scalar sequences, then if Π_z is the permutation matrix that puts z in order and Δ is a matrix that find the first difference (scaled by the distance between adjacent elements of z) then

$$Q = \Pi_z^* \Delta^* \Delta \Pi_z$$

A dispersion function that uses more general interpolants than piecewise linear can be obtained by using concepts from approximation theory [27]. Choose $(\lambda_n)_{n=1}^{\infty}$ as a sequence of nonnegative real numbers such that $\sum_{n=1}^{\infty} \lambda_n < \infty$. Then define the set of potential interpolants $\mathbb{K}$ as

$$\mathbb{K} = \left\{ \mathscr{I} : \mathscr{I}(z) = \sum_{n=1}^{\infty} a_n \phi_n(z), \sum_{n=1}^{\infty} \frac{|a_n|^2}{\lambda_n} < \infty \right\}$$

where $\phi_n(z) = \exp(-2\pi jnz)$. That is, the interpolants have a Fourier series expansion with coefficients $\{a_n\}_{n=1}^{\infty}$ that must decay to zero since $\lambda_n \to 0$. This in effect enforces the functions $\mathscr{I} \in \mathbb{K}$ to possess a smoothness property by limiting the influence of the higher frequency components.

For any two functions $\mathscr{I}_1(\cdot) = \sum_{n=1}^{\infty} a_n \phi_n(\cdot) \in \mathbb{K}$ and $\mathscr{I}_2(\cdot) = \sum_{n=1}^{\infty} b_n \phi_n(\cdot) \in \mathbb{K}$, we can define an inner product

$$\langle \mathscr{I}_1, \mathscr{I}_2 \rangle_K = \sum_{n=1}^{\infty} \frac{a_n^* b_m}{\lambda_n}$$

Using the norm induced by this inner product, we can rewrite $\mathbb{K}$ as

$$\mathbb{K} = \left\{ \mathscr{I} : \mathscr{I}(z) = \sum_{n=1}^{\infty} a_n \phi(z), \|\mathscr{I}\|_K^2 < \infty \right\} \tag{8.13}$$

The norm $\|\cdot\|_K$ clearly penalizes variation. If the points $\{z_k, w_k\}_{k=1}^{L}$ induce high-frequency oscillations in the interpolant, the "cost" of interpolating is higher.

Now, given a point cloud (z, w), the dispersion function is defined to be the norm $\|\cdot\|_K$ of the smoothest function that interpolates the points.

Definition 8.6.

$$\mathscr{D}(z,w) = \begin{array}{ll} \min_{\mathscr{I}\in\mathbb{K}} & \|\mathscr{I}\|_K^2 \\ \text{s.t.} & w_k = \mathscr{I}(z_k) \end{array}$$

As defined, it appears that we need to compute the Fourier coefficients of a particular interpolant to determine its cost. However, it is possible to form a so-called kernel function that can capture the equivalent information. Define the kernel function associated with $\mathbb{K}$ as

$$K(z_1, z_2) = \sum_{n=1}^{\infty} \lambda_n \phi_n(z_1)^* \phi_n(z_2) \tag{8.14}$$

Since $\phi_n(z_2 - z_1) = \phi_n(z_1)^* \phi_n(z_2)$, we have $K(z_1, z_2) = K(z_2 - z_1) = \sum_{n=1}^{\infty} \lambda_n \phi_n(z_2 - z_1)$, so the kernel function has only one argument.

With $\mathbb{K}$ and K defined as in (8.13) and (8.14), it turns out (see [27]) that $\mathscr{D}(z, w)$ is a quadratic function in w with the explicit form

$$\mathscr{D}(z, w) = (w)^* Q w \tag{8.15}$$

where

$$Q = \begin{bmatrix} K(z_1 - z_1) & \cdots & K(z_L - z_1) \\ \vdots & \ddots & \vdots \\ K(z_1 - z_L) & \cdots & K(z_L - z_L) \end{bmatrix}^{-1}$$

We see from above there is an equivalence between choosing $(\lambda_n)_{n=1}^{\infty}$ and choosing a kernel K. In fact, all calculations can be done using only K.

A common kernel is the Gaussian kernel

$$K(z_1, z_2) = \frac{1}{\sqrt{2\pi\sigma^2}} \exp\left(-\frac{(z_1 - z_2)^2}{2\sigma^2} \right)$$

but other examples can be found in [27].

Now armed with a dispersion function, we can return to the estimation problem. In the case when the true function is known to satisfy the constraint $\mathscr{D}(\mathscr{N}(z), z) < \delta$, the identification problem is solved via the solution to the optimization problem

$$\begin{array}{ll} \min_{e,w} & \|e\|^2 \\ \text{subject to} & \mathscr{D}(z, w) \leq \delta \\ & y = \mathscr{L}_{yu} u + \mathscr{L}_{ye} e + \mathscr{L}_{yw} w \end{array}$$

Note that the factorization discussed in section 8.5.1 can also be used in cases when $\mathscr{L}_y$ is unstable. An equivalent formulation for a specific choice of γ would be

$$\begin{array}{ll} \min_{e,w} & \|e\|^2 + \gamma \mathscr{D}(z, w) \\ \text{subject to} & y = \mathscr{L}_{yu} u + \mathscr{L}_{ye} e + \mathscr{L}_{yw} w \end{array}$$

which has a very similar form as for the standard estimation problem (8.12), except that a different cost is used to evaluate w. This formulation could also be used when δ

is unknown by selecting γ using a cross-validation procedure [34]. In the case when z is measurable, all of the dispersion functions discussed above are quadratic in w, and thus these optimization problems are convex, indeed quadratic, and can be easily solved.

Convergence results in the nonparametric case are also available; see [9] for details.

8.6 Identification example

As an illustration, we consider the identification of an automobile suspension model from a set of experimental data. The data was measured on a car equipped with semi-active controlled dampers and a CDC-Skyhook (continuous damping control) system, positioned on a four-poster test bench. The CDC-Skyhook system is characterized by several control settings. In this example, a setting called “Soft” has been considered, in which the dampers are controlled with current modulated by the CDC-Skyhook control unit according to a comfort calibration. A symmetric (left = right) random road profile has been used for data acquisition. The vehicle has been assumed to travel at a constant speed of $V = 60$ km/h. Since the considered road profiles are symmetric with respect to the longitudinal axis of the vehicle, a half-car model structure has been considered (in particular, the one described in section 8.3 and shown in Figures 8.5 and 8.6). The data set, consisting of 6000 data acquired with a sampling time $\tau = 1/512$ s, has been partitioned as follows:

Identification set: First 4000 data
Validation set: Data from 4001 to 6000

The model for the block $\mathscr{L}_{\mathsf{CE}}$ (representing the behavior of the chassis and engine) has been obtained from first principles, assuming that the rigid chassis interacts with the engine through a linear spring and dumper. The model parameters have been provided by the car manufacturer. The models for the blocks $\mathscr{L}_{\mathsf{WT}_*}$ (representing the overall behavior of front and rear wheels and tires) have been obtained by standard linear identification methods, using the measurements of p_*, a_*, $\Delta p_*, \Delta\dot{p}_*$ of the identification set (we recall that $*$ stands for f or r, i.e., for front or rear). In particular, linear output error (OE) models of $\mathscr{L}_{\mathsf{WT}_*}$ have been identified by means of the MATLAB® System Identification Toolbox [15]. Several orders have been considered, and the best results have been obtained by two $OE(6,4,0)$ models. In Figure 8.15, the measured values of $\Delta\dot{p}_f$ are compared to a simulation of the identified model on a subset of the validation set. These linear models match the validation data very well. The blocks S_f and S_r, representing the front and rear suspensions, have been identified by means of the algorithm described in section 8.5.1.

The LFT-half-car model structure (8.4) has been considered, where the nonlinearities $\mathscr{N}_*$ in (8.4) have been parametrized by a finite set of basis functions:

$$\mathscr{N}_*(z_*) = \sum_{k=1}^{N_*} \theta_{*k}\phi_*^{[k]}(z_*)$$

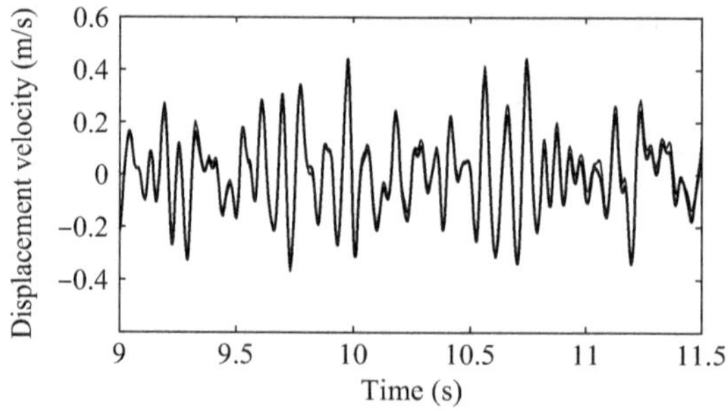

Figure 8.15 Front displacement velocity. Bold line, measured; thin line, simulated by $\mathscr{L}_{\mathsf{WT}_f}$

where $*$ stands for f or r, $\phi_*^{[k]}(z_*) = [\sigma(\beta_{*1}z_* - \lambda_{*1}), \sigma(\beta_{*2}z_* - \lambda_{*2}), \ldots, 1]$, $\lambda_{*k} \in \mathbb{R}$, $\beta_{*k} \in \mathbb{R}^{n_*}$, $n_* = \dim(z_*)$ and $\sigma(h) = 2/(1 + e^{-2h}) - 1$ is a sigmoidal basis function. The parameters β_{*k}, λ_{*k} have been randomly chosen in ranges such that the basis functions $\sigma(\beta_{*k}z_* - \lambda_{*k})$ are nonconstant for the measured z_{*t}. Several values of n_1 and n_2 (see (8.3)) have been considered in the range $[1, 10]$ and several values of N_f and N_r in the range $[1, 50]$. The model simulation accuracy on the identification set and the model complexity have been used as criteria for choosing the orders n_1 and n_2 and the numbers of basis functions N_f and N_r. A good compromise between simulation accuracy and complexity has been obtained by means of the following values $n_1 = 5, n_2 = 1, N_f = N_r = 25$. After choosing the model orders and the numbers of basis functions, the parameter vector $\theta = [\theta_f' \quad \theta_r']' = [\theta_{f1} \cdots \theta_{fN_f} \quad \theta_{r1} \cdots \theta_{rN_r}]'$ has been identified by means of the algorithm described in section 8.5.1 using the identification data set.

The overall model, called M1 and composed of the linear systems $\mathscr{L}_{\mathsf{CE}}$, $\mathscr{L}_{\mathsf{WT}_f}$, $\mathscr{L}_{\mathsf{WT}_r}$, D_f, D_r and of the identified static nonlinearities $\mathscr{N}_f$ and $\mathscr{N}_r$, provided quite satisfactory simulation results on the validation set. Indeed, the root mean square simulation error (*RMSE*) between the simulated front and rear accelerations and the validation data resulted to be $RMSE_f = 0.74$ and $RMSE_r = 0.53$, respectively. These values are quite small compared to the acceleration standard deviations. The simulation of model M1 can be observed in Figures 8.16 and 8.17, where the measured and simulated data are shown for a subset of the validation set. Similar results have been obtained for the entire data set.

In the first part of this example, the LTI system $\mathscr{L}$ has been supposed to be known: a model based on mechanical laws has been used for the chassis–engine block and black-box models have been used for the wheel–tire blocks. The latter are very accurate models in comparison with the measured data (see Figure 8.15). However, no particular information on the quality of the chassis–engine model is available. This model is only known to work well in closed loop with the wheel–tire models and the identified nonlinear suspensions.

In order to check if better results could be obtained, the iterative identification algorithm of [19] has been applied. In this algorithm, the LTI system $\mathscr{L}$ is not assumed

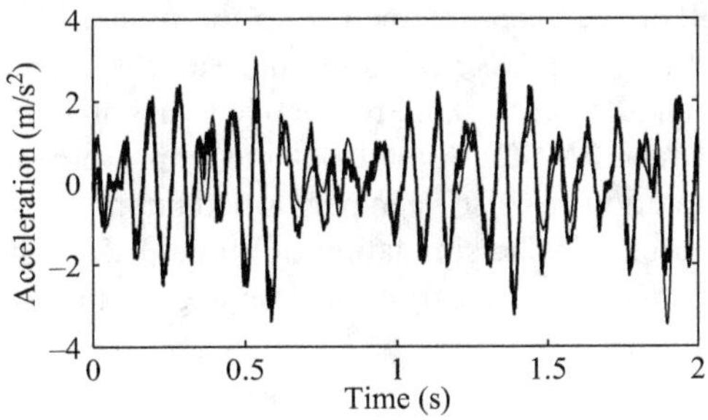

Figure 8.16 Front chassis acceleration. Bold line, measured; thin line, simulated by model M1

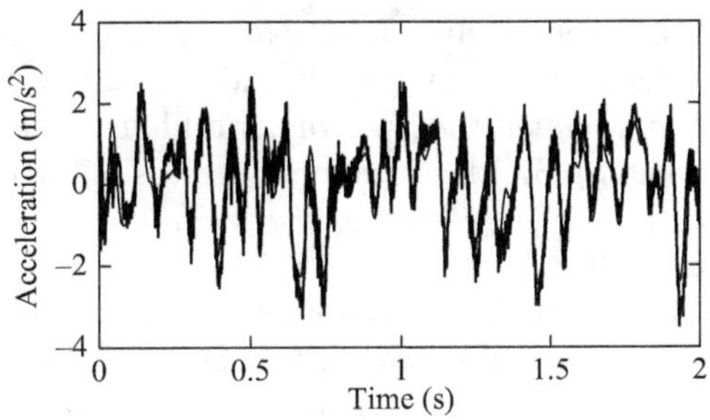

Figure 8.17 Rear chassis acceleration. Bold line, measured; thin line, simulated by model M1

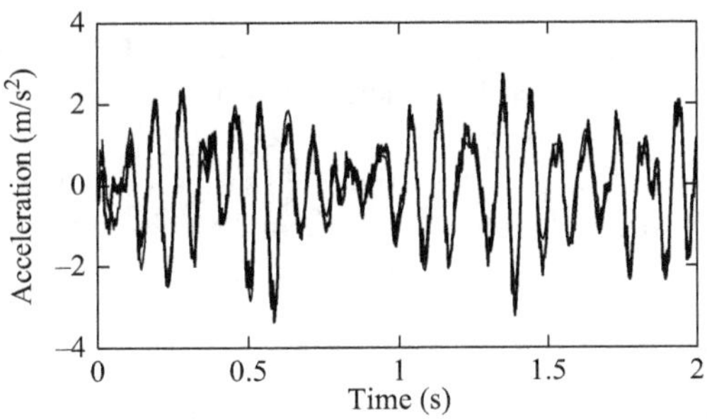

Figure 8.18 Front chassis acceleration. Bold line, measured; thin line, simulated by model M2

known but is identified at each step of the algorithm. Starting from the model M1, a second iteration has been performed. Using the forces generated from the identified suspensions as inputs and the measured accelerations as outputs, a new model of the chassis–engine block has been identified. Then, using the new chassis–engine model and the wheel–tire models of the first identification, the algorithm of section 8.5.1 has been applied to reidentify the suspensions.

Interesting results have been obtained from the model identified at the second iteration and called M2: the RMSE of front and rear chassis accelerations resulted to be $RMSE_f = 0.48$ and $RMSE_r = 0.52$, respectively. In comparison with model M1, the improvement of rear RMSE is small but the improvement of front RMSE is significant. A third iteration has also been performed but no further improvements of RMSEs have been observed. The simulation of model M2 is shown in Figure 8.18, where the measured and simulated front chassis accelerations are compared for a subset of the validation set.

References

[1] dos Santos, P.L., Perdicoúlis, T.A., Novara, C., Ramos, J., Rivera, D. (Eds.): Linear parameter-varying system identification. In: *Advanced Series in Electrical and Computer Engineering*. World Scientific (2011)

[2] Billings, S., Fakhouri, S.: Identification of a class of nonlinear systems using correlation analysis. *Proceedings of the Institution of Electrical Engineers* **125**(7), 691–697 (1978)

[3] Billings, S., Fakhouri, S.: Identification of systems containing linear dynamic and static nonlinear elements. *Automatica* **18**(1), 15–26 (1982)

[4] Brown, R.G., Hwang, P.Y.C.: *Introduction to Random Signals and Applied Kalman Filtering*, 4th edn. Wiley, New York, NY (2012)

[5] Chang, F., Luus, R.: A noniterative method for identification using Hammerstein model. *IEEE Transactions on Automatic Control* **16**(5), 464–468 (1971)

[6] Falck, T., Suykens, J.A., Schoukens, J., De Moor, B.: Nuclear norm regularization for overparametrized Hammerstein systems. In: *Proc. IEEE Conference on Decision and Control*, pp. 7202–7207 (2010)

[7] Greblicki, W.: Non-parametric orthogonal series identification of Hammerstein systems. *International Journal of Systems Science* **20**(12), 2355–2367 (1989)

[8] Hogberg, M., Bewley, T., Henningson, D.: Relaminarization of $re_\tau = 100$ turbulence using gain scheduling and linear state-feedback control. *Physics of Fluids* **15**(11), 3572–3575 (2003)

[9] Hsu, K., Poolla, K., Vincent, T.L.: Identification of structured nonlinear systems. *IEEE Transactions on Automatic Control* **53**(11), 2497–2513 (2008)

[10] Hsu, K., Vincent, T.L., Poolla, K.: Nonparametric methods for the identification of linear parameter varying systems. In: *Proc. IEEE International Conference on Computer-Aided Control Systems*, pp. 846–851 (2008)

[11] Kaminer, I., Pascoal, A.M., Khargonekar, P.P., Coleman, E.E.: A velocity algorithm for the implementation of gain-scheduled controllers. *Automatica* **31**(8), 1185–1191 (1995)

[12] Krtolica, R., Hrovat, D.: Optimal active suspension control based on a half-car model: An analytical solution. *IEEE Transactions on Automatic Control* **37**(4), 2238–2243 (1992)

[13] Lee, L.H., Poolla, K.: Identification of linear parameter-varying systems via LFTs. In: *Decision and Control, 1996, Proceedings of the 35th IEEE*, vol. 2, pp. 1545–1550. IEEE (1996)

[14] Ljung, L.: *System Identification: Theory for the User*. Wiley, New York, NY (1999)

[15] Ljung, L.: *System Identification Toolbox for Use with MATLAB*. The Mathworks, Natick, MA (2007)

[16] Ljung, L.: Perspectives on system identification. *Annual Reviews in Control* **34**(1), 1–12 (2010)

[17] Lovera, M., Novara, C., dos Santos, P.L., Rivera, D.: Guest editorial special issue on applied LPV modeling and identification. *IEEE Transactions on Control Systems Technology* **19**(1), 1 –4 (2011). doi 10.1109/TCST.2010.2090416

[18] Lu, J., DePoyster, M.: Multiobjective optimal suspension control to achieve integrated ride and handling perfomance. *IEEE Transactions on Control Systems Technology* **10**(6), 807–821 (2002)

[19] Milanese, M., Novara, C.: Structured set membership identification of nonlinear systems with application to vehicles with controlled suspension. *Control Engineering Practice* **15/1**, 1–16 (2007)

[20] Narendra, K., Gallman, P.: An iterative method for the identification of nonlinear systems using a Hammerstein model. *IEEE Transactions on Automatic Control* **11**(3), 546–550 (1966)

[21] Navarro-Lopez, E.: What makes the control of discontinuous dynamical systems so complex?. In: S. Sivasundaram (ed.). *Mathematical Analysis and Applications in Engineering Aerospace and Sciences*. Cambridge Scientific Publishers, UK (2013)

[22] Novara, C., Vincent, T., Hsu, K., Milanese, M., Poolla, K.: Parametric identification of structured nonlinear systems. *Automatica* **47**(4), 711–721 (2011)

[23] Pepona, E., Paoletti, S., Garulli, A., Date, P: Identification of piecewise affine LFR models of interconnected systems. *IEEE Transactions on Control Systems Technology* **19**(1), 148–155 (2011)

[24] Rabitz, H., Aliş, Ö.F., Shorter, J., Shim, K.: Efficient input-output model representations. *Computer Physics Communications* **117**(1), 11–20 (1999)

[25] Rugh, W., Shamma, J.: Research on gain scheduling. *Automatica* **36**, 1401–1425 (2000)

[26] Sanandaji, B.M., Vincent, T.L., Colclasure, A.M., Kee, R.J.: Modeling and control of tubular solid-oxide fuel cell systems: II. Nonlinear model reduction and model predictive control. *Journal of Power Sources* **196**(1), 208–217 (2011)

[27] Schölkopf, B., Smola, A.J.: *Learning with Kernels: Support Vector Machines, Regularization, Optimization and Beyond*. MIT Press, Cambridge, MA (2002)

[28] Sjöberg, J., Zhang, Q., Ljung, L., Benveniste, A., Delyon, B., Glorennec, P.Y., Hjalmarsson, H., Juditsky, A.: Nonlinear black-box modeling in system identification: a unified overview. *Automatica* **31**(12), 1691–1724 (1995)

[29] Suykens, J.A., Vandewalle, J.: *Nonlinear Modeling: Advanced Black-Box Techniques*. Kluswer Academic Publishers, Dordrecht, The Netherlands (1998)

[30] Van Mulders, A., Schoukens, J., Vanbeylen, L.: Identification of systems with localised nonlinearity: from state-space to block-structured models. *Automatica* **49**(5), 1392–1396 (2013)

[31] Vanbeylen, L.: From two frequency response measurements to the powerful nonlinear lfr model. In: *Proc. International Instrumentation and Measurement Technology Conference (I2MTC)*, pp. 2109–2113. IEEE (2012)

[32] Vandersteen, G., Schoukens, J.: Measurement and identification of nonlinear systems consisting of linear dynamic blocks and one static nonlinearity. *IEEE Transactions on Automatic Control* **44**(6), 1266–1271 (1999)

[33] Vapnik, V.: *The Nature of Statistical Learning Theory*. Springer, New York, NY (1999)

[34] Wahba, G.: *Spline Models for Observational Data* (vol. 59). SIAM (1990)

[35] Wei, X., del Re, L.: Gain scheduled H_∞ control for air path systems of diesel engines using LPV techniques. *IEEE Transactions on Control Systems Technology* **15**(3), 406–415, Philadelphia, PA (2007). doi 10.1109/TCST.2007.894633

Chapter 9
Linear fractional LPV model identification from local experiments using an H_∞-based glocal approach

Daniel Vizer[1], *Guillaume Mercère*[2], *Edouard Laroche*[3] *and Olivier Prot*[4]

Abstract

In this chapter a new identification technique is introduced to estimate a linear fractional representation (LFR) of a linear parameter-varying (LPV) system from local experiments using a dedicated non-smooth optimization procedure. Having access to a reliable set of local models, this technique consists more specifically in optimizing an H_∞-norm-based cost function measuring the fit between the local information (represented by the locally estimated LTI models) and the local behavior of a parameterized global LPV model. The method presented in this chapter results directly in an LPV model whose parametric matrices can be rational functions of the scheduling variables without any interpolation step (required usually by the local approach) and without writing the local fully parameterized LTI state-space models with respect to a coherent basis. On top of that, specific attention is paid to parameterized LPV models satisfying a fully parameterized or a physically structured linear fractional representation. This identification procedure is tested and validated with a simulation example.

9.1 Introduction

The idea of approximating the dynamic behavior of a non-linear system with a linear parameter-varying (LPV) model, i.e., a linear model the parameters of which depend

[1]University of Technology and Economics of Budapest, Budapest, Hungary and University of Poitiers, Poitiers Cedex, France, e-mail: vizer@iit.bme.hu
[2]University of Poitiers, Poitiers Cedex, France, e-mail: guillaume.mercere@univ-poitiers.fr
[3]University of Strasbourg, Illkirch Cedex, France, e-mail: laroche@unistra.fr
[4]University of Limoges, Limoges Cedex, France, e-mail: olivier.prot@unilim.fr

on measurable time-varying signals called the scheduling variables, has attracted a lot of attention from last two decades (see, e.g., [32, 21] and the references therein for an historical and recent overview). Most of the techniques developed to identify LPV models can be sorted into two main families according to the assumptions made on the excitability of the scheduling variables involved in the LPV representation of the system to identify. Historically, the first developments (the so-called global approach) assumed that one global experiment can be performed for which the control inputs as well as the scheduling variables can be both persistently excited (see, e.g., [18, 35]). By construction, these techniques assume that the scheduling variables $p \in \mathbb{R}^{n_p}$ are not only measurable but also totally controllable and excitable. By holding these assumptions, the global methods are limited to specific systems. Indeed, as it can be seen in Section 9.3 as well as in Chapter 16 of this book, for several practical cases, such a global experiment is not conceivable mainly for financial or safety reasons. In order to circumvent this problem, the so-called local approach can be considered instead. By local approach, it is meant that a multistep procedure is undertaken. More precisely,

1. a finite set of scheduling variable values p_i, $i \in \{1, \ldots, N_{op}\}$, $N_{op} \in \mathbb{N}^*$, is selected instead of the continuous excitation of $p \in \mathbb{R}^{n_p}$,
2. experimental data are gathered for each operating point determined by the fixed values of the scheduling variables p_i, $i \in \{1, \ldots, N_{op}\}$,
3. N_{op} local LTI (state-space) models are estimated based on the measured data gathered locally,
4. the estimated local LTI models are interpolated in order to obtain a global p-dependent model.

Such an approach is considered among others in [31, 12, 22, 24, 25] as well as in the following of this chapter where LPV state-space representations are handled exclusively.

As pointed out first in [33], the challenging step of the local approach is mainly the interpolation of the local LTI models when state-space forms are tackled. As shown in the aforementioned reference, the standard local techniques can indeed lead to a global LPV model unable to mimic the dynamic behavior of the system to identify even if the local models are accurately estimated. As shown in [33, 32], the description of the local models in coherent bases before the interpolation step is a crucial stage in the local identification procedure. This difficulty can be explained as follows. On a real system, the matrices composing the LPV state-space representation of the system may be dynamic functions[1] of $p(t)$, i.e., they may depend on the time derivatives or time-shifted versions of the scheduling variables. When the real system satisfies a dynamic scheduling dependency, a mis-specification of the mapping relating the state-space matrices of the model with respect to p can lead to a bad input-output dynamic behavior of the global LPV model over the entire operating regime of the identified process even if the model fits the local operating points. This difficulty is especially encountered in the black-box model based framework, i.e., when no information about the structure

[1] This type of parametrization can be well-described by following a behavioral approach as in [34].

of the real system which generates the data is available. As highlighted in [33, 17], all the local state-space models should be identified in a dependent way in order to ensure that the local models, which are determined up to a similarity transformation, are coherent to each others. Such a constraint is really difficult to satisfy in a general black-box framework. In this chapter the challenging problems described above are tackled. A gray-box LPV model-based technique is more precisely developed to reach this goal.

In short, the main idea of the method suggested hereafter consists in avoiding the standard interpolation step by resorting to a "glocal[2] method" able to yield directly a global LPV model in the so-called linear fractional representation (LFR) [18, 23] from the available local experiments. Since the identification procedure described herein is based on local experiments but the final LPV model is obtained by optimizing one global cost function, it circumvents the standard problems related to the interpolation of local models such as [22, 32]

- the numerical issues if an ill-conditioned parameterization for the model class is chosen (e.g., use of canonical forms with coefficients with huge magnitude variations (see [31] for an illustration of this problem)),
- the challenge of realizing all the models with respect to the same state variables which comes into play when dealing with the problem of interpolating black-box local state-space models.

A glocal approach, which also avoids the standard interpolation step, is described in [24] where a global affine LPV model is determined from local experiments by using a meta-heuristic optimization technique such as a differential evolution algorithm [29]. In this chapter, inspired by the discussions available in [28, 5], the frequency responses of the local LTI state-space models and the local frozen[3] LPV-LFR of the system are compared to each other at each corresponding operating point in order to build a global cost function using local experiments. In order to circumvent the difficulties encountered in the black-box framework, the required LPV-LFR is derived by taking the advantage of an initial study of the dynamic equations governing the behavior of the non-linear system. Notice indeed that

- the glocal method developed in [24] is only able to yield affine LPV models,
- the approach presented in [28] is based on H_2-norm,
- the H_∞-based technique described in [5] is restricted to LTI, linear time-periodic and affine LPV models.

Based on local experiments (corresponding each to almost constant scheduling variable values), the technique presented hereafter involves the following steps:

[2] The basic idea of realizing global functions by local actions is inspired from S. Hara's work dedicated to the design of "glocal controllers" [14]. For this reason, the identification procedure described hereafter will be also called "glocal" in the sequel.

[3] The term "frozen LPV" is borrowed from [32, 34]. A frozen LPV model is defined hereafter as the dynamic representation of an LPV model for which the scheduling variable trajectory is constant, i.e., $p(t) = \bar{p}$ for all $t \in \mathbb{R}$ or $\mathbb{Z}$.

1. determination of an LPV-LFR model of the system with a decent dimension that can be structured by using the available nonlinear equations describing the behavior of the system under study and standard linearization techniques [32],
2. estimation of local black-box LTI state-space models based on the available local experimental data,
3. minimization of a (non-smooth and non-convex) cost function comparing the frequency responses of the local black-box state-space models with the frozen LPV-LFR at each corresponding operating point in order to estimate the unknown parameters of the LPV linear fractional transformation (LFT) description of the system to identify.

As illustrated in Section 9.3, then in Chapter 16, such a procedure bypasses the challenging issues encountered when a standard local technique is used.

This chapter is organized as follows. Section 9.2 introduces the developed LPV-LFR identification method based on a dedicated structured H_∞ synthesis algorithm relying on a specific non-smooth optimization criterion. In Section 9.3, this identification procedure is tested and validated with a simulation example. Eventually, Section 9.4 concludes this chapter.

9.2 Identification method

9.2.1 Problem formulation, definitions, and notations

In many fields of application such as energetics, robotics, or aeronautics, the systems to identify have nonstationary and/or nonlinear behaviors and, often, their dynamics depend on external measurable variables. For such real systems, LPV models can be very attractive as far as model identification is concerned [32]. Keeping this idea in mind, we focus in what follows on LPV state-space models in the form

$$\dot{x}(t) = A(p(t),\Theta)x(t) + B(p(t),\Theta)u(t) \tag{9.1a}$$

$$y(t) = C(p(t),\Theta)x(t) + D(p(t),\Theta)u(t) \tag{9.1b}$$

where $x(t) \in \mathbb{R}^{n_x}$ is the state vector, $u(t) \in \mathbb{R}^{n_u}$ is the input vector, $y(t) \in \mathbb{R}^{n_y}$ is the output vector. The state matrices $(A(p(t),\Theta), B(p(t),\Theta), C(p(t),\Theta), D(p(t),\Theta))$ are functions of measurable time-varying signals gathered into a vector $p(t) \in \mathbb{K} \subseteq \mathbb{R}^{n_p}$ and used hereafter as the scheduling variable vector where $t \in \mathbb{R}$ and where $\mathbb{K}$ is a compact set called the scheduling space [34]. Notice that the state matrices depend on the instantaneous time value of $p(t)$, which is called a static-dependence [34]. The vector $\Theta \in \mathbb{R}^N$ contains the unknown parameters to identify.

In this chapter, a specific attention is paid to the linear fractional LPV state-space description [37]. Such a parameter dependence is indeed more general than the standard affine dependence usually encountered in system identification [21] and is

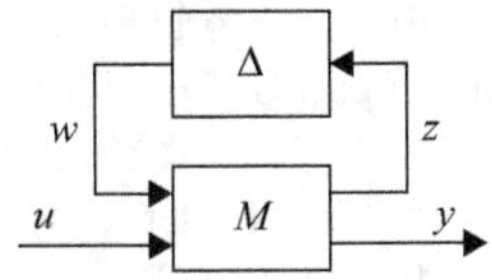

Figure 9.1 LFR (or LFT description) of the system to identify as an LTI dynamic system with a partial feedback connection composed of a time-varying static block $\Delta(p(t))$

often used in the robust control literature. The block diagram of the LFT description can be seen in Figure 9.1 where Δ is a $p(t)$-dependent matrix defined as

$$\Delta(p(t)) = \text{diag}(p_1(t)I_{r_1}, \ldots, p_{n_p}(t)I_{r_{n_p}}) \tag{9.2}$$

The signals $w(t) \in \mathbb{R}^r$ and $z(t) \in \mathbb{R}^r$ are inner auxiliary signals with $r = \sum_{k=1}^{n_p} r_k$ linked by the equation $w(t) = \Delta(p(t))z(t)$ and M is an LTI system with the following state-space representation

$$\begin{bmatrix} \dot{x}(t) \\ z(t) \\ y(t) \end{bmatrix} = \underbrace{\begin{bmatrix} A_0(\Theta) & B_w(\Theta) & B_0(\Theta) \\ C_z(\Theta) & D_{zw}(\Theta) & D_{zu}(\Theta) \\ C_0(\Theta) & D_{yw}(\Theta) & D_{yu}(\Theta) \end{bmatrix}}_{M(\Theta)} \begin{bmatrix} x(t) \\ w(t) \\ u(t) \end{bmatrix} \tag{9.3}$$

The matrices $(A_0(\Theta), B_w(\Theta), \ldots, D_{yu}(\Theta))$ composing the matrix $M(\Theta)$ are time-invariant matrices with appropriate dimensions. From this LFR, the direct state-space representation (9.1) can be obtained with the following state matrices [37]

$$A(p(t),\Theta) = A_0(\Theta) + B_w(\Theta)\Delta(p(t))(I - D_{zw}(\Theta)\Delta(p(t)))^{-1}C_z(\Theta) \tag{9.4a}$$

$$B(p(t),\Theta) = B_0(\Theta) + B_w(\Theta)\Delta(p(t))(I - D_{zw}(\Theta)\Delta(p(t)))^{-1}D_{zu}(\Theta) \tag{9.4b}$$

$$C(p(t),\Theta) = C_0(\Theta) + D_{yw}(\Theta)\Delta(p(t))(I - D_{zw}(\Theta)\Delta(p(t)))^{-1}C_z(\Theta) \tag{9.4c}$$

$$D(p(t),\Theta) = D_0(\Theta) + D_{yw}(\Theta)\Delta(p(t))(I - D_{zw}(\Theta)\Delta(p(t)))^{-1}D_{zu}(\Theta) \tag{9.4d}$$

It can be seen that the elements of the system matrices in (9.4) are

1. rational functions of the scheduling variable vector when $D_{zw}(\Theta) \neq 0$,
2. affine functions when $D_{zw}(\Theta) = 0$.

As it was explained previously, the method developed to identify the parameter vector Θ of the afore-introduced LPV-LFR is composed of three steps which are described thoroughly in this sub-section.

1. The first step aims at determining a linear fractional (LF) structure of the global LPV model to identify. Such an LF structure can be obtained in different manners according to the prior knowledge available on the system behavior. Hence, when

no prior information is available, using a fully parameterized linear fractional LPV state-space representation can be considered as the trivial choice. On the contrary, a structured gray-box LPV-LFR should be favored when the nonlinear equations governing the behavior of the system are known *a priori*. The methods aiming at deriving structured LPV-LFR from the nonlinear equations governing the behavior of the system mainly resorts to extensions of the familiar notions of linearization, such as the Jacobian linearization, the state transformation technique or the function substitution method (see [32, Chap. 7] and [8] for an overview of the main analytic developments). All these black-box and gray-box modeling considerations are detailed in Sub-section 9.2.2. Notice right now that the number of unknown parameters to estimate as well as the structure of the system given by (9.3) depend on the major initial study of the nonlinear equations governing the system behavior.

2. The second step consists in choosing a set of constant scheduling variables values for $N_{op} \in \mathbb{N}^*$ local working points so that all the working range of the system is covered (i.e., N_{op} is large enough) and the "distance" between two constant working points is small enough.[4] Then, local experimental data sets are generated for each working point satisfying the local conditions of excitation (see [19, 36] for more details about the consistence of excitation conditions) followed by the estimation of local black-box LTI state-space models by using standard and well-known techniques which can be found, e.g., in [20]. By denoting these i local black-box LTI state-space models with the quadruplet $(\mathfrak{A}_i, \mathfrak{B}_i, \mathfrak{C}_i, \mathfrak{D}_i)$, $i \in \{1, \ldots, N_{op}\}$, we can easily define N_{op} local LTI transfer functions as
$$\mathfrak{G}_i(s) = \mathfrak{D}_i + \mathfrak{C}_i(sI - \mathfrak{A}_i)^{-1}\mathfrak{B}_i \tag{9.5}$$
where s is the Laplace variable.
3. The third step focuses on the estimation of the vector Θ gathering the unknown parameters of the LPV state-space matrices (9.4) from the local information $\mathfrak{G}_i(s)$, $i \in \{1, \ldots, N_{op}\}$. In this chapter, such an estimation problem is solved by resorting to the minimization of the cost function
$$\min_{\Theta} \max_{i \in \{1,\ldots,N_{op}\}} \|\mathfrak{G}_i(s) - \mathscr{G}(s, \Delta(p_i), \Theta)\|_{\diamond} \tag{9.6}$$
where $\diamond$ stands for any arbitrary norm on systems and where $\mathscr{G}(s, \Delta(p_i), \Theta)$, $i \in \{1, \ldots, N_{op}\}$, is the local frozen linear fractional description of the system to identify written by applying the same frozen values of the scheduling variable vector p_i as the ones used for the identification of $(\mathfrak{A}_i, \mathfrak{B}_i, \mathfrak{C}_i, \mathfrak{D}_i)$, $i \in \{1, \ldots, N_{op}\}$, i.e.,
$$\mathscr{G}(s, \Delta(p_i), \Theta) = D(p_i, \Theta) + C(p_i, \Theta)(sI - A(p_i, \Theta))^{-1}B(p_i, \Theta) \tag{9.7}$$
where the matrices $(A(p_i, \Theta), B(p_i, \Theta), C(p_i, \Theta), D(p_i, \Theta))$ satisfy the structure of the matrices defined in (9.4) for $p(t) = p_i$, $i \in \{1, \ldots, N_{op}\}$. Roughly speaking,

[4] Notice that in [15], in order to maximize the accuracy of the interpolated LPV model, the optimal selection of the local operating points is examined. The solution developed in [15] is unfortunately restricted to a SISO-ARX LPV model and cannot be extended to linear fractional LPV models directly.

this cost function is used hereafter to match the LPV-LFR with the available local LTI realizations. In what follows, a specific attention will be paid to the H_∞-norm that compares the frequency responses of the local black-box state-space models with the frozen structured LPV-LFR at each corresponding operating point.

Because second step is no more a problem since many efficient techniques are now available in the literature [19, 36] to get consistent local fully parameterized state-space black-box models, this step is not detailed herein anymore. Hence, the next two sub-sections focus on first step and third step which constitute the main contributions of this chapter.

9.2.2 Determination of the structure of $\mathcal{G}(s, \Delta(p_i), \Theta)$

The first step of the identification procedure suggested in this chapter consists in determining the structure of the matrix $\Delta(p)$ as well as the matrix $M(\Theta)$ composed of the state-space matrices given in (9.4). As claimed previously, the structure of the matrices $\Delta(p)$ and $M(\Theta)$ composing the sought LPV-LFR highly depends on the available prior knowledge about the system to identify. In practice, the matrix $\Delta(p)$ can be defined with a diagonal structure, each parameter appearing once and being repeated several times according to the complexity of the final LPV-LFR. The main challenging step concerns the choice of $M(\Theta)$. As shown hereafter, three different cases, i.e., three different matrices $M(\Theta)$, can be encountered. For the sake of conciseness, in order to illustrate the derivation of the LPV-LFR, these three cases are explicitly introduced by using a well-known system example: the standard mass-spring-damper system (see [11]) where, for this specific application, the mass $m(t)$ is assumed to be time-varying and measurable as required when LPV systems are handled. A state-space representation of the system is given below:

$$\begin{bmatrix} \dot{x}(t) \\ \ddot{x}(t) \end{bmatrix} = \begin{bmatrix} 0 & 1 \\ -\dfrac{k}{m(t)} & -\dfrac{f}{m(t)} \end{bmatrix} \begin{bmatrix} x(t) \\ \dot{x}(t) \end{bmatrix} + \begin{bmatrix} 0 \\ \dfrac{1}{m(t)} \end{bmatrix} u(t) \tag{9.8a}$$

$$y(t) = \begin{bmatrix} 0 & 1 \end{bmatrix} \begin{bmatrix} x(t) \\ \dot{x}(t) \end{bmatrix} \tag{9.8b}$$

In the following, the mass is assumed to satisfy the relation $m(t) = m_0 + m_1(t)$ and $m_1(t)$ will be considered as the scheduling variable $p(t)$. For this specific case, the matrix $\Delta(p)$ is defined with a diagonal structure, the scheduling variable p being repeated once. As far as the matrix $M(\Theta)$ is concerned, more precisely its matrix-components, three cases can be considered according to the available prior information.

1. The first case corresponds to the black-box experimental modeling framework where $M_1(\Theta_1)$ is a fully parameterized state-space representation and $\Theta_1 \in \mathbb{R}^{N_1}$.

In this case, only the input-output behavior of the system is of interest. In this case, the matrix $M_1(\Theta_1)$ can be written for the system defined by (9.8) as follows:

$$M_1(\Theta_1) = \begin{bmatrix} A_0(\Theta_1) & B_w(\Theta_1) & B_0(\Theta_1) \\ C_z(\Theta_1) & D_{zw}(\Theta_1) & D_{zu}(\Theta_1) \\ C_0(\Theta_1) & D_{yw}(\Theta_1) & D_0(\Theta_1) \end{bmatrix} = \begin{bmatrix} \vartheta_1^1 & \vartheta_2^1 & \vartheta_3^1 & \vartheta_4^1 \\ \vartheta_5^1 & \vartheta_6^1 & \vartheta_7^1 & \vartheta_8^1 \\ \vartheta_9^1 & \vartheta_{10}^1 & \vartheta_{11}^1 & \vartheta_{12}^1 \\ \vartheta_{13}^1 & \vartheta_{14}^1 & \vartheta_{15}^1 & \vartheta_{16}^1 \end{bmatrix} \quad (9.9)$$

where $\Theta_1 = \begin{bmatrix} \vartheta_1^1 & \dots & \vartheta_{16}^1 \end{bmatrix}^\top \in \mathbb{R}^{16}$. Then $\mathcal{G}_1(s, \Delta(p_i), \Theta_1)$ can be calculated by using the expression given by (9.7) for fixed values of m_{1_i} and $\Delta(p_i) = m_{1_i}$, $i \in \{1, \dots, N_{op}\}$.

2. The second case deals with gray-box models obtained by the application of an analytic modeling technique, as suggested in [32, Chap. 7]. In this particular case, an initial study of the nonlinear equations governing the behavior of the system is taken into account. This consideration results in a fixed structure of the global LPV-LFR. By fixed structure, it is meant that some entries of the state-space representation can be fixed to be equal to 0, 1, or other known constant values. Thus, only the unknown parameters are stacked into the parameter vector $\Theta_2 \in \mathbb{R}^{N_2}$. For the mass-spring-damper system introduced previously,

$$M_2(\Theta_2) = \begin{bmatrix} A_0(\Theta_2) & B_w(\Theta_2) & B_0(\Theta_2) \\ C_z(\Theta_2) & D_{zw}(\Theta_2) & D_{zu}(\Theta_2) \\ C_0(\Theta_2) & D_{yw}(\Theta_2) & D_0(\Theta_2) \end{bmatrix} = \begin{bmatrix} 0 & 1 & 0 & 0 \\ \vartheta_1^2 & \vartheta_2^2 & \vartheta_1^2 & \vartheta_3^2 \\ \vartheta_1^2 & \vartheta_2^2 & \vartheta_3^2 & \vartheta_3^2 \\ 0 & 1 & 0 & 0 \end{bmatrix} \quad (9.10)$$

with $\Theta_2 = \begin{bmatrix} \vartheta_1^2 & \vartheta_2^2 & \vartheta_3^2 \end{bmatrix}^\top \in \mathbb{R}^3$ and $\Delta(p_i) = m_{1_i}$, $i \in \{1, \dots, N_{op}\}$.

3. The third case is related to the previous one by taking into account that Θ_2 can be written as a function of physical parameters $\theta \in \mathbb{R}^{N_\theta}$. This extension is possible because the analytical modeling allows us not only to fix the structure of the global LPV model but also to derive it as a function of the physical parameters. In order to emphasize this dependency, the unknown parameter vector is written hereafter as $\Theta_2(\theta)$. Again, for the mass-spring-damper system, we get

$$M_2(\Theta_2(\theta)) = \begin{bmatrix} A_0(\Theta_2(\theta)) & B_w(\Theta_2(\theta)) & B_0(\Theta_2(\theta)) \\ C_z(\Theta_2(\theta)) & D_{zw}(\Theta_2(\theta)) & D_{zu}(\Theta_2(\theta)) \\ C_0(\Theta_2(\theta)) & D_{yw}(\Theta_2(\theta)) & D_0(\Theta_2(\theta)) \end{bmatrix}$$

$$= \begin{bmatrix} 0 & 1 & 0 & 0 \\ -\dfrac{\theta_1}{\theta_3} & -\dfrac{\theta_2}{\theta_3} & -\dfrac{\theta_1}{\theta_3} & -\dfrac{1}{\theta_3} \\ -\dfrac{\theta_1}{\theta_3} & -\dfrac{\theta_2}{\theta_3} & -\dfrac{1}{\theta_3} & -\dfrac{1}{\theta_3} \\ 0 & 1 & 0 & 0 \end{bmatrix} \quad (9.11)$$

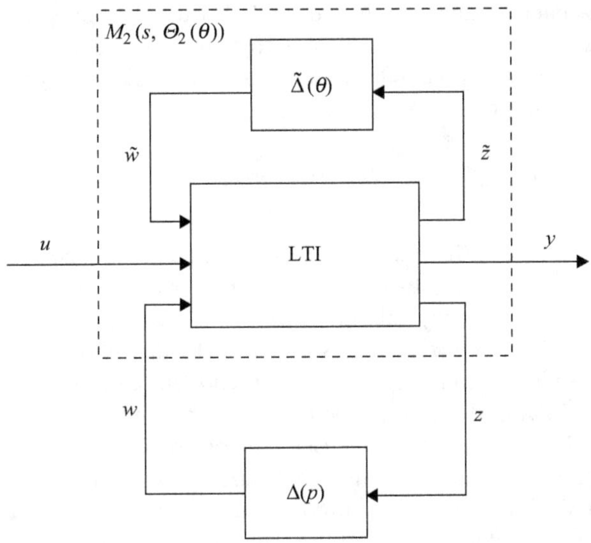

Figure 9.2 LFT description of a system with a specific focus on unknown physical parameters θ

with $\theta = \begin{bmatrix}\theta_1 & \theta_2 & \theta_3\end{bmatrix}^\top = \begin{bmatrix}k & f & m_0\end{bmatrix}^\top \in \mathbb{R}^3$ as the vector containing the physical parameters. Thus, the redefined unknown parameter vector can be written as

$$\Theta_2(\theta) = \begin{bmatrix}-\frac{\theta_1}{\theta_3} & -\frac{\theta_2}{\theta_3} & -\frac{1}{\theta_3}\end{bmatrix}^\top \qquad (9.12)$$

It is important to point out that, from this new dependency, as shown in Figure 9.2, a second partial feedback connection can be drawn from the initial LFT description by introducing a matrix $\tilde{\Delta}(\theta)$ as a diagonal matrix composed of the physical parameters θ as well as an LTI dynamic system containing known parameters (mainly zeros and ones). Thus, the global LPV model can be determined as a new LFR where the Δ matrix includes both the scheduling variables and the parameters to be estimated.

Notice here that $N_2 < N_1$ but N_θ is not always smaller than N_2 as it can be seen through the applied example as well. This latter property of course depends on the structure of the system under study.

9.2.3 H_∞-based optimization technique

As introduced in Sub-section 9.2.1, in order to estimate the parameter vector Θ, it is suggested optimizing a cost function comparing the frequency responses of the local black-box state-space models and the frozen LPV-LFR at each corresponding operating point. The H_∞-norm is selected herein to be used as the distance measurement between the frequency responses because of its beneficial properties [37]. The cost function given by (9.6) can then be written as

$$\min_{\Theta} \max_{i \in \{1,\ldots,N_{op}\}} \|\mathfrak{G}_i(s) - \mathcal{G}(s, \Delta(p_i), \Theta)\|_\infty \tag{9.13}$$

Because the H_∞-norm is the maximal singular values of the complex gain matrix over all the frequencies for multi-input multi-output (MIMO) systems (which simplifies to the maximum frequency gain in the case of single-input single-output (SISO) systems (see e.g., [37] for further details)), it holds by definition that, if the maximal value of the H_∞-norm found in (9.13) is small enough (e.g., much smaller than 1), then the maximum amplification of the difference between the involved frequency responses is small, so the distance between the considered systems is small as well. Said differently, if the maximal value of the H_∞-norm found in (9.13) is small enough, the LPV-LFR is a decent approximation of all the local LTI state-space models for all the considered values of the scheduling variables. Moreover, it should provide a good approximation of the system for any trajectory of the scheduling variables.

In the following, it is shown how the identification problem stated beforehand can be recast into a structured H_∞-synthesis problem. Let us denote, at each operating point $i \in \{1,\ldots,N_{op}\}$, the following difference:

$$G_i(s,\Theta) = \mathfrak{G}_i(s) - \mathcal{G}(s, \Delta(p_i), \Theta) \tag{9.14}$$

Remark 9.1. In order to achieve a more accurate fit over some specific frequency range chosen by the user, the transfer $G_i(s,\Theta)$ can be tuned with specific input and output weighting functions $W_O(s)$ and $W_I(s)$ as follows:

$$W_O(s)(\mathfrak{G}_i(s) - \mathcal{G}(s, \Delta(p_i), \Theta))W_I(s) \tag{9.15}$$

This specific feature, interesting from a system identification point of view, is possible thanks to the use of an H_∞-synthesis problem. Thus, in the following, the notation $G_i(s,\Theta)$ will be used for the representations (9.14) and (9.15) equivalently.

Let us then introduce the block-diagonal term $\breve{G}(s,\overline{p},\Theta)$ having the form

$$\breve{G}(s,\overline{p},\Theta) = \text{blockdiag}\big(G_1(s,\Theta),\ldots,G_{N_{op}}(s,\Theta)\big) \tag{9.16}$$

where $\overline{p} = [p_1,\ldots,p_{N_{op}}]^\top$ is a vector containing the scheduling variables. As a result of the block diagonal structure, it can be written that

$$\|\breve{G}(s,\overline{p},\Theta)\|_\infty = \max_i \|G_i(s,\Theta)\|_\infty \tag{9.17}$$

Hence, the cost function given by (9.13) can be recast as

$$\min_{\Theta} \|\breve{G}(s,\overline{p},\Theta)\|_{\infty} \tag{9.18}$$

The optimization problem given by (9.18) is a structured H_∞-design problem. At its beginning, i.e., in the 90's, H_∞-synthesis was restricted to the synthesis of nonstructured, full-order systems resulting in a convex optimization problem [37]. However, recent developments in non-smooth and non-convex optimization have given birth to new methods for structured H_∞-synthesis with applications to controller and observer synthesis. The application of these methods for model identification remains a promising research field. It is worth highlighting that there is a significant difference between the identification problem studied in this chapter and the synthesis of control laws mentioned previously. Indeed, in the control law synthesis case, a local optimum may be a satisfying solution, while in our context, the global optimum is sought, particularly in the gray-box model identification framework, in order to get the true parameter vector $\Theta_\bullet$.

Let us summarize the main advantages of the proposed technique:

1. the LPV-LFR is more general than the affine LPV representation [5],
2. such a representation can be directly estimated, thus avoiding the standard interpolation step,
3. the black-box local LTI state-space models do neither have to be written in a coherent basis nor to have the same order,
4. physical parameters can also be estimated based on the available prior information of the structure of the system.

Notice that the challenging problems of the local approaches e.g., the problem of the basis coherence [33] and the inaccuracy of the dynamic behavior of the final LPV model caused by the interpolation step [33, 32] are circumvented by the proposed method.

Remark 9.2. It is interesting to point out that the following developments are not restricted to the LFT description. For instance, as soon as the transfer $\mathscr{G}(s,\Delta(p_i),\Theta)$ is differentiable with respect to Θ, the approach suggested in this chapter can be extended straightforwardly.

In order to minimize such a cost function given by (9.18), several methods are available. A detailed discussion about these different possibilities and the chosen method for this chapter can be found in the next sub-section.

9.2.4 Computing the H_∞-norm

In order to solve (9.18), i.e., in order to minimize the H_∞-norm of $\breve{G}(s,\overline{p},\Theta)$, the first step consists in computing the value of this function efficiently. Clearly,

$$\left\|\breve{G}(\jmath\omega,\overline{p},\Theta)\right\|_{\infty}^{2} = \sup_{\omega\in[0,\infty]} \lambda_1(\breve{G}^H(\jmath\omega,\overline{p},\Theta)\breve{G}(\jmath\omega,\overline{p},\Theta)) \tag{9.19}$$

Thus, solving (9.18) is equivalent to the following optimization problem:

$$\min_{\Theta} \sup_{\omega \in [0,\infty]} \lambda_1(\breve{G}^H(\jmath\omega,\overline{p},\Theta)\breve{G}(\jmath\omega,\overline{p},\Theta)) \tag{9.20}$$

where $\lambda_1(\bullet)$ is the maximum eigenvalue function which is a convex but a non-smooth[5] function [2]. From (9.20), the computation of the objective function seems to be a hard task because, by construction, the global optimum of a non-smooth and non-convex function over $[0, \infty]$ must be estimated. Fortunately, when $\breve{G}$ has a state-space representation, this step can be performed efficiently by using a bisection algorithm based on a hamiltonian calculus [3]. Thus, under the assumption that $\breve{G}$ can be realized by a state-space representation, the map $\omega \mapsto \breve{G}(\jmath\omega,\overline{p},\Theta)$ is a rational matrix function which is smooth for all $\omega \in [0, \infty]$. The bisection algorithm used to compute the value of the cost function involved in problem (9.20) moreover permits the computation of the set of active frequencies $\Omega(\overline{p},\Theta)$ defined as follows:

$$\Omega(\overline{p},\Theta) = \left\{\omega \in [0,\infty] : \lambda_1(\breve{G}^H(\jmath\omega,\overline{p},\Theta)\breve{G}(\jmath\omega,\overline{p},\Theta)) = \left\|\breve{G}(\bullet,\overline{p},\Theta)\right\|_\infty^2\right\} \tag{9.21}$$

This set is of prime importance for the following non-smooth minimization algorithm. It is indeed used to define as well as to compute the sub-gradients of the underlying cost function. As shown in [3], the number of frequencies in $\Omega(\overline{p},\Theta)$ is either finite, or $\Omega(\overline{p},\Theta) = [0,\infty]$. In practice, the set $\Omega(\overline{p},\Theta)$ is too small and is replaced by an extended set of active frequencies [3], which is defined by

$$\Omega_e(\overline{p},\Theta) = \Omega(\overline{p},\Theta) \cup \tilde{\Omega}(\overline{p},\Theta) \tag{9.22}$$

where $\tilde{\Omega}(\overline{p},\Theta)$ is a finite set of frequencies chosen adequately. See [3] for some hints about this choice.

9.2.5 Minimizing the H_∞-norm

During the last decade, many algorithms have been developed for the minimization of non-convex and non-smooth functions [9, 10, 16, 26]. These algorithms have been successfully applied in the control framework, for example, in order to synthesize a stabilizing controller or in the case of H_∞-synthesis [1]. While this kind of controller design problems are known to be NP-hard, the techniques developed in this non-smooth framework have allowed to synthesize efficient structured controllers, even for systems with a great number of states.

[5] In this chapter, a function is non-smooth if it is a Lipschitz continuous function which is not continuously differentiable.

As shown by the substantial literature dedicated to non-convex and non-smooth functions optimization [9, 10, 16, 26], many techniques are now available to optimize the cost function (9.20) used in the identification framework addressed in this chapter. As explained in [30], numerical techniques in non-smooth optimization can be divided roughly into two categories: the direct and the indirect techniques. Herein, it has been chosen to use a direct method because of its interesting numerical properties (see [1, 27] for details). For this specific class of techniques, local properties of the function to optimize are obtained through variational analysis. At this point, we may recall that, for a non-smooth function f, the notion of gradient of f at a point x is no longer valid and must be replaced by the sub-differential $\partial f(x)$. Herein, the Clarke sub-differential of a function will be considered [9]. The sub-differential $\partial f(x)$ is a set, the elements of which are called the sub-gradients. The sub-gradients of f at point x give some first order information about the geometry of f at x. Aware of this specificity, three main techniques can be used to minimize the non-smooth function (9.20). A first approach consists in using a gradient sampling method, where the information about the sub-differential at point x is built by choosing some points around x randomly and by computing a sub-gradient at these points [7]. For instance, the software `HIFOO` is an implementation of the gradient sampling algorithm. This software has been successfully used in many applications [13, 4]. `HIFOO` works well in practice but, by construction, is not a deterministic method. It is indeed based on a random sampling algorithm. Moreover, the random sampling method may be quite slow because a good description of the sub-differential at point x may require a large number of samples. Thus, a second technique (see [1] for details) consists in building the information on the sub-differential at point x from the extended set of active frequencies $\Omega_e(\overline{p}, \Theta)$, then generating a descent direction from a local model [1], finally using a line-search technique in order to decrease the value of the cost function value. This approach has been implemented in the `hinfstruct` function available in the Robust Control Toolbox of MATLAB®. This specific but standard function is used hereafter to optimize the cost function (9.20). Notice however that the proximity control algorithm, described in [27, 3] could also have been utilized to perform the minimization of the non-smooth function (9.20). Originally, this algorithm is an improvement of the method described in [1], with better convergence properties. The main idea of the algorithm consists in using a local convex model ϕ of the cost function as a cutting plane generator from which a polyhedral model of the objective is built and used to generate trial points in order to minimize the cost function value. Because of the inherent instability of the cutting plane method [6], a proximity control term is introduced into the model in order to stabilize the iterations. This algorithm is a first order method and is thus quite slow to converge. It is however possible to add up a second order term into the model [27] but this is only a theoretical possibility. No theoretical studies have been performed yet to see if a second order term improves the convergence properties of the algorithm. The interested reader can read [1] for more information concerning the cutting plane method as well as the sub-gradient computation.

9.3 Identification results

9.3.1 System description

In order to illustrate the performance of the proposed method, a two-mass system in translation (see Figure 9.3) is considered as a simulation example. The first vehicle of mass m_1 and linear speed v_1 is actuated with a force F. The second vehicle of mass m_2 and linear speed v_2 suffers from a dissipative force $-f_2 v_2$. These two vehicles are linked with a flexible transmission of strength k and friction ratio f. Denoting δ the distance between the two vehicles, the force produced on the first vehicle is $-k\delta - f(v_1 - v_2)$ and the opposite on the second vehicle. The driving vehicle is equipped with a linear speed controller $F = K(u - v_1)$ where u is the speed reference and K is the control gain. The first vehicle speed is considered as the output ($y = v_1$). The second vehicle is loaded with a mass m_c which is considered as variable, i.e., $m_c(t)$ is time-varying. By considering $x = \begin{bmatrix} m_1 v_1 & m_2 v_2 & \delta \end{bmatrix}^\top$ as the state vector, an LPV model structure similar to the one described in (9.1) can be obtained with $p(t) = m_c(t)$ and where its state matrices satisfy

$$A(m_c(t)) = \begin{bmatrix} -\dfrac{f+K}{m_1} & \dfrac{f}{m_2+m_c(t)} & -k \\ \dfrac{f}{m_1} & -\dfrac{f+f_2}{m_2+m_c(t)} & k \\ \dfrac{1}{m_1} & -\dfrac{1}{m_2+m_c(t)} & 0 \end{bmatrix} \quad B(m_c(t)) = \begin{bmatrix} -K \\ 0 \\ 0 \end{bmatrix}$$
$$C(m_c(t)) = \begin{bmatrix} 0 \\ \dfrac{1}{m_2+m_c(t)} \\ 0 \end{bmatrix}^\top \tag{9.23}$$

and $D(m_c(t)) = 0$. Starting from this LPV state-space representation, a black-box fully parameterized linear fractional LPV model can be derived by following the procedure introduced in Sub-section 9.2.2. Such a black-box LFR is characterized by a matrix $\Delta(p(t))$ and a matrix $M_1(\Theta_1)$ defined as follows:

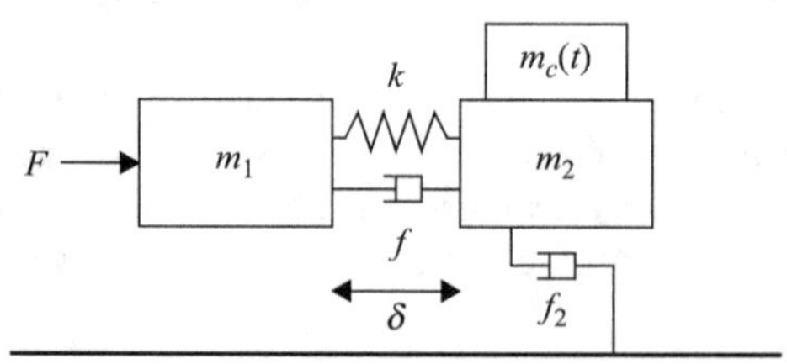

Figure 9.3 The translational two-mass system used for illustration

$$\Delta(p(t)) = p(t) = m_c(t) \tag{9.24a}$$

$$M_1(\Theta_1) = \begin{bmatrix} A_0 & B_w & B_0 \\ C_z & D_{zw} & D_{zu} \\ C_0 & D_{yw} & D_0 \end{bmatrix} = \begin{bmatrix} \vartheta_1 & \vartheta_2 & \vartheta_3 & \vartheta_4 & \vartheta_5 \\ \vartheta_6 & \vartheta_7 & \vartheta_8 & \vartheta_9 & \vartheta_{10} \\ \vartheta_{11} & \vartheta_{12} & \vartheta_{13} & \vartheta_{14} & \vartheta_{15} \\ \vartheta_{16} & \vartheta_{17} & \vartheta_{18} & \vartheta_{19} & \vartheta_{20} \\ \vartheta_{21} & \vartheta_{22} & \vartheta_{23} & \vartheta_{24} & \vartheta_{25} \end{bmatrix} \tag{9.24b}$$

Similarly, a gray-box LFR of the system focusing on the physical parameters $\begin{bmatrix} m_1 & m_2 & k & K & f & f_2 \end{bmatrix}$ can be determined with a matrix $\Delta(p(t))$ and a structured matrix $M_2(\Theta_2(\theta))$ defined as follows:

$$\Delta(p(t)) = p(t) = m_c(t) \tag{9.25a}$$

$$M_2(\Theta_2(\theta)) = \begin{bmatrix} A_0 & B_w & B_0 \\ C_z & D_{zw} & D_{zu} \\ C_0 & D_{yw} & D_0 \end{bmatrix} = \begin{bmatrix} -\dfrac{f}{m_1} & \dfrac{f}{m_2} & -k & -\dfrac{f}{m_2} & -K \\ \dfrac{f}{m_1} & -\dfrac{f+f_2}{m_2} & k & \dfrac{f}{m_2} & 0 \\ \dfrac{1}{m_1} & -\dfrac{1}{m_2} & 0 & \dfrac{1}{m_2} & 0 \\ 0 & \dfrac{1}{m_2} & 0 & -\dfrac{1}{m_2} & 0 \\ \dfrac{1}{m_1} & 0 & 0 & 0 & 0 \end{bmatrix} \tag{9.25b}$$

where θ gathers the unknown physical parameter to estimate, i.e., $\theta = \begin{bmatrix} m_1 & m_2 & k & K & f & f_2 \end{bmatrix}^\top$.

9.3.2 Linear fractional LPV model identification

9.3.2.1 Identification of the local LTI state-space models

The first step of the method presented in this chapter consists in selecting N_{op} local operating points, then determining N_{op} local fully parametrized LTI state-space models $\mathfrak{G}_i(s) = (\mathfrak{A}_i, \mathfrak{B}_i, \mathfrak{C}_i, \mathfrak{D}_i)$, $i \in \{1, \ldots, N_{op}\}$, from the locally acquired data sets. For this simulation example, the range of p is chosen equal to $[0 : 0.05 : 0.8]$ with a 0.05 step between the operating points. Thus, $N_{op} = 17$. For each value of p_i, $i \in \{1, \ldots, N_{op}\}$, a noise-free I/O data sequence is generated from an input signal chosen as a pseudo-random binary sequence (PRBS) built so that the local dynamic of the system is well-excited. The local black-box fully parametrized LTI state-space models are estimated by resorting to the well-known ordinary[6] multivariable output-error state space (MOESP) subspace-based identification method [36]. In order to get local continuous-time models, the discrete-to-continuous-time domain conversion is performed by using the bilinear Tustin approximation to the derivative (e.g., the function `d2cm` available in MATLAB). Once these N_{op} local black-box fully parametrized LTI state-space models are estimated from these N_{op} local data sets, they are easily validated by comparing their local frequency responses with the frequency responses of their frozen analytic counterparts, i.e., the simulated local LTI models satisfying in (9.23) for constant values of m_c. As shown in Figure 9.4, the local estimated models are able to picture the local system dynamics accurately.

The second step of the procedure aims at determining the parameters of the global LFT description of the system from the available local information, i.e., from $\mathfrak{G}_i(s) = (\mathfrak{A}_i, \mathfrak{B}_i, \mathfrak{C}_i, \mathfrak{D}_i)$, $i \in \{1, \ldots, N_{op}\}$. According to the prior knowledge available about the system dynamics, two complimentary structures $M_1(\Theta_1)$ and $M_2(\Theta_2(\theta))$ can be considered.

9.3.2.2 Focus on $M_1(\Theta_1)$

First of all, a fully parameterized LFT description is considered, i.e., the unknown parameters composing $M_1(\Theta_1)$ are sought. In this case, $M_1(\Theta_1)$ has the structure given by (9.24b) and $\Theta_1 \in \mathbb{R}^{25}$. The main difficulty with such a fully parameterized model class is twofold. First, because of the lack of prior knowledge about the system in this black-box context, determining an initial parameter vector for the optimization algorithm is far from being easy while the final model estimate can be very sensitive to this initialization step. Second, because such a fully parameterized model is not identifiable [19], several estimated parameter vectors Θ_1 can lead to indistinguishable input-output behaviors. This last characteristic can make the selection of the "good" parameter vector difficult. Hereafter, the final value of $\|\breve{G}(s, \overline{p}, \Theta)\|_\infty$ is used as an efficient and straightforward selection criterion. As far as the first difficulty is concerned, in order to test the robustness of the H_∞-based optimization technique with respect to the initialization, a Monte Carlo simulation of size 100 is carried out. More precisely, for each run, 25 different and uniformly distributed random values generated in the range $[0, 1]$ are used as initial values of the 25 unknown matrix entries. Such a Monte Carlo simulation leads to figures gathered in Table 9.1. These figures show that the optimization algorithm developed for this specific case is rather robust with respect to the initialization because 100% of the runs lead to reliable estimated fully parameterized LFR corresponding to a small value of $\|\breve{G}(s, \overline{p}, \Theta)\|_\infty$

[6] The ordinary MOESP algorithm is sufficient for this simulation example because noise-free data are handled. The Authors have chosen to deal with noise-free data because proving that the PI MOESP or PO MOESP algorithms are able to deal with different noise characteristics is not the goal of this simulation example.

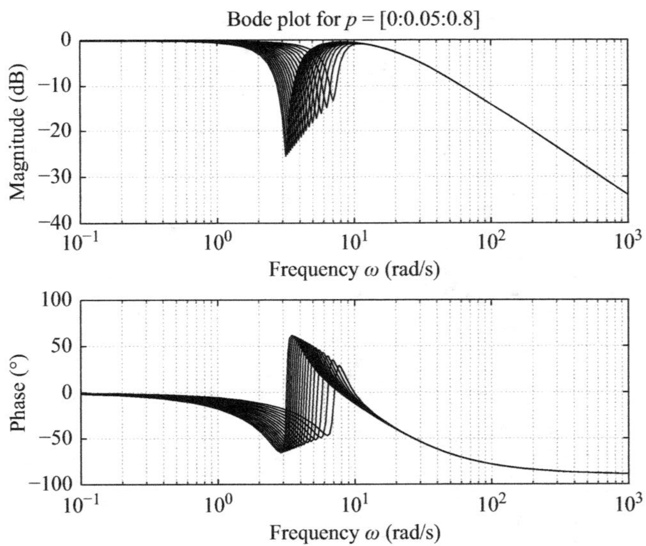

Bode diagram of the local black-box fully parametrized LTI state-space models w.r.t. p.

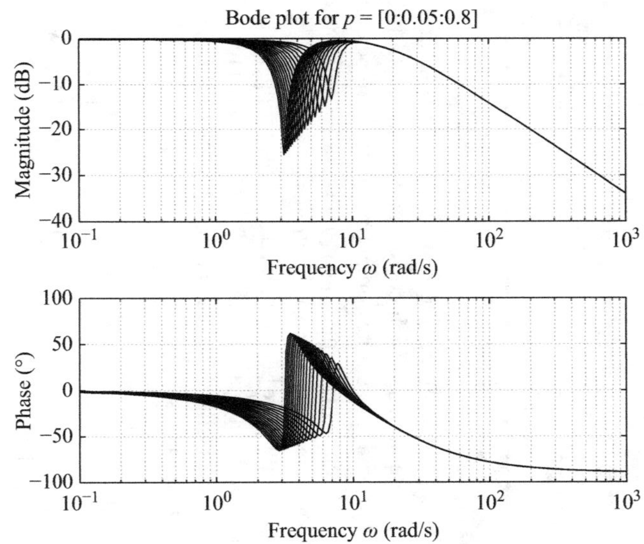

Bode diagram of the local analytic models w.r.t. p.

Figure 9.4 Frequency response validation of the estimated local LTI state-space models

Table 9.1 Evaluation of the sensitivity of the optimization algorithm w.r.t. the initialization. Case of $M_1(\Theta_1)$

	Best	Worst	Mean
$\|\breve{G}(s,\bar{p},\Theta)\|_\infty$	1.17e-7	0.0043	8.65e-5

(0.0043 at worst). Notice that, on average, the final values of the estimated parameters are obtained after 365 iterations, which corresponds to about 15 minutes on a recent laptop computer. Although these time computing figures can be considered as significant, the reader must keep in mind that 25 parameters are sought when $M_1(\Theta_1)$ is handled which leads to a quite complex cost function shape.

Notice that these good results have been obtained by considering, for each value of $i \in \{1,\ldots,N_{op}\}$, different weighting matrices W_I and W_O (see (9.15) for a definition). More precisely, it has been chosen to use, for each value of $i \in \{1,\ldots,N_{op}\}$, (see Figure 9.5 for their respective shapes)

$$W_I(s) = W_O(s) = \mathfrak{G}_i(s) \tag{9.26}$$

This information is indeed the best local filter we can have access to when no prior information is available about the system.

9.3.2.3 Focus on $M_2(\Theta_2(\theta))$

Second, the prior information about the system under study is taken into account in order to determine the physical parameters, namely, m_1, m_2, k, K, f and f_2, i.e., $\theta \in \mathbb{R}^7$. In this particular case, the matrix $M_2(\Theta_2(\theta))$ has the form given by (9.25b). Again, the initialization of the H_∞-based algorithm used in this chapter with a reliable initial parameter vector is essential to ensure the convergence toward the global optimum of the cost function (9.20). By assuming that we have access to prior information about the unknown parameters such as a range of evolution, the initialization is performed herein randomly and independently for each parameter θ_i, $i \in \{1,\ldots,7\}$, by using the following form:

$$\theta_{i_{init}} = \theta_{i_{real}}(1 + 4(r_i - 0.5)) \tag{9.27}$$

where $i = \{1,\ldots,7\}$, $\theta_{i_{real}}$ is the real value of the sought parameter and r_i is a random number uniformly distributed on the open interval $[0, 1]$. Said differently, according to (9.27), the initial values of the physical parameters may vary randomly in a $\pm 200\%$ wide range around their real values. As far as W_I and W_O are concerned, the same experimental framework as for $M_1(\Theta_1)$ is considered, i.e., $W_I(s) = W_O(s) = \mathfrak{G}_i(s)$, $i \in \{1,\ldots,N_{op}\}$.

In order to examine the effectiveness of the proposed method, a Monte Carlo simulation of size 100 is carried out again. For this specific framework, the optimization takes in average about 90 seconds per trial (which corresponds to 60 iterations).

Table 9.2 Evaluation of the sensitivity of the optimization algorithm w.r.t. the initialization. Case of $M_2(\Theta_2(\theta))$

Physical parameter	**Real value**	**Mean value**	**Best value of $\|\breve{G}(s,\overline{p},\Theta)\|_\infty$**	**Worst value of $\|\breve{G}(s,\overline{p},\Theta)\|_\infty$**
m_1 (kg)	0.2	0.24	0.25	0.034
m_2 (kg)	0.2	0.19	0.18	2e-11
K (N.s/m)	4	4.24	4.49	1.25
f (N.s/m)	0.1	0.21	0.064	7.46
f_2 (N.s/m)	0.04	0.056	0.02	1.14
k (N/m)	10	9.84	9.65	0.14

This computation reduction (in comparison with the fully parameterized case) can be explained by the fact that, now, only six unknown parameters are estimated.

In Table 9.2, the real physical parameters, their estimated mean values, the estimated values for the best and the worst values of the fit measurement $\|\breve{G}(s,\overline{p},\Theta)\|_\infty$ obtained during these 100 runs are gathered. The physical parameters are quite well-estimated except for the friction terms. An heuristic test[7] of sensitivity shows that these friction parameters have less effect on the output of the system, which can explain this estimation difficulty. This figure proves however the efficiency of the developed technique quantitatively.

9.3.3 Validation

Once the estimated fully parameterized and structured linear fractional LPV models are available, the last step of the identification procedure consists in validating theses estimated models. Two techniques are suggested in what follows.

First, the magnitude of the frequency responses of the estimated LPV models are calculated at each operating point, then compared to the local magnitude of the frequency responses of the real system computed for fixed values of $m_c(t)$, i.e., $m_c = [0 : 0.05 : 0.8]$. For both cases, the best LPV-LFR, i.e., the model having the smallest value of $\|\breve{G}(s,\overline{p},\Theta)\|_\infty$, is selected for this validation step. The magnitude versus frequency plots gathered in Figure 9.5 prove that the selected LPV-LFT descriptions are able to capture the local behavior of the system to identify efficiently on a quite large frequency range. Notice indeed that the estimated fully parameterized LPV-LFR knows difficulties to reproduce the high frequency responses of the locally generated real system accurately. This misfit indeed only occurs for $\omega > 100$ rad/s. Better results are obtained with the estimated structured LPV-LFR which is able to mimic the system frequency response on the whole frequency range. This observation leads to the conclusion that, when it is possible, using prior knowledge for the

[7] This sensitivity analysis is not explained in detail herein because the necessary and sufficient condition linked to this sensitivity test is not proved so far.

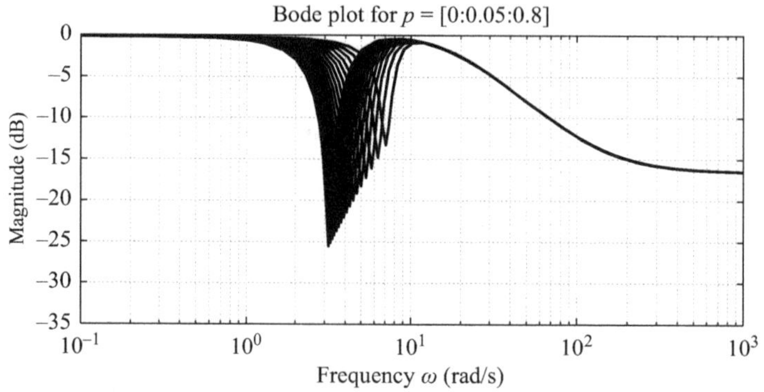

Bode magnitude diagram of the frozen fully parameterized LPV-LFR w.r.t. p.

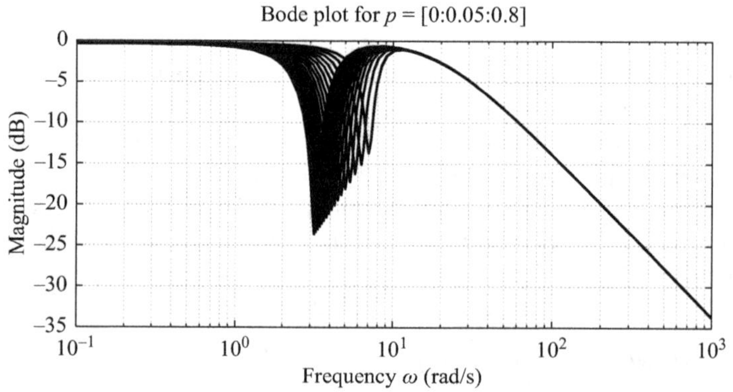

Bode magnitude diagram of the frozen structured LPV-LFR w.r.t. p.

Figure 9.5 Frequency response validation of the frozen estimated LPV-LFR

construction of the structure of the LFR and focusing on the estimation of physical parameters improve the final LPV model response, at least locally.

The second and last step of the validation procedure of the estimated LFT models consists in comparing the outputs of both estimated LPV-LFR with the output of the real system when the scheduling variable is excited continuously in the range $[0, 0.8]$. To reach this goal, the scheduling variable $m_c(t)$ is designed as a chirp signal characterized by frequencies that vary linearly with respect to time in the range $[0 - 15]$ rad/s over the time interval of 750 s (see Figure 9.6) while the evolution of the excitation signal u is a PRBS built so that the whole dynamic of the system is well-excited. The comparison of the output of the real system with the output of the fully parameterized and the structured LPV-LFR can be seen in Figures 9.7 and 9.8 respectively. These curves show that the obtained LPV-LFR are adequate approximations of the real system. The following fit measurements (see (9.28)) are also introduced in order to quantify the model quality on validation data (where y stands for the real

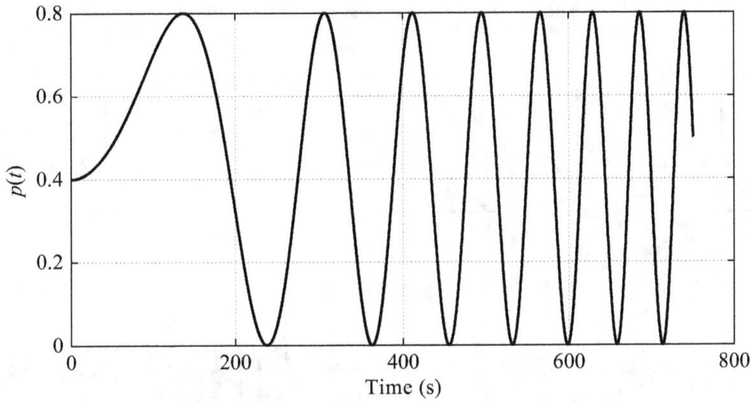

Figure 9.6 Evolution of the scheduling variable $p(t) = m_c(t)$ during the LPV model validation step

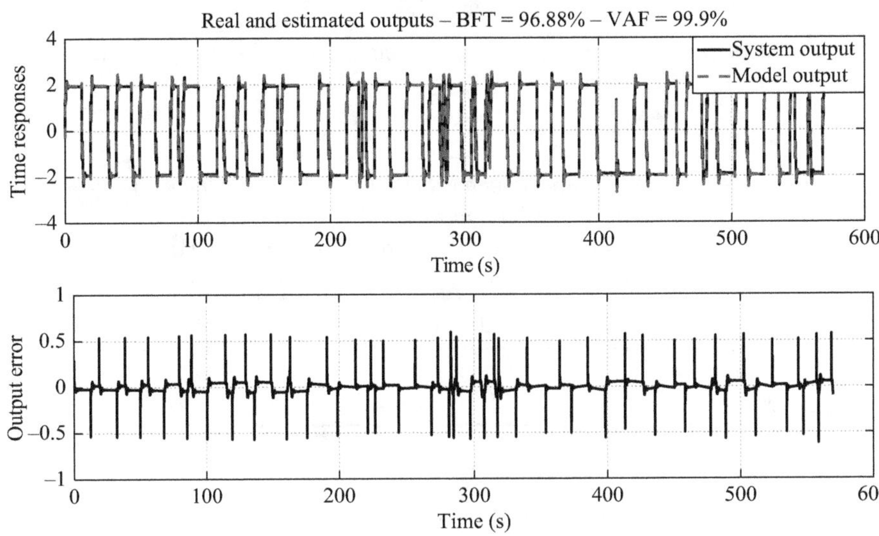

Figure 9.7 Comparison of the time responses of the real system (-) and the fully parameterized LPV-LFR (- -)

system output and $\hat{y}$ for the identified LPV model output and var(•) is the variance of •):

$$\text{BFT} = 100 \times \max\left(1 - \frac{\|y - \hat{y}\|_2}{\|y - \text{mean}(y)\|_2}, 0\right) \tag{9.28a}$$

$$\text{VAF} = 100 \times \max\left(1 - \frac{\text{var}(y - \hat{y})}{\text{var}(y)}, 0\right) \tag{9.28b}$$

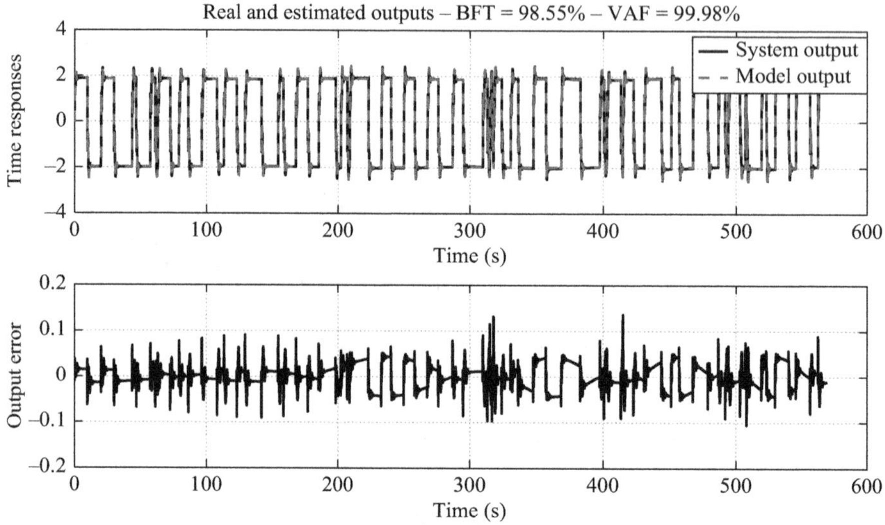

Figure 9.8 Comparison of the outputs of the real system (-) and the structured LPV-LFR (- -)

Table 9.3 Performance metrics (BFT and VAF) for the estimated (fully parameterized (fullyp.) and structured (struc.)) LPV/LFR on validation data

	BFT (%)	**VAF (%)**
y (fullyp.)	96.8	99.9
y (struc.)	98.5	99.9

These figures (see Table 9.3) prove that the proposed method gives access to accurate and reliable linear fractional LPV models capable to capture the behavior of a quite complex system. Notice that, even if the fully parameterized LPV-LFR is less efficient as far as the local frequency responses are concerned (see Figure 9.5), its reliability is proved from an I/O response point of view.

9.4 Conclusions

In this Chapter, through the study of linear fractional LPV state-space representations, some challenging problems and new solutions to the standard local approach for LPV system identification have been illustrated. More specifically, by starting from reliable locally estimated fully parameterized state-space forms, a new algorithm based on a specific H_∞-based model matching cost function has been developed to give access to a consistent global LPV-LFT description of the system to identify without requiring any interpolation step (usually required by the local approach). Such a technique has many advantages which can be listed as follows. First, the final LPV-LFR may be considered as fully parametrized or structured according to the available prior information about the system under study. Second, this method can be used to identify LPV models satisfying an affine or a rational dependency in a similar way. Last, but not least, local models with different state orders can be involved without any restriction. The efficiency of the aforementioned technique has been illustrated via a specific numerical example built so that black-box and gray-box state-space LPV-LFR identification can be tested. Thanks to its implementation user-friendliness and the availability of efficient H_∞-synthesis tools such as the ones presented in [1, 3], such a method can be considered as a promising solution to the challenging problem of coordinate bases coherence encountered by most of the local techniques.

References

[1] P. Apkarian and D. Noll. Nonsmooth H_∞ synthesis. *IEEE Transactions on Automatic Control*, 51:71–86, 2006.

[2] P. Apkarian, D. Noll, and O. Prot. A trust region spectral bundle method for non-convex eigenvalue optimization. *SIAM Journal of Optimization*, 19:281–306, 2008.

[3] P. Apkarian, D. Noll, and O. Prot. A proximity control algorithm to minimize nonsmooth and nonconvex semi-infinite maximum eigenvalue functions. *Journal of Convex Analysis*, 16:641–666, 2009.

[4] D. Arzelier, G. Deaconu, S. Gumussoy, and D. Henrion. H2 for HIFOO. In *Proceedings of the International Conference on Control and Optimization with Industrial Applications*, Ankara, Turkey, August 2011.

[5] M. Bergamasco. *Continuous-time model identification with applications to rotorcraft dynamics*. PhD thesis, Politecnico di Milano, Milan, Italy, 2013.

[6] J. Bonnans, J. Gilbert, C. Lemaréchal, and C. Sagastizábal. *Numerical optimization*. Springer-Verlag, Berlin Heidelberg New York 2006.

[7] J. Burke, A. Lewis, and M. Overton. A robust gradient sampling algorithm for nonsmooth, nonconvex optimization. *SIAM Journal on Optimization*, 25:751–779, 2005.

[8] F. Casella and M. Lovera. LPV/LFT modelling and identification: overview, synergies and a case study. In *Proceedings of the IEEE Multi-Conference on Systems and Control*, San Antonio, Texas, USA, September 2008.

[9] F. Clarke. *Nonsmooth Analysis*. Wiley, New York, 1983.

[10] F. Clarke. *Optimization and nonsmooth analysis*. Society for Industrial and Applied Mathematics, SIAM, Philadelphia, 1990.

[11] R. Dorf and R. Bishop. *Modern control systems*. Prentice Hall, eleventh edition, Prentice Hall, Upper Saddle River, New Jersey, 2008.

[12] W. Groot Wassink, M. van de Wal, C. Scherer, and O. Bosgra. LPV control for a wafer stage: beyond the theoretical solution. *Control Engineering Practice*, 13:231–245, 2005.

[13] S. Gumussoy, D. Henrion, M. Millstone, and M. Overton. Multiobjective robust control with HIFOO 2.0. In *Proceedings of the IFAC Symposium on Robust Control Design*, Haifa, Israel, June 2009.

[14] S. Hara. Workshop on "glocal control." In *Proceedings of the IEEE Multi-Conference on Systems and Control*, Yokohama, Japan, September 2010.

[15] A. Khalate, X. Bombois, R. Tóth, and R. Babuska. Optimal experimental design for LPV identification using a local approach. In *Proceedings of the IFAC Symposium on System Identification*, Saint-malo, France, July 2009.

[16] D. Klatte and B. Kummer. *Nonsmooth equations in optimization: regularity, calculus, methods and applications*. Springer, Dordrecht, 2002.

[17] B. Kulcsár and R. Tóth. On the similarity state transformation for linear parameter-varying systems. In *Proceedings of the IFAC World Congress*, Milano, Italy, August 2011.

[18] L. Lee and K. Poolla. Identification of linear parameter varying systems using non linear programming. *Journal of Dynamic Systems, Measurements and Control*, 121:71–78, 1999.

[19] L. Ljung. *System identification. Theory for the user*. Prentice Hall, Upper Saddle River, 2nd edition, 1999.

[20] L. Ljung. *System identification toolbox - for use with MATLAB*. Mathworks, 5th edition, Mathworks, Natick, 2000.

[21] P. Lopes dos Santos, T. Azevedo Perdicoúlis, C. Novara, J. Ramos, and D. Rivera. *Linear Parameter-Varying System Identification: new developments and trends*. Advanced Series in Electrical and Computer Engineering, World Scientific Publishing, Singapore, 2011.

[22] M. Lovera and G. Mercère. Identification for gain scheduling: a balanced subspace approach. In *Proceedings of the American Control Conference*, New York, USA, July 2007.

[23] J. Magni. Linear fractional representation with a toolbox for use with MATLAB. Technical report, ONERA, Toulouse, 2011.

[24] G. Mercère, E. Laroche, and M. Lovera. Identification of a flexible robot manipulator using a linear parameter-varying descriptor state-space structure. In *Proceedings of the IEEE Conference on Decision and Control and European Control Conference*, Orlando, Florida, USA, December 2011.

[25] G. Mercère, E. Laroche, and O. Prot. Analytical modelling and grey-box identification of a flexible arm using a linear parameter-varying model. In *Proceedings of the IFAC Symposium on System Identification*, Brussels, Belgium, July 2012.

[26] J. Nocedal and S. Wright. *Numerical Optimization*. Springer-Verlag, New York, 2006.

[27] D. Noll, O. Prot, and A. Rondepierre. A proximity control algorithm to minimize nonsmooth and nonconvex functions. *Pacific Journal of Optimization*, 4:571–604, 2008.

[28] D. Petersson. Nonlinear optimization approaches to H2-Norm based LPV modelling and control. Master's thesis, Linköping University, Linköping, Sweden, 2010.

[29] K. Price, R. Storn, and J. Lampinen. *Differential evolution: a practical approach to global optimization*. Springer, Berlin Heidelberg, 2005.

[30] T. Rockafellar. Nonsmooth optimization. *Mathematical programming: state of the art*, University of Michigan Press, Ann Arbor, pp. 248–258, 1994.

[31] M. Steinbuch, R. van de Molengraft, and A. van der Voort. Experimental modelling and LPV control of a motion system. In *Proceedings of the American Control Conference*, Denver, Colorado, USA, June 2003.

[32] R. Tóth. *Identification and Modeling of Linear Parameter-Varying Systems*. Springer Verlag. Lecture Notes in Control and Information Sciences 403, Springer-Verlag, Berlin, 2010.

[33] R. Tóth, F. Felici, P. Heuberger, and P. Van den Hof. Discrete-time LPV I/O and state-space representations, differences of behavior and pitfalls of interpolation. In *Proceedings of the European Control Conference*, Kos, Greece, July 2007.

[34] R. Tóth, J. Willems, P. Heuberger, and P. Van den Hof. The behavioral approach to linear parameter-varying systems. *IEEE Transactions on Automatic Control*, 56:2499–2514, 2011.

[35] J. W. van Wingerden. *Control of wind turbines with smart rotors: proof of concept and LPV subspace identification*. PhD thesis, Delft University of Technology, Delft, The Netherlands, 2008.

[36] M. Verhaegen and V. Verdult. *Filtering and system identification: a least squares approach*. Cambridge University Press, New York, 2007.

[37] K. Zhou, J. Doyle, and K. Glover. *Robust and optimal control*. Prentice Hall, Englewood Cliffs, New Jersey, 1996.

Chapter 10
Object-oriented modelling of spacecraft dynamics: tools and case studies

Marco Lovera[1] *and Francesco Casella*[2]

Abstract

The development process for spacecraft control systems relies heavily on modelling and simulation tools for spacecraft dynamics. For this reason, there is a strong need for adequate design tools in order to cope efficiently with tight budget and time constraints for space missions. In this chapter the main issues related to the modelling and simulation of satellite dynamics for control purposes are first discussed and an object-oriented modelling framework, implemented as a Modelica library, is then presented. The proposed tools enable a unified approach to a range of problems spanning from initial mission design and actuator sizing phases, down to detailed closed-loop simulation of the control system, including realistic models of sensors and actuators. It also promotes the reuse of modelling knowledge among similar missions, thus minimizing the design effort for any new project. The proposed framework and the Modelica library are demonstrated by means of two case studies.

10.1 Introduction

The safe and satisfactory operation of a satellite, in terms of its mission objectives, is strongly related to the performance level of its on-board attitude and orbit control systems, which provide the ability to maintain a desired orientation in space (or, e.g., carry out predefined attitude manoeuvres) and track a desired, nominal orbit in spite of the presence of external disturbances. Furthermore, the trend towards missions based on constellations or formations of small satellites has led to the formulation of even more complex control problems, dealing with the simultaneous relative motion (in terms of both attitude and position) of multiple spacecraft. This has resulted in an increasing need for efficient design tools in every domain involved in spacecraft design, particularly in the area of control-oriented modelling and simulation.

[1] Politecnico di Milano, Milano, Italy, e-mail: marco.lovera@polimi.it
[2] Politecnico di Milano, Milano, Italy, e-mail: francesco.casella@polimi.it

Specific tools are required for the design of both the system architecture and the Attitude and Orbit Control System (AOCS), bearing in mind the principles of reusability, flexibility and modularity.

The main goal of such tools should be to provide the designer with a *unified environment*, satisfying the following requirements:

- To enable the rapid development, testing and integration of new space systems in a unified, multi-physics framework
- To allow the support and management of a variety of spacecraft configurations and space environment descriptions covering all the phases of mission development
- To permit the reuse and customization of models across various missions
- To provide reusable models the interconnection of which can represent complex space systems
- To handle multi-spacecraft simulations
- To ensure visibility into the simulator models and state variables

Many commercial tools are available to support one or more of the above-mentioned requirements in the development of AOCS subsystems; however, none of them seems capable to provide the complete coverage of the whole development cycle in a sufficiently flexible way.

The systematic approach to modelling and simulation offered by modern acausal, equation-based object-oriented languages such as Modelica (see Chapter 2 in this book and [11, 4, 3]) allows to develop efficient spacecraft simulation tools. Note, in passing, that there is an increasing interest for multi-domain problems in the spacecraft control design community (see, e.g., [23]), an area that would benefit from the availability of simulation tools based on the object-oriented approach.

The development of simulation tools for satellite attitude and orbit dynamics within the object-oriented paradigm has been the subject of previous work (see [28], where an overview of the existing tools for AOCS modelling is presented). The use of Modelica for aerospace applications has recently led to the development of a library for flight dynamics (see [13]), while some preliminary results in the development of a Modelica spacecraft modelling library have been presented in [7]. More recently, the model components presented in the cited reference have been revised in order to take advantage of the Modelica MultiBody library (see [14]) which turns out to be extremely suitable to serve as a basis for the development of models for the mechanical parts of spacecraft models. In addition, exploiting Modelica's advanced features (see, e.g., [16]) it has been possible to redesign the library [7, 20] in order to improve the flexibility and the usability of the simulation tools. Furthermore, a recent extension of the MultiBody library (see [25]) is proving specially beneficial for the simulation of spacecraft with flexible appendages (see also [24]), while some related results on the development of high accuracy methods for the simulation of orbit dynamics have been presented in [1].

In view of the above discussion, this chapter aims at presenting a unified modelling approach to spacecraft dynamics modelling and simulation for control purposes based on object-oriented modelling concepts, and the Modelica library Space Flight

Dynamics that has been developed to implement it. The capabilities of the proposed approach are then demonstrated by presenting two case studies corresponding to applied research work developed in the last few years.

The chapter is organized as follows: the proposed approach to spacecraft modelling is described in sections 10.2 and 10.3, together with the new models and components developed specifically to implement this framework; the results obtained in three case studies are presented and discussed in section 10.4. More precisely, in section 10.4.1 the problem of determining disturbance torques using dynamic inversion is discussed, while in section 10.4.2 the development of a control-oriented simulation model for the tuning of a magnetic detumbling attitude control mode is presented.

10.2 The Modelica Space Flight Dynamics library

As already pointed out in [7, 20], the specific features of the Modelica modelling language make it a very natural candidate for developing flexible modelling and simulation tools for spacecraft dynamics, for a number of reasons.

- Spacecraft dynamics can be modelled by extending suitable classes available in the standard Modelica MultiBody library [15]: rigid bodies, rotational, translational, and prismatic joints, fixed displacements, elastic elements etc.
- In particular, in the framework of the MultiBody library, coordinate frames correspond to mechanical connectors, and kinematic transformations are easily and intuitively defined by means of suitable library elements.
- The Modelica language provides dedicated constructs to represent physical fields and environmental quantities. As one of the main issues in the modelling of spacecraft dynamics is describing the interaction between the satellite and the space environment (gravitational field, geomagnetic field, the Earth's atmosphere etc.), this feature turns out to be extremely useful. In particular, with a suitable choice of the environment interfaces, a whole range of models of increasing complexity can be implemented for each of the relevant environmental fields.
- As will be described in greater detail in the following, control systems components such as sensors and actuators can also be easily represented in the Modelica paradigm. As an example, a component for the simulation of magnetic torquers is modelled in terms of interaction with the geomagnetic field, while the momentum exchange between spacecraft and wheels is modelled via a simple mechanical connector allowing one rotational degree of freedom.[1]
- Packages containing data sheets for each class can be constructed and components easily modified within each spacecraft model, using Modelica's advanced features (see, e.g., [16]).

[1] Mounting errors, which may give rise to inter-axis coupling and vibrations, can be easily accounted for, whenever needed.

- Modelica models of the spacecraft hardware can easily be coupled with procedural control algorithms written as, e.g., *C* or *MATLAB*® code, allowing the designer to reuse a wide range of available specific algorithms and routines he/she is confident with, without going through all the trouble of re-implementing them as Modelica code.
- Finally, as the basic components of the library are independent from each other, one can exploit this flexibility in order to build simulation models of increasing complexity and accuracy according to the needs associated with each phase of the AOCS development process.

In addition to the basic models available within the Modelica MultiBody Library, recent extensions to the library allow the modelling and simulation of multibody systems with flexible elements (see [2, 25]) making it possible to describe spacecraft with flexible appendages such as gravity gradient booms, antennae or solar panel arrays.

The Space Flight Dynamics Library described in this chapter encompasses all the necessary utilities to set up a reliable and quick-to-use scenario for a generic space mission, providing a wide choice of most commonly used models for AOCS sensors, actuators and controls. The Space Flight Dynamics Library's model reusability is such that, as new missions are conceived, the library can be used as a base upon which a simulator can be readily and easily built. This goal can be achieved simply by interconnecting the standard Space Flight Dynamics Library objects, possibly with new components purposely designed to cope with specific mission requirements, regardless of the specific space mission scenario, in terms of mission environment (e.g. planet Earth and Mars and the solar system), spacecraft configuration, or embarked on-board systems (e.g. sensors, actuators, controls).

The generic spacecraft simulator contains one extended **World** model and one (or more) **Spacecraft** models. Multiple spacecraft can be used for, e.g., formation flying studies.

The extended **World** model extends the standard **World** model of the MultiBody library. It provides all the additional functions needed for a complete representation of the space environment as seen by an orbiting spacecraft:

- definition of the relevant inertial reference frame;
- definition of the base for simulation time (the so-called *epoch*);
- gravitational field model, possibly including the Sun and Moon third-body effects;
- geomagnetic field model;
- atmospheric model;
- solar radiation and eclipse model;
- models for Sun and Moon ephemeris.

These additional functions play a major role in the realistic simulation of the dynamics of a spacecraft, as the linear and angular motion of a satellite are significantly influenced by its interaction with the space environment.

The **Spacecraft** model is a completely reconfigurable model of a single spacecraft, obtained by the connection of the following components:

1. **SpacecraftDynamics:** This model is obtained by connecting a Body model from the standard MultiBody library, which represents a rigid body with six degrees of freedom, with models that describe the interaction between the spacecraft and the space environment, and that allow special initialization of the spacecraft position and velocity based on orbital data. Data for custom orbits and spacecraft inertial properties and geometry (influencing both aerodynamic and solar radiation behaviour) are stored in dedicated library packages. It is also possible to consider multibody spacecraft, e.g. satellites with flexible appendages, by just connecting the extra mechanical elements to the rigid-body model of the spacecraft bulk (see [24] for details).
2. **SensorBlock:** This *replaceable* model consists in a reconfigurable set of attitude sensors to be chosen among custom Space Flight Dynamics Library **SensorBlock** implementations. Replaceable models are a key object-oriented feature of the Modelica language, which allow to substitute a model with another one, provided that they have the same interface, i.e. the same connectors to the outer world. This allows to redeclare the basic **Spacecraft** model by choosing, e.g., a specific star tracker sensor, corresponding to a specified supplier's serial number; its configuration, defined by star tracker number, location and orientation with respect to the spacecraft's reference frame; and the level of detail of the model, ranging from an ideal angle measurement to a measurement corrupted with white noise and bias, up to a detailed sampled-time digital implementation, with computational delay and digital information about measurement availability. The Space Flight Dynamics Library encompasses mathematical models of different degree of complexity for star sensors, gyroscopes, magnetometers and GPS receivers.
3. **ActuatorBlock:** This *replaceable* model consists in a reconfigurable set of attitude control actuators to be chosen among custom implementations, in the same spirit of the SensorBlock model. Mathematical models with different degrees of detail for commonly used actuators and actuator sets have been implemented in the library, including momentum and reaction wheels, magnetic torquers and cold gas thrusters. Currently implemented sensor and actuator models have been developed within a control-oriented framework, i.e. they are functional models, rather than explicit physical models, of the measurement process. This is usually the best choice for control system studies. However, if necessary, more advanced physical models could be developed within the considered model architecture.
4. **ControlBlock:** This *replaceable* model implements the spacecraft Attitude Control System (ACS), including blocks supervising the basic attitude determination, attitude control and control allocation functions. This block can either contain a Modelica description of the controller, or an interface to an external description of the controller, e.g. a Simulink® block diagram, or the C-code of a digital controller.

A more detailed bottom-up description of the key library elements is given in the following section.

10.3 Structure of the spacecraft simulation models

10.3.1 Extended World model

This model extends the standard **World** model of the MultiBody library by fully describing the space environment in which the spacecraft(s) fly. More precisely, the user can define the initial date and time of the simulation, and the planet the spacecraft will orbit around. It is then possible to choose among the available models for the gravity field (J_2, J_4 or the more general JGM-3 model for gravitational potential; see [12]), for the magnetic field (dipole, quadrupole and the IGRF model; see [30]), for the Sun and Moon ephemeris tables, and for the atmospheric density model. The Space Flight Dynamics Library contains the coefficients for Earth, Moon, Jupiter, and the Sun, and more (e.g. Mars) may be added in the future.

As is well known (see, e.g. [30, 12]), the gravitational potential of a planet U_g may be described by the function

$$\begin{aligned} U_g(r,\theta,\lambda) = -\frac{\mu}{r}\Bigg\{ & 1+\sum_{n=2}^{\infty}\left(\frac{R_e}{r}\right)^n J_n P_n(\cos(\theta)) \\ & +\sum_{n=2}^{\infty}\sum_{m=1}^{n}\left(\frac{R_e}{r}\right)^n P_n^m(\cos(\theta))(C_n^m\cos(m\lambda)+S_n^m\sin(m\lambda))\Bigg\} \end{aligned} \quad (10.1)$$

where P_n^m are the Legendre polynomials

$$\begin{aligned} P_n(x) &= \frac{1}{2^n n!}\frac{d^n}{dx^n}(x^2-1)^n \\ P_n^m(x) &= (1-x^2)^{m/2}\frac{d^m P_n(x)}{dx^m} \end{aligned}$$

R_e is the mean equatorial planet radius, r, θ and λ are the point's spherical coordinates and coefficients J_n, C_n^m, S_n^m are the zonal, sectoral and tesseral coefficients.

Depending on the mission characteristics and on the purpose of attitude control simulations, a satisfactory approximation can be obtained by choosing the order of the expansion in a suitable way. The gravitational field components (expressed in spherical coordinates) are then given by

$$g = -\nabla U_g = -\left\{\frac{\partial U_g}{\partial r}, \frac{1}{r}\frac{\partial U_g}{\partial \theta}, \frac{1}{r\sin(\theta)}\frac{\partial U_g}{\partial \lambda}\right\} \quad (10.2)$$

This function supplements the gravitational field functions which are already available in the standard World model of the MultiBody library, i.e. no gravity, constant gravity and point-mass gravity.

Similarly, the geomagnetic potential U_m is described by the function

$$U_m(r,\theta,\lambda) = \frac{R_e}{\mu}\sum_{n=0}^{\infty}\sum_{m=0}^{n}\left(\frac{R_e}{r}\right)^{n+1} P_{nm}(\cos(\theta))(g_n^m\cos(m\lambda) + h_n^m\sin(m\lambda)) \quad (10.3)$$

where g_n^m and h_n^m are the Gauss coefficients appropriate to the Schmidt polynomials P_{nm}

$$P_{n,0}(x) = P_n^0(x)$$

$$P_{n,m}(x) = \left(\frac{2(n-m)!}{(n+m)!}\right)^{1/2} P_n^m(x)$$

The coefficients for the geomagnetic potential adopted in the simulation environment correspond to the so-called International Geomagnetic Reference Field (IGRF) model for the Earth's magnetic field (see [30]). The components of the geomagnetic field (expressed in spherical coordinates) are then given by

$$B = -\nabla U_m = -\left\{\frac{\partial U_m}{\partial r}, \frac{1}{r}\frac{\partial U_m}{\partial \theta}, \frac{1}{r\sin(\theta)}\frac{\partial U_m}{\partial \lambda}\right\} \quad (10.4)$$

Models for the atmospheric density and the Sun and Moon positions have also been implemented, according to [12, 29]; the details are omitted here for the sake of conciseness. The Sun position can be used to compute the radiation pressure on the spacecraft, using suitable eclipse algorithms, while both the Sun and Moon positions can be used to include third-body effects into the gravitational field.

10.3.2 SpacecraftDynamics model

This component accounts for the spacecraft inertia, and for its interaction with the space environment. It is obtained by the connection of different subcomponents through 3D mechanical connectors.

The 3D translational and rotational dynamics of the spacecraft is described by the standard **Body** model from the MultiBody library; this model basically contains Newton's and Euler's equations of motion.

The effect of the gravity field is accounted for using the standard design of the Multibody library: the **Body** model accesses the function to compute the gravity field contained in the **World** model, through the `inner/outer` construct of the Modelica language. This construct is precisely introduced to allow the propagation of field effects to multiple-component models without the need of explicit connections.

With respect to the gravity forces, the standard **World** model has been extended to include more accurate field models, as explained in section 10.3.1.

Besides gravity, several other forces and torques are applied onto the spacecraft due to its interaction with the space environment:

- gravity gradient torques;
- magnetic torques, arising from the presence of a non-zero spacecraft's residual magnetic dipole;
- aerodynamic forces and torques, produced by the interaction with the planet's atmosphere;
- forces and torques originated by the solar radiation pressure.

Each of these effects is described by a corresponding interaction model, connected to the Body. Each interaction model gets the spacecraft position and orientation from the mechanical connector, computes the forces and torques generated by the interaction with the environment using the extended World functions (via the `inner/outer` construct), and then applies them to the body through the mechanical connector. Set-up parameters are available to selectively include only the effects that are interesting for a particular simulation. An additional model allows to initialize the spacecraft position, velocity and orientation as a function of the classical orbital parameters.

The solar radiation pressure model is now described in some detail. Solar radiation pressure arises from the interaction between the spacecraft geometry, defined as an assembly of planar and possibly cylindrical surfaces, and the average solar radiation pressure Φ. When the spacecraft is fully illuminated, the force acting upon a single exposed surface is given by the momentum exchange law

$$\frac{dq_{sc}}{dt} = p_{\odot}\frac{(1\,AU)^2}{\|R\|^2}\sum_{i=1}^{\#surf} A_i\cos(\theta_i)[(1-\varepsilon_i)(-\hat{R}) + 2\varepsilon_i\cos(\theta_i)(-n_{ext,i})] \quad (10.5)$$

where q_{sc} is the spacecraft's momentum, $p_{\odot} = 4.56e^{-6}$ J m^{-2} is the mean solar radiation pressure at 1 astronomical unit (AU), R is the relative position vector from the spacecraft centre of mass to the Sun, A_i, ε_i and $n_{ext,i}$ are the single surface area, reflectivity coefficient and external surface unit vector respectively. Finally, $\theta_i = a\cos(\hat{R}\cdot n_{ext,i})$.

When a spherical body interposes between the spacecraft and the Sun, the latter is partially or totally eclipsed, and the force reduces accordingly by a factor[2]:

$$\nu(r_{sun}, r_{s/c}) = 1 - \left(\frac{\alpha a^2 + \beta b^2 - ac\sin\alpha}{\pi b^2}\right) \quad (10.6)$$

[2] The shadow function $\nu(r_{sun}, r_{s/c})$ is derived under the assumption of occulting body infinitely far from the spacecraft.

where

$$\alpha = \arcsin\left(\frac{b\sin\beta}{a}\right), \quad \beta = \arccos\left(\frac{c^2+b^2-a^2}{2bc}\right) \tag{10.7}$$

a, b are the Sun and occulting body apparent radii respectively, and c is the apparent distance between the geometrical centres of the Sun and occulting body.

Thus, the overall force acting on the spacecraft is

$$\frac{dq_{sc}}{dt} = \nu \cdot p_{\odot}\frac{(1\,AU)^2}{\|R\|^2}\sum_{i=1}^{\#surf} A_i\cos(\theta_i)\cdot[(1-\varepsilon_i)(-\hat{R}) + 2\varepsilon_i\cos(\theta_i)(-n_{ext,i})] \tag{10.8}$$

while the associated torque is computed accordingly, once the centre of pressure of each and every surface composing the spacecraft geometry is defined.

The spacecraft data (i.e. inertial properties, geometry, surface reflectivity, material etc.) can be easily retrieved from appropriate datasheet records. Moreover, to cope with classic space mission requirements, two initialization options are allowed:

- standard initialization of the **Modelica** MultiBody library: cartesian position and velocity, Euler angles and angular velocity;
- custom initialization, based on current simulation Universal Time (set within the extended World model), nominal orbit (specified by six orbital parameters, retrieved from appropriate records), orientation and angular rate relative to the orbital reference frame.

The mechanical connector of the spacecraft body, corresponding to the spacecraft frame of reference, is also made available as an external connector of the **Spacecraft-Dynamics** model itself. This allows to connect the model to the sensor and actuator models, which will provide the additional forces and torques commanded by the control system.

Mulitibody spacecraft can easily be accommodated within this framework, by connecting, e.g., the flexible appendage model to the spacecraft body connector.

10.3.3 Spacecraft model

One level up in the the model hierarchy, the structure of the **Spacecraft** model can be described as follows. The **SpacecraftDynamics** model has a mechanical port, which is connected to the mechanical ports of the **SensorBlock** and **ActuatorBlock** models, thus describing a rigid mechanical connection between the spacecraft body model (including the interaction with the environment) and the sensor and actuator models. These two latter models are then connected to the **ControlBlock** by *expandable bus connectors*, which carry all the input and output control signals.

As explained above, all of these models are *replaceable*. This means that the basic Spacecraft model contains default implementation for these blocks, but it is then possible to define and instantiate a whole range of Spacecraft models by *redeclaring* the sensor and actuator blocks with other implementations, which may differ by configuration (e.g. control moment gyros rather than momentum wheels) and/or by level of detail of the model. The only constraint is that the default model and its replacement have the same interface, i.e. the same connectors. Modelica tools with graphical user interfaces allow to do this intuitively, e.g. through drop-down menus: the *choicesAllMatching* annotation tells the GUI to include in the menu all the models that are compatible with the interface of the replaceable models. This also means that whenever a new model is added to the library, all these menus are implicitly updated without further intervention, which is very convenient from the library developer's perspective.

As an example, the default choice for the replaceable model **SensorBlock**, comprising three star trackers, one GPS receiver and one magnetometer, is depicted in Figure 10.1. As can be seen from the figure, models for each of the on-board sensors are included; in particular, each sensor is characterized by a mechanical interface, corresponding to the physical mounting of the instrument on the satellite body (taking into account the definition of the local sensor reference frame via a suitable change

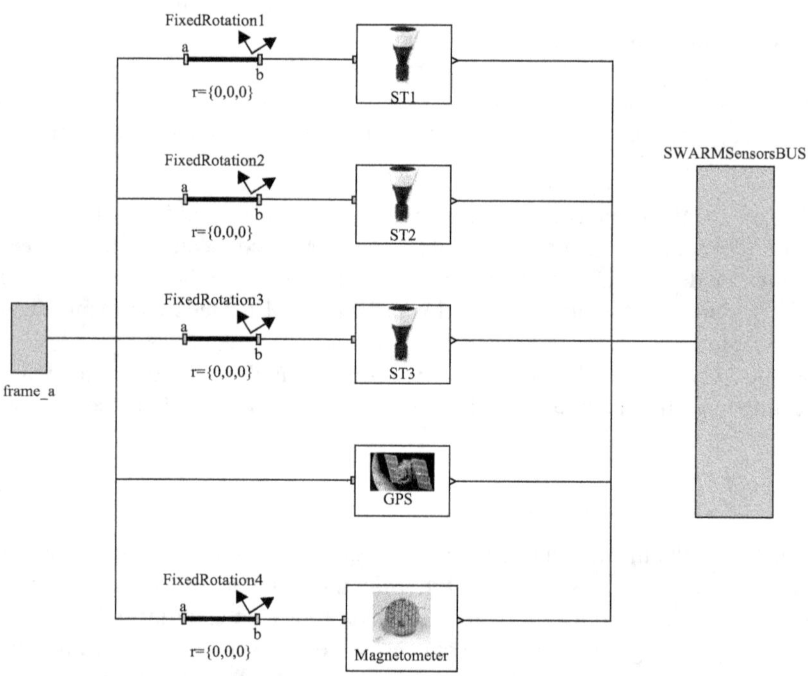

*Figure 10.1 Default choice for replaceable model **SensorBlock**, comprising three star trackers, one GPS receiver and one magnetometer*

of coordinates) and by a signal interface. The sensor data bus is therefore defined by the collection of output signals coming from each of the available sensors, by using Modelica *expandable connectors* (see [27]).

The sensor and actuator models exploit the *encapsulation* features of the Modelica language. The core definition for each sensor/actuator model is at the interface level: a mechanical port and signal bus port. The sensor/actuator model can then be connected to the **SpacecraftDynamics** model, no matter what the actual implementation of the model is.

Let us once again point out that the Spacecraft model is completely customizable for what concerns actuators, sensors and controls, which can be selected among standard library models via dedicated popup menus using graphical user interfaces. The possible customization are virtually countless: limiting the discussion only to possible sensors customization, the present Space Flight Dynamics Library implementations allow for choosing among all possible combinations arising from star tracker models and configurations, models for GPS receivers, magnetometers models, gyroscope models and sensors block instances.

The following code fragment shows, as an example, how it is possible to instantiate a spacecraft model with a specific implementation of **SensorBlock**, via the *redeclare* construct. The model *Example* contains an instance *world* of the **World** model, and an instance *spacecraft* of the **SpacecraftBase**, where a specific *SensorBlock* implementation has been chosen, containing a particular kind of start tracker assembly, where the single star tracker datasheet is specified by an appropriate record.

```
model Example
  import SpacecraftDynamics.Spacecraft.*;
  inner Environment.World world;
  Implementations.SpacecraftBase spacecraft(
      redeclare model SensorBlock =
        Sensors.Implementations.GPS_StarTracker_MagField
        (redeclare model StarTrackerConf =
            Sensors.Components.StarTrackers.Assemblies.SingleST_conf
            (redeclare model StarTracker =
                Sensors.Components.StarTrackers.StarTrackerBase
                (data=Sensors.Components.StarTrackers.Datasheets.ESA2006_prj))));
end Example;
```

This operation is of course inconvenient to perform in textual form by the end user, but it can be supported by the graphical user interface of the Modelica tool that is used to compile the model.

The advanced object-oriented features of the Modelica language (in particular the *redeclare/replaceable* constructs) have thus been used to design a highly flexible, reconfigurable and reusable modelling framework; on the other hand, this framework can be exploited in an intuitive way through graphical user interfaces by end users, which need not be familiar with the intricacies of an object-oriented modelling language.

10.4 Case studies

The Spacecraft Dynamics Library has been already applied in a number of spacecraft modelling and simulation studies (see [19, 20] and the references therein) associated with various attitude control system design problems. In the following sections, two representative examples of the obtained results are presented. More precisely, in section 10.4.1 the application of the library to the analysis of the external forces and torques acting on the spacecraft along a specified trajectory is demonstrated, while section 10.4.2 deals with the application of the library to the problem of assessing the performance of a magnetic detumbling control law for a LEO small satellite. The interested reader can check, e.g., [7, 20] for additional case studies, dealing with modelling and simulation problems arising when considering spacecraft with large flexible appendages.

10.4.1 Assessing external disturbances via dynamic inversion

In this section, the focus will be on the application of the library in the AOCS preliminary design stage, i.e. when the spacecraft architecture is not yet completely defined and different options are being evaluated, depending on the specific mission profile. At this stage, it is indeed convenient to maintain for the spacecraft a *higher level* structure (i.e. one where only requirements are specified, not specific equipments), to evaluate its interaction with the space environment for different mission profiles and to be concerned only afterwards about the choice of specific equipment to be embarked. One of the main tasks in this stage is to evaluate the external disturbance forces and torques acting on the spacecraft, depending on the mission profile.

This task can be in principle performed in many different ways. The simplest approach would be to rely on simple worst case formulae, such as the ones given in [5], for the maximum amplitude of external disturbances. Such formulae provide a very simple way of determining (usually conservative) estimates of the perturbations the satellite will experience during its operation.

On the other hand, one could think of reducing the conservatism of the previous approach by running a closed-loop simulation using a simple attitude control algorithm to maintain the satellite near its nominal operating conditions (e.g. the Earth-pointing attitude) and computing the external disturbances as outputs of the simulation. Clearly, the latter approach will provide a much more accurate result; however, it requires a preliminary design of the attitude control law, which may be time-consuming.

A more efficient way of dealing with this task can be devised by taking advantage of the a-causal nature of Modelica models: since the Modelica representation of spacecraft dynamics does not prescribe a priori which variables are to be considered as inputs for the model and which constitute the outputs, one can proceed by assigning a specified orbit and attitude trajectory to the satellite and computing by means of an *inverse simulation* the disturbance forces and torques the satellite will experience in such an operating condition, or, as an alternative, the control forces and torques that an ideal control system should exert on the spacecraft in order to maintain the desired trajectory. In such a way, relying on the Modelica tool for symbolic dynamic inversion, it is possible to achieve a fast preliminary assessment of the disturbance forces and torques acting upon the spacecraft in its nominal orbit and attitude.

Taking advantage of the Space Flight Dynamics Library features, it is an easy task to derive a customization of the base **Spacecraft** model which can be used to perform this preliminary analysis: it is sufficient to *assemble* a new spacecraft model with neither **ControlBlock** nor **ActuatorBlock**, to define an unknown *control torque* to be applied as input torque to the spacecraft and assign the desired spacecraft's initial attitude and angular rate time-history. The Modelica tool then automatically takes care of solving the resulting system of non-linear equations to derive the control forces and torques necessary to keep the spacecraft in its nominal orbit and attitude.

As an example, a satellite operating on a circular, near-polar orbit ($i = 86.9$ degree inclination) with an altitude of 450 km (and a corresponding orbital period of 5614.8 seconds) has been considered. The satellite inertial properties are as follows:

- Satellite mass $m = 500$ [kg]
- Satellite inertia matrix: $I = \begin{bmatrix} 30 & 2 & -18 \\ 2 & 1080 & -0.1 \\ -18 & -0.1 & 1070 \end{bmatrix}$ kg m^2

For aerodynamic modelling purposes, a default cubic geometry was assumed for the satellite, comprising six surfaces, each with 1 m^2 surface area, reflectivity coefficient $\varepsilon = 0.02$ and centre of pressure located at the surface geometric centre. As far as the interaction with the environment is concerned, aerodynamic drag, solar radiation pressure and a residual magnetic dipole of 1 A m^2 upon each spacecraft body axis were considered as sources of disturbance torques, while default choices were selected for the geomagnetic and gravity fields (i.e. the dipole model for the geomagnetic field and the J_2 model for the gravitational field), Sun ephemeris and atmosphere model (i.e. the Harris-Priester air density model; see [12]). The simulation was initialized at GMT 1200, March 21, 2007. Finally, the results are referred to the following axes:

- Earth-centred inertial (ECI) reference frame. The origin of these axes is in the Earth's centre. The x-axis is parallel to the line of nodes (i.e. the intersection of the ecliptic with the Earth's equatorial plane). The z-axis is parallel to the Earth's geographic North-South axis and pointing North. The y-axis completes the right-handed orthogonal triad.
- Orbital reference frame. The origin is located in the spacecraft centre of gravity. The z-axis points towards nadir. The y-axis is perpendicular to the orbital plane, and positive towards negative orbit normal. The x-axis completes the right-handed orthogonal triad.
- Satellite body axes (roll-pitch-yaw). The origin is located in the spacecraft centre of gravity. The x-axis (roll) is aligned with nominal flight direction. The z-axis (yaw), perpendicular to flight direction, points towards nadir. The y-axis (pitch) completes the right-handed orthogonal triad.

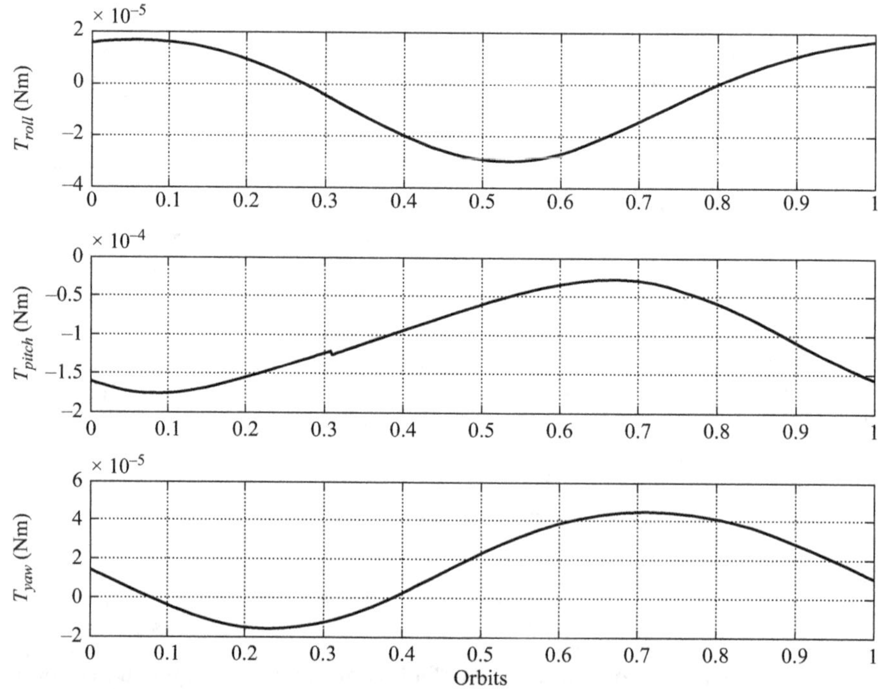

Figure 10.2 Overall control torques applied by dynamic inversion

Figures 10.2–10.6 show the computed disturbance torques experienced by the Earth-pointing satellite assuming that it is perfectly aligned with its orbital reference frame.

In particular, it is apparent from Figure 10.4 that the solar radiation disturbance torque experiences a sudden drop to zero when the Earth interposes between the spacecraft and the Sun, and takes back a non-zero value as soon as the spacecraft gets full Sun illumination again. The effect of the eclipse period is visible, although in a less evident way, also in the time history of the aerodynamic torques given in Figure 10.3: indeed, during the eclipse the satellite experiences a significant variation in the atmospheric density, which leads to a smaller variation of the aerodynamic torques. An accurate assessment of the overall disturbance torques acting on the satellite constitutes a valuable input for the sizing of the control actuators; furthermore, repeating the illustrated inverse simulation for different values of the mission parameters can easily lead to a worst case assessment of the environmental conditions for the spacecraft throughout its operation.

10.4.2 Magnetic detumbling for small satellite attitude control

The library's flexibility can be exploited to quickly evaluate and compare a number of different approaches to the design of digital attitude controllers. A tradeoff between

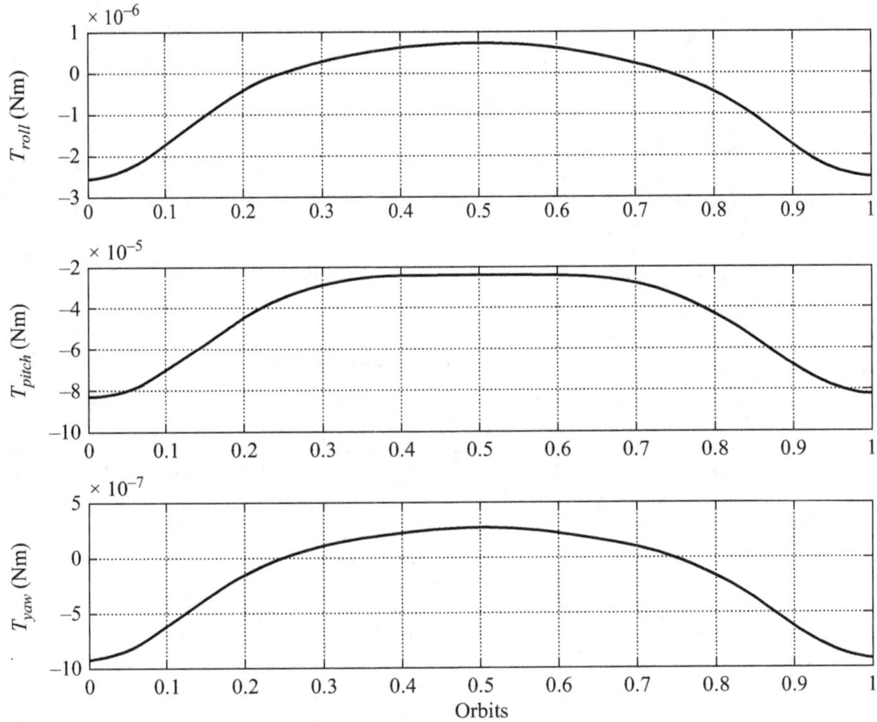

Figure 10.3 Aerodynamic disturbance torques experienced by the satellite in its nominal orbit and attitude

control performance and required on-board hardware, architectural simplicity and spacecraft requirements (i.e. mass, size etc.) can then be made, leading to the final spacecraft configuration. Magnetic coils have been used since the very beginning of the space age as a simple and efficient torque generation mechanism for attitude and momentum control [26, 8, 9, 31, 33, 21, 22]. As is well known, the operation of such actuators is based on the interaction between the magnetic field generated by the current-driven coils and the magnetic field of the Earth and therefore provides a simple and low-cost solution to the problem of generating torques on-board a satellite. More precisely, magnetic torquers can be used either as main actuators for attitude control in momentum biased or gravity gradient attitude control architectures or as secondary actuators for momentum management tasks in zero momentum reaction wheel-based configurations.

The main issue associated with magnetic coils, however, is the fact that magnetic torques are instantaneously constrained to lie in the plane orthogonal to the local magnetic field vector, so that the angular dynamics of a magnetically actuated spacecraft is completely controllable only in the time-varying sense (thanks to the variability of the geomagnetic field). This leads to a number of difficulties in the design of the attitude control law, which have been a very active research topic in the last few

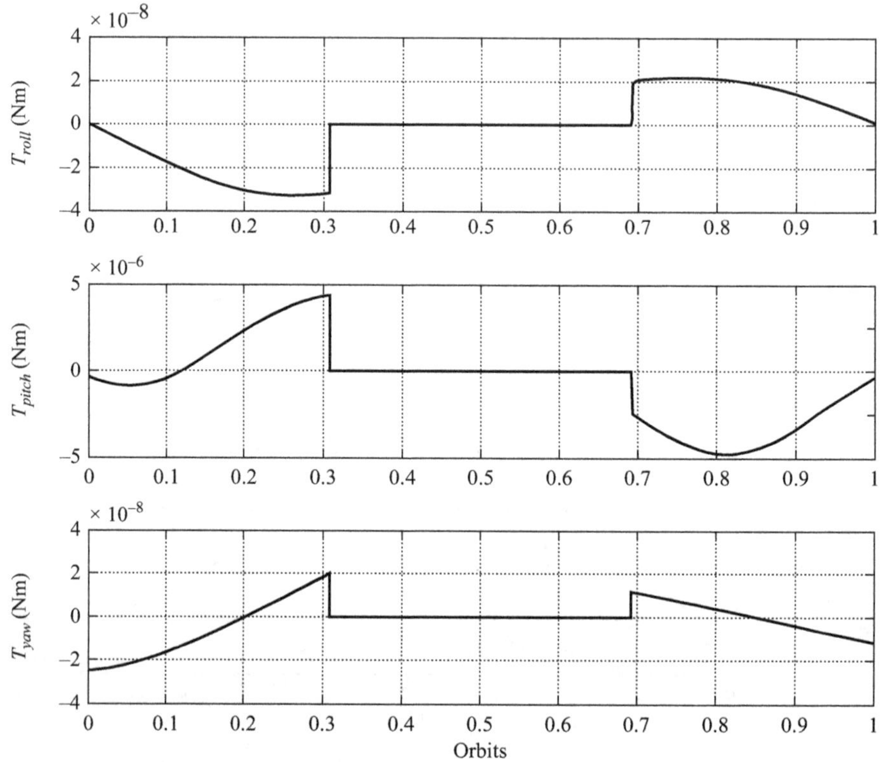

Figure 10.4 Solar radiation disturbance torques experienced by the satellite in its nominal orbit and attitude

years. In particular, as far as the design of nominal attitude control laws is concerned (i.e. design problems based on linearized models) the recent work has focused on the use of optimal and robust periodic control theory for the design of state and output feedback regulators for a number of different ACS configurations, in which magnetic torquers are assisted in their operation with *passive* means, namely gravity gradient spacecraft [32, 18], momentum-biased spacecraft [17, 10, 6, 26], or satellites exploiting passive aerodynamic properties [33]. However, in spite of this extensive activity, little or no attention has been dedicated to the development of design techniques for the (frequently adopted) architecture in which magnetic coils are aided in an *active* way by a suitable propulsion system (e.g. cold gas thrusters). In this case, the design problem has specific emphasis on achieving a closed loop behaviour that requires only a *restricted* support from the propulsion system, while taking into account its non-linearities (such as saturation and dead zone) which, combined with the time-periodicity of the magnetic coils, give rise to a class of very challenging design problems.

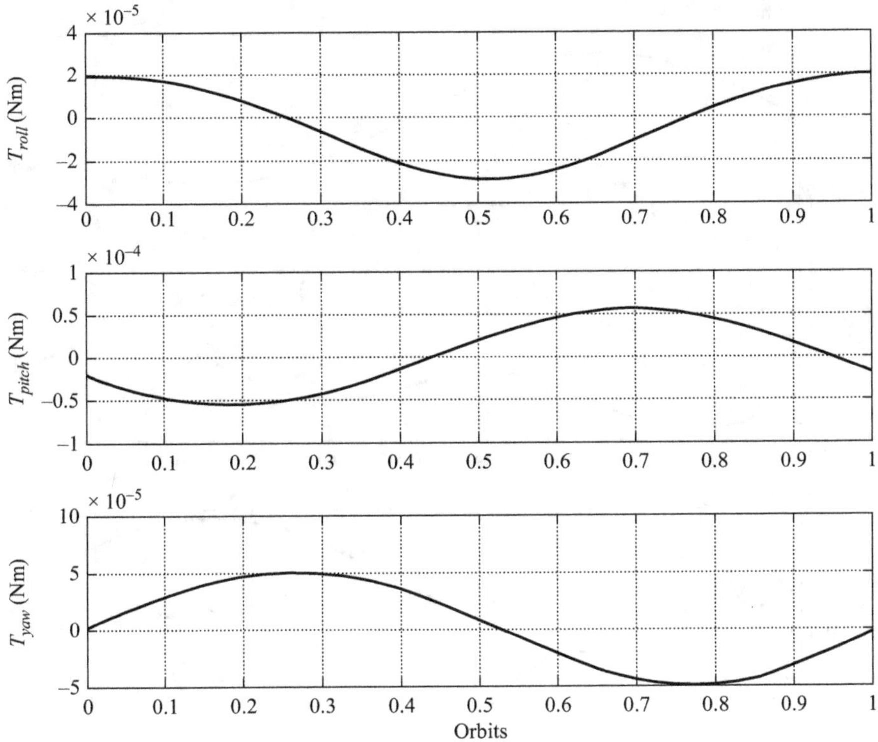

Figure 10.5 Magnetic disturbance torques experienced by the satellite in its nominal orbit and attitude

The first task the ACS must perform after separation of the spacecraft from the launcher is to detumble it, i.e. to bring it from an initial condition possibly characterized by a large angular momentum to a final one that, while remaining essentially arbitrary in terms of direction of the angular rate vector, is compatible with the operation of subsequent ACS modes capable of achieving also (coarse) pointing requirements.

Detumbling must be performed by the spacecraft in a fully autonomous way. Therefore, in order to minimize the risk of failure in the course of this operation, tight requirements on the reliability of the involved actuators and sensors and on the simplicity of the adopted control law are usually driving the design of the detumbling mode.

As is well known since the 1960s, magnetic control is an excellent candidate for this task, as it allows to effectively detumble a spacecraft using only magnetic torquers as actuators and magnetometers as sensors for the spacecraft angular rate. As already mentioned, magnetic torquers are extremely reliable, as they do not involve moving parts and are therefore less sensitive to the large vibratory loads during launch than,

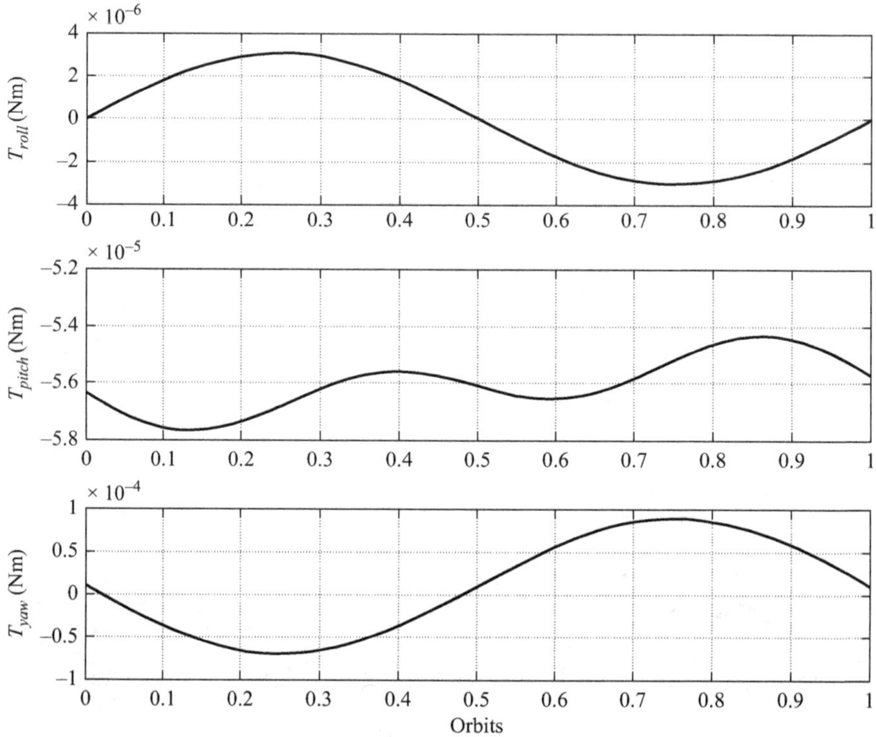

Figure 10.6 Gravity gradient disturbance torques experienced by the satellite in its nominal orbit and attitude

e.g., inertial actuators. Similarly, magnetometers are reliable, relatively inexpensive equipment, for which redundancy can be easily achieved if necessary.

The technology enabling magnetic detumbling is the celebrated b-dot control law (in all its variants developed and flown through the years), which is presented and analysed in the following sections. Subsequently, a simulation-based approach to the tuning and performance verification of the b-dot algorithm is illustrated.

10.4.2.1 Spacecraft model

The model of a rigid spacecraft with magnetic actuation can be described in various reference frames [30]. For the purpose of the present analysis, the following reference systems are adopted.

- Earth-centred inertial (ECI) reference axes. The origin of these axes is in the Earth's centre. The x-axis is parallel to the line of nodes, i.e. the intersection between the Earth's equatorial plane and the plane of the ecliptic, and is positive in the Vernal equinox direction (Aries point). The z-axis is defined as being

parallel to the Earth's geographic North-South axis and pointing north. The y-axis completes the right-handed orthogonal triad.
- Satellite body axes. The origin of these axes is in the satellite centre of mass; the axes are assumed to coincide with the body's principal inertia axes.

The attitude dynamics can be expressed by the well-known Euler's equations [30]

$$I\dot{\omega} = S(\omega)I\omega + T_{coils} + T_{dist} \tag{10.9}$$

where $\omega = \begin{bmatrix}\omega_x & \omega_y & \omega_z\end{bmatrix}^T \in \mathbb{R}^3$ is the vector of spacecraft angular rates, expressed in body frame, $I \in \mathbb{R}^{3\times 3}$ is the inertia matrix, $S(\omega)$ is given by

$$S(\omega) = \begin{bmatrix} 0 & \omega_z & -\omega_y \\ -\omega_z & 0 & \omega_x \\ \omega_y & -\omega_x & 0 \end{bmatrix} \tag{10.10}$$

$T_{coils} \in \mathbb{R}^3$ is the vector of external torques induced by the magnetic coils and $T_{dist} \in \mathbb{R}^3$ is the vector of external disturbance torques.

In turn, the attitude kinematics can be described either by means of the cosine matrix A or by means of the four Euler parameters (or quaternions), which lead to the following representation

$$\dot{\boldsymbol{q}} = W(\boldsymbol{q})\omega \tag{10.11}$$

where $\boldsymbol{q} = \begin{bmatrix}q_1 & q_2 & q_3 & q_4\end{bmatrix}^T = \begin{bmatrix}\mathbf{q}^T & q_4\end{bmatrix}^T$ and

$$W(\boldsymbol{q}) = \frac{1}{2}\begin{bmatrix} q_4 & -q_3 & q_2 \\ q_3 & q_4 & -q_1 \\ -q_2 & q_1 & q_4 \\ -q_1 & -q_2 & -q_3 \end{bmatrix} \tag{10.12}$$

Note that the attitude of inertially pointing spacecraft is usually referred to the ECI reference frame.

The magnetic attitude control torques are generated by a set of three magnetic coils, aligned with the spacecraft principal inertia axes, which generate torques according to the law

$$T_{coils} = m \times \tilde{b}(t) = S(b(t))m \tag{10.13}$$

where $\times$ denotes the vector cross product, $m \in \mathbb{R}^3$ is the vector of magnetic dipoles for the three coils, $b(t) \in \mathbb{R}^3$ is the vector formed with the components of the Earth's magnetic field in the body frame of reference. Note that the vector $b(t)$ can be expressed in terms of the attitude matrix $A(\boldsymbol{q})$ [30] and of the magnetic field vector expressed in the inertial coordinates, namely $b_0(t)$, as

$$b(t) = A(\boldsymbol{q})b_0(t) \tag{10.14}$$

and that the orthogonality of $A(\boldsymbol{q})$ implies $\|b(t)\| = \|b_0(t)\|$. Since $S(b(t))$ is structurally singular, as mentioned in section 10.1, magnetic actuators do not provide full controllability of the system at each time instant. In particular, it is easy to see that rank $(S(b(t))) = 2$ (since $\|b_0(t)\| \neq 0$ along all orbits of practical interest for magnetic control) and that the kernel of $S(b(t))$ is given by the vector $b(t)$ itself, i.e., at each time instant it is *not* possible to apply a control torque along the direction of $b(t)$.

10.4.2.2 Magnetic detumbling

Consider the problem of designing a feedback law for the dynamics of a rigid body subject to magnetic actuation

$$\dot{A} = S(\omega)A \tag{10.15}$$

$$I\dot{\omega} = S(\omega)I\omega + S(b)m \tag{10.16}$$

with the objective of reducing the angular rate of the spacecraft from the initial value $\omega(0) = \omega_o$ to, ideally, zero, using only feedback provided by a triaxial magnetometer:

$$b = Ab_0 \tag{10.17}$$

Clearly, magnetic field measurements can provide attitude information (in the form of a scalar vector measurement), but cannot provide rate information to a feedback controller. Magnetometer output, however, becomes relevant to the solution of the above-stated problem if one considers the time derivative of the measured geomagnetic field, as computed from the above measurement equation:

$$\dot{b} = \dot{\widehat{Ab_0}} = \dot{A}b_0 + A\dot{b}_0 \tag{10.18}$$

which, recalling (10.15), can be equivalently written as

$$\dot{b} = S(\omega)Ab_0 + A\dot{b}_0 = S(\omega)b + A\dot{b}_0 \tag{10.19}$$

Finally, by commuting the factors in the cross-product appearing at the right-hand side and recalling that $S^T(\cdot) = -S(\cdot)$, one gets

$$\dot{b} = S^T(b)\omega + A\dot{b}_0 \tag{10.20}$$

With the derivative of the measured geomagnetic field ($\dot{b}$, hence the name b-dot for the magnetic detumbling controller) written in this form, it is apparent that the first term is proportional to the body angular rate of the spacecraft, through a (singular) matrix gain ($S^T(b)$) that is a function of the measured magnetic field itself. Furthermore, the presence of the second term, which is proportional to the rate of variation of the geomagnetic field in the orbital frame (as seen from the body frame), can be interpreted as a limit to the precision of the angular rate information that $\dot{b}$ can provide, so that $\dot{b}$ can be considered a meaningful measure of the spacecraft angular rate provided that its magnitude is significantly larger than the orbital angular rate ω_0. As this is typically the case for the initial condition of a detumbling mode, $\dot{b}$ can be considered a valuable source of feedback for magnetic detumbling.

Many choices for the design of a magnetic detumbling controller based on $\dot{b}$ have been considered in the literature. The simplest one, which neglects actuator saturation and therefore suits the purpose of illustrating the main concept, is given by simple static linear feedback of the form

$$m = -k\dot{b} \tag{10.21}$$

where $k > 0$ is a scalar control gain. Now, substituting (10.21) in (10.15) and (10.16), the closed-loop dynamics of the magnetically controlled spacecraft can be written as

$$\dot{A} = S(\omega)A \tag{10.22}$$

$$I\dot{\omega} = S(\omega)I\omega - kS(b)S^T(b)\omega - kA\dot{b}_0 \tag{10.23}$$

As can be seen from (10.23), up to the effect of the non-vanishing perturbation $-kA\dot{b}_0$, the effect of the b-dot feedback on the dynamics of the spacecraft is to introduce an angular rate feedback with a gain given by $-kS(b)S^T(b)$, which is negative semi-definite in view of the fact that $S(b)S^T(b) \geq 0$. In other words, letting $T(\omega) = \frac{1}{2}\omega^T I\omega$ the angular kinetic energy of the spacecraft, the computation of the derivative along the trajectories of the closed-loop system ((10.22) and (10.23)) of $T(\omega)$ gives

$$\dot{T}(\omega) = -k\omega^T S(b)S^T(b)\omega \tag{10.24}$$

and since $\dot{T}(\omega) \leq 0$, from an engineering perspective it can be claimed that the b-dot law can effectively remove angular momentum from the spacecraft whenever ω is not parallel to b, while leaving it unchanged when $\omega^T b = 0$.

10.4.2.3 Case study: detumbling a small LEO satellite

A simulation case study is presented, based on a specific mission scenario related to a typical LEO small satellite mission.

The considered satellite has a geometry of the 'box' type; within the Modelica Space Flight Dynamics Library paradigm, the geometry can be implemented as the collection of six planar surfaces. For the considered study, the satellite operates in an Earth-pointing attitude, along a near polar, circular orbit ($i = 89°$, $e = 0$) with an altitude of 450 km and a corresponding orbital period of 5614.8 s.

As in the previous case, a Modelica Space Flight Dynamics Library model of the satellite was easily obtained as a specialization of the base **SpacecraftBase** class. The following model components were used:

- Model **GPSStarTrackerMagField**. It describes a sensor configuration including a GPS receiver, a magnetometer and a generic assembly of star trackers.

- Model **MagnetotorquersActuatorsBlock**. It includes a set of three mutually orthogonal magnetic coils.
- Model **SSACS**. It executes the attitude determination and control functions of the considered small satellite.

For the b-dot detumbling algorithm a Monte Carlo study was carried out to define a robust tuning with respect to

- uncertainty on the initial angular rate: ±50% of an initial reference value chosen as 50 times the orbital angular rate, per axis;
- uncertainty on the attitude: ±1 on all components of the attitude quaternion;
- uncertainty on the moments of inertia: ±10% on each of the three principal moments.

More precisely, 50 simulations were carried out for different choices of the gain k of the b-dot law to assess the settling time of the norm of the angular rate vector of the spacecraft with respect to a threshold equal to twice the orbital angular rate. As a result, the following statistics were computed: mean, standard deviation, maximum and minimum values. The numerical values are depicted in Figure 10.7, from which, in view of a conservative, worst-case tuning, a value of $k = 2 \times 10^{-3}$ is chosen, which

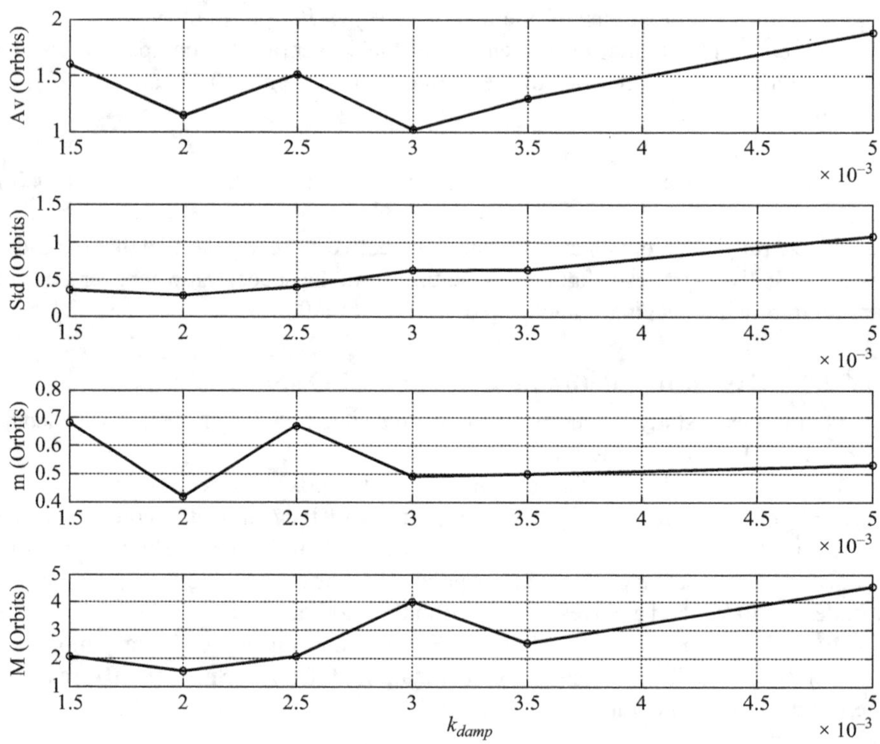

Figure 10.7 Monte-Carlo controller tuning statistics

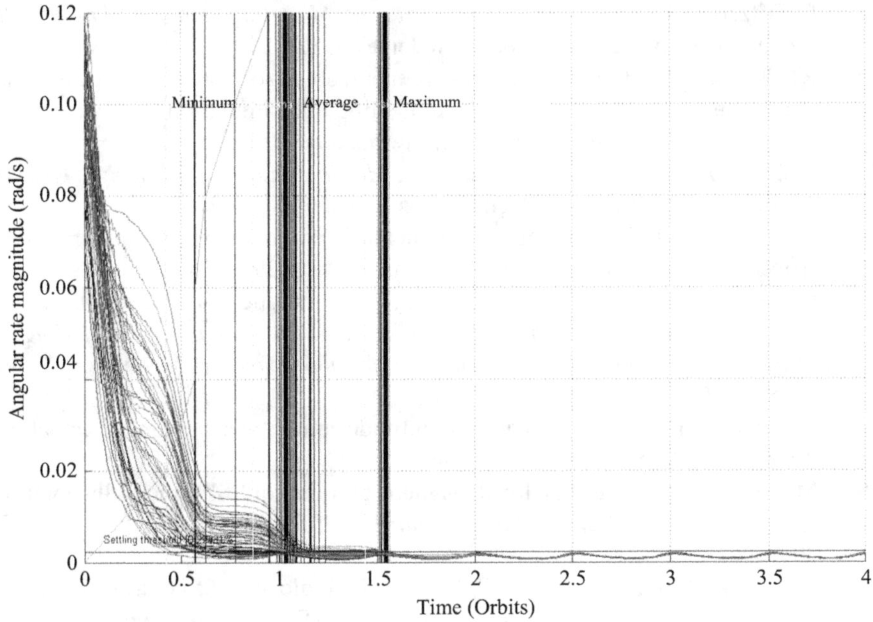

Figure 10.8 Closed-loop evolution of angular rate magnitude for the Monte Carlo study

guarantees a worst-case settling time of about 1.5 orbits. Finally, in Figure 10.8 a time-domain representation of the simulation results is provided.

10.5 Concluding remarks

In this chapter the main issues related to the modelling and simulation of spacecraft dynamics are described, and the results obtained so far in developing a Modelica-based modelling framework for spacecraft simulation are presented. Finally, two case studies demonstrating the usefulness of the proposed approach are discussed.

References

[1] F. Casella and M. Lovera. High accuracy simulation of orbital dynamics: an object-oriented approach. *Simulation Modelling Practice and Theory*, 16(8):1040–1054, 2008.

[2] G. Ferretti, F. Schiavo, and L. Viganò. Object-oriented modelling and simulation of flexible multibody thin beams in Modelica with the finite element method. In *4th Modelica Conference*, Hamburg-Harburg, Germany, March 7 and 8, 2005.

[3] P. Fritzson. *Principles of Object-Oriented Modeling and Simulation with Modelica 2.1*. Wiley, Hoboken, New Jersey, USA, 2003.

[4] P. Fritzson and P. Bunus. Modelica – a general object-oriented language for continuous and discrete-event system modelling and simulation. In *Proceedings of the 35th IEEE Annual Simulation Symposium*, 2002.

[5] W.J. Larson and J.R. Wertz (eds.). *Space Mission Analysis and Design*. Kluwer Academic, Dordrecht, The Netherlands, 1992.

[6] M. Lovera. Optimal magnetic momentum control for inertially pointing spacecraft. *European Journal of Control*, 7(1):30–39, 2001.

[7] M. Lovera. Control-oriented modelling and simulation of spacecraft attitude and orbit dynamics. *Journal of Mathematical and Computer Modelling of Dynamical Systems, Special issue on Modular Physical Modelling*, 12(1): 73–88, 2006.

[8] M. Lovera and A. Astolfi. Spacecraft attitude control using magnetic actuators. *Automatica*, 40(8):1405–1414, 2004.

[9] M. Lovera and A. Astolfi. Global magnetic attitude control of inertially pointing spacecraft. *Journal of Guidance, Control and Dynamics*, 28(5):1065–1072, 2005.

[10] M. Lovera, E. De Marchi, and S. Bittanti. Periodic attitude control techniques for small satellites with magnetic actuators. *IEEE Transactions on Control Systems Technology*, 10(1):90–95, 2002.

[11] The Modelica Association. *Modelica – A Unified Object-Oriented Language for Physical Systems Modeling* – Language Specification Version 3.0. September 5, 2007. URL: http://www.modelica.org/news_items/documents/ModelicaSpec30.pdf

[12] O. Montenbruck and E. Gill. *Satellite Orbits: Models, Methods, Applications*. Springer, London, UK, 2000.

[13] D. Moorman and G. Looye. The Modelica flight dynamics library. In *Proceedings of the 2nd International Modelica Conference*, Oberpfaffenhofen, Germany, 2002.

[14] M. Otter, H. Elmqvist, and S. E. Mattsson. The new Modelica Multi-Body library. In *Proceedings of the 3rd International Modelica Conference*, Linköping, Sweden, 2003.

[15] M. Otter, H. Elmqvist, and S. E. Mattsson. The new Modelica Multi-Body library. In *Proceedings of the 3rd International Modelica Conference*, pp. 311–330, Linköping, Sweden, November 3–4, 2003. URL: http://www.modelica.org/events/Conference2003/papers/h37_Otter_ multibody.pdf

[16] M. Otter and H. Olsson. New features in Modelica 2.0. In *Proceedings of the 2nd International Modelica Conference*, 2002.

[17] M. Pittelkau. Optimal periodic control for spacecraft pointing and attitude determination. *Journal of Guidance, Control and Dynamics*, 16(6): 1078–1084, 1993.

[18] M. Psiaki. Magnetic torquer attitude control via asymptotic periodic linear quadratic regulation. *Journal of Guidance, Control and Dynamics*, 24(2): 386–394, 2001.

[19] T. Pulecchi. *Advanced Techniques for Satellites Modeling and Attitude Control*. Ph.D. thesis, Politecnico di Milano, 2008.

[20] T. Pulecchi, F. Casella, and M. Lovera. Object-oriented modelling for spacecraft dynamics: tools and applications. *Simulation Modelling Practice and Theory*, 18(1):63–86, 2010.

[21] T. Pulecchi and M. Lovera. Attitude control of spacecraft with partially magnetic actuation. In *17th IFAC Symposium on Automatic Control in Aerospace*, Toulouse, France, 2007.

[22] T. Pulecchi, M. Lovera, and A. Varga. Optimal discrete-time design of magnetic attitude control laws. In *Proceedings of the 6th International ESA Conference on Guidance, Navigation and Control Systems*, Loutraki, Greece, 2005.

[23] C. Roithmayr, C. Karlgaard, R. Kumar, and D. Bose. Integrated power and attitude control with spacecraft flywheels and control moment gyroscopes. *Journal of Guidance, Control, and Dynamics*, 27(5):859–873, 2004.

[24] F. Schiavo and M. Lovera. Modelling, simulation and control of spacecraft with flexible appendages. In *Proc. of the 5th International Symposium on Mathematical Modelling*, Vienna, Austria, 2006.

[25] F. Schiavo, L. Viganò, and G. Ferretti. Modular modelling of flexible beams for multibody systems. *Multibody Systems Dynamics*, 15(3):263–286, 2006.

[26] E. Silani and M. Lovera. Magnetic spacecraft attitude control: a survey and some new results. *Control Engineering Practice*, 13(3):357–371, 2005.

[27] The Modelica Association. *Modelica – A Unified Object-Oriented Language for Physical Systems Modeling*. Language Specification Version 2.2, 2005. URL: www.modelica.org.

[28] A. Turner. *An Open-Source, Extensible Spacecraft Simulation and Modeling Environment Framework*. Master's thesis, Virginia Polytechnic Institute and State University, 2003.

[29] D. Vallado. *Fundamentals of Astrodynamics and Applications*. Microcosm Press/Kluwer Academic Press, Dordrecht, The Netherlands, 2001.

[30] J. Wertz. *Spacecraft Attitude Determination and Control*. D. Reidel, Dordrecht, The Netherlands, 1978.

[31] R. Wisniewski. *Satellite Attitude Control Using Only Electromagnetic Actuation*. Ph.D. thesis, Aalborg University, Denmark, 1996.

[32] R. Wisniewski and L.M. Markley. Optimal magnetic attitude control. In *14th IFAC World Congress*, Beijing, China, 1999.

[33] M. Wood, W.-H. Chen, and D. Fertin. Model predictive control of low Earth orbiting spacecraft with magneto-torquers. In *IEEE International Conference on Control Applications*, Munich, Germany, 2006.

Chapter 11

Control-oriented aeroelastic BizJet low-order LFT modeling

Charles Poussot-Vassal[1], *Clement Roos*[2], *Pierre Vuillemin*[3], *Olivier Cantinaud*[4] *and Jean-Patrick Lacoste*[5]

Abstract

To accelerate aircraft conception and reduce development costs, computer-aided preliminary design is widely used (e.g., for controller design, performance analysis etc.). The computer-based approach has been rendered possible thanks to advances in modeling tools, allowing to faithfully reproduce complex physical phenomena with limited – expensive – experimental tests. Due to these economical considerations and technological advances, aeronautical control engineers can rely and work with a considerable amount of very accurate models. The counterpart of this accuracy is the resulting numerical complexity which leads (*i*) to a prohibitively large number of variables to manage, rendering the control design task very complex (i.e., numerical tools become nearly inefficient), and (*ii*) to models with an accuracy level too high for the control synthesis purpose (indeed modern control techniques usually require low-order representations). In this chapter, based on multiple initial large-scale linear time invariant models, the problem of constructing a suitable low-order parameter dependent model, appropriate to the control design purpose, is addressed. The proposed solution is illustrated on a complex generic Business Jet (BizJet) aeroelastic control-design problem.

11.1 Introduction

11.1.1 Foreword on the Dassault-Aviation BizJet models

Aeroservoelastic generic Business Jet (BizJet) aircraft behavior is often modeled at frozen flight conditions and mass center of gravity position configurations using both

[1]ONERA, Toulouse, France, e-mail: charles.poussot-vassal@onera.fr
[2]ONERA, Toulouse, France, e-mail: clement.roos@onera.fr
[3]ONERA, Toulouse, France, e-mail: pierre.vuillemin@onera.fr
[4]Dassault Aviation, Saint Cloud, France, e-mail: olivier.cantinaud@dassault-aviation.com
[5]Dassault Aviation, Saint Cloud, France, e-mail: jean-patrick.lacoste@dassault-aviation.com

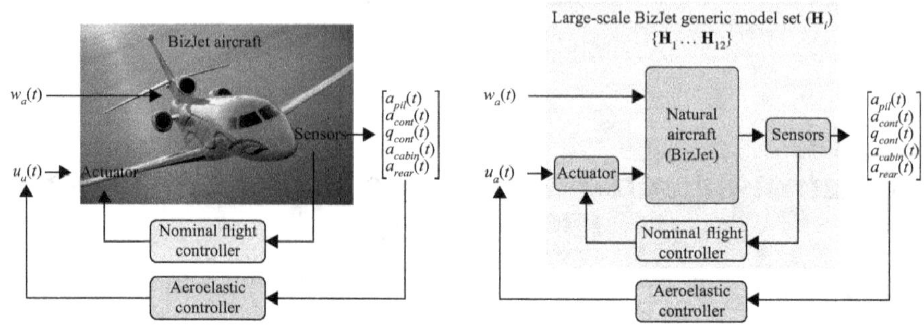

Figure 11.1 A Dassault-Aviation generic BizJet aircraft (left) and its equivalent multi-LTI block diagram representation (right) with nominal and aeroelastic controllers

high fidelity numerical tools (e.g., finite elements, modal identification, etc.) and flight dynamics experts knowledge. Aeroservoelastic models take into account the physics involved in the aeroelastic phenomena (structural loads, unsteady aerodynamic loads, etc.) and the flight control system behavior (actuator dynamics, sensors bandwidth, measurement delays, etc.).

This faithful modeling usually results in a set of large-scale linear time invariant (LTI) models. In the considered application, the set of LTI models provided by Dassault-Aviation describes the aeroservoelastic BizJet aircraft and its nominal flight control laws[1] at:

- 3 Mach numbers (0.8, 0.85, and 0.9), and
- 4 flight altitudes (35,000 ft, 40,000 ft, 45,000 ft, and 51,000 ft).

Thus, twelve LTI state-space model realizations, denoted $\mathbf{H}_i$ ($i = 1, \dots, 12$), describing the BizJet aircraft behavior at frozen flight configurations, are considered here. Each realization $\mathbf{H}_i$ has a state vector with dimension around 650 (see Figure 11.1).

Note that the LTI models $\mathbf{H}_i$ also include the nominal flight control laws designed by Dassault-Aviation control engineers. In this study, the objective is to build a suitable model for the design and the analysis of an additional aeroelastic controller with objective to attenuate undesirable vibrations over a bounded frequency range (see next subsection).

11.1.2 The BizJet aircraft aeroelatic control problem

The nominal flight control laws[1] (see Figure 11.1) provide good flight dynamics and load performance at low frequencies. However, some aerodynamic perturbations are likely to occur and can generate an undesirable response of some medium frequency modes on some parts of the aircraft within a specific frequency range. As illustrated

[1] The nominal flight control laws are designed with the aim to ensure good handling qualities (internal stability of the platform, fast, smooth, and accurate responses to pilot commands) and a rejection of the aeroelastic modes on control.

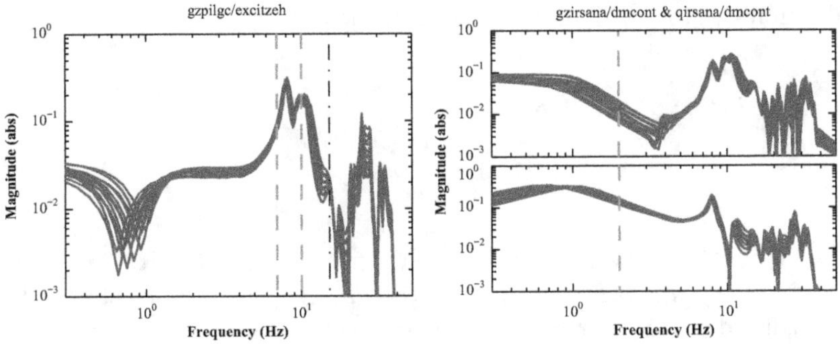

Figure 11.2 Normalized frequency responses from aerodynamical disturbance $w_a(t)$ to pilot cabin vertical acceleration $a_{pil}(t)$ (left) and from control sensors to control input (right), for all considered large-scale models

on the left frame of Figure 11.2, perturbations on the rear part of the aircraft (A/C) can generate vibrations in the cockpit or in the cabin between, e.g., 5 Hz and 15 Hz. This is why an additional aeroelastic controller (lower block on Figure 11.1) is introduced to attenuate these vibrations by imposing suitable deflections of the control surfaces (e.g., elevators located on the rear part of the A/C). These deflections are calculated using information from sensors such as accelerometers and gyrometers (see Figure 11.1). The aeroelastic control law aims then at attenuating undesirable behaviors until an upper frequency, generally limited by the actuator bandwidth (additional details can be found in [19]). Finally, the aeroelastic control law must be robust to parametric changes. As an illustration, Figure 11.2 illustrates the frequency responses of the comfort (left) and constraint (right) transfers for all the parametric configuration of the considered generic BizJet system.

It is important to note that, while the low frequency nominal flight controller can be designed using classical flight dynamics equations, the aeroelastic one has to account for medium and high frequencies, which makes it necessary to use accurate large-scale models. As it is not the purpose of this chapter, the aeroelastic control strategy design is not presented. However, for the sake of completeness, it is important to note that the aeroelastic control law rolls off after 15 Hz (the interested reader can refer to [19] for additional informations).

11.1.3 Mathematical problem formulation

Let us consider n_s stable, n_u inputs n_y outputs MIMO LTI continuous-time dynamical systems of order n_i $(i = 1 \dots n_s)$ described by $H_i : \mathbb{C} \rightarrow \mathbb{C}^{n_y \times n_u}$, a $\mathcal{H}_p^{n_y \times n_u}$ complex-valued matrix function[2] defined as

$$H_i(s) = C_i(sI_{n_i} - A_i)^{-1}B_i + D_i \in \mathbb{C}^{n_y \times n_u}, \quad i = 1, \dots, n_s \tag{11.1}$$

[2] The $\mathcal{H}_p^{n_y \times n_u}$ $(p > 0)$ denotes the set of $n_y \times n_u$ complex-valued matrix functions $H(s)$ with components $h_{xy}(s)$ $(x = 1, \dots, n_y$ and $y = 1, \dots, n_u)$ that are analytic in the open right half-plane $\mathbb{C}_+$.

A realization $\mathbf{H}_i : (A_i, B_i, C_i, D_i)$ of $H_i(s)$ is defined as

$$\mathbf{H}_i : \begin{cases} \dot{x}_i(t) = A_i x_i(t) + B_i u(t) \\ y(t) = C_i x_i(t) + D_i u(t) \end{cases}, \quad i = 1, \ldots, n_s \tag{11.2}$$

where $A_i \in \mathbb{R}^{n_i \times n_i}$ (Hurwitz), $B_i \in \mathbb{R}^{n_i \times n_u}$, $C_i \in \mathbb{R}^{n_y \times n_i}$ and $D_i \in \mathbb{R}^{n_y \times n_u}$. Moreover, $x(t) \in \mathbb{R}^{n_i} := \mathscr{X}$, $u(t) \in \mathbb{R}^{n_u} := \mathscr{U}$ and $y(t) \in \mathbb{R}^{n_y} := \mathscr{Y}$. The objective is to find $\hat{\mathbf{H}}(\delta)$, a reduced-order parametrized model of order $r \ll n_i$:

$$\hat{\mathbf{H}}(\delta) : \begin{cases} \dot{\hat{x}}(t) = \hat{A}(\delta)\hat{x}(t) + \hat{B}(\delta)u(t) \\ y(t) = \hat{C}(\delta)\hat{x}(t) + \hat{D}(\delta)u(t) \end{cases} \tag{11.3}$$

where $\hat{A}(\delta) \in \mathbb{R}^{r \times r}$, $\hat{B}(\delta) \in \mathbb{R}^{r \times n_u}$, $\hat{C}(\delta) \in \mathbb{R}^{n_y \times r}$, and $\hat{D}(\delta) \in \mathbb{R}^{n_y \times n_u}$. The model $\hat{\mathbf{H}}(\delta)$ should well approximate $\mathbf{H}_i$ when $\delta = \delta_i$ (the frozen parametric configurations), and its eigenvalues and frequency responses should evolve as smoothly as possible whatever the variation of δ inside the whole considered parametric domain. Other requirements are that $\hat{A}(\delta)$, $\hat{B}(\delta)$, $\hat{C}(\delta)$, and $\hat{D}(\delta)$ depend polynomially or rationally on $\delta \in \mathbb{R}^l$ to allow the generation of a suitable LFR (as depicted in Figure 11.3), and that $\hat{A}(\delta)$ remains Hurwitz over the δ continuum (A_i is initially Hurwitz). Finally, the size m of the block-diagonal matrix Δ must be as low as possible to ensure that analysis and design tools can be applied with sufficient accuracy and reasonable computational cost.

More precisely, in the considered BizJet aircraft, $n_s = 12$ models $(\mathbf{H}_i)_{i \in [1,12]}$ are available. The two components $\delta(1) = \delta_{mach}$ and $\delta(2) = \delta_{alti}$ of δ correspond to the Mach number and the A/C altitude, respectively. Each model has an order $n_i \approx 650$ states, $n_u = 2$ inputs (representing the commanded elevator deflection u_a and the aerodynamical disturbance w_a) and $n_y = 5$ outputs (representing the vertical acceleration a_{cont} and the pitch rate q_{cont} at the center of gravity of the aircraft, as well as the vertical accelerations a_{pil}, a_{cabin}, and a_{rear} at the pilot seat, the middle of the cabin and the rear of the fuselage respectively). The input u_a and the outputs a_{cont}, q_{cont}, a_{pil}

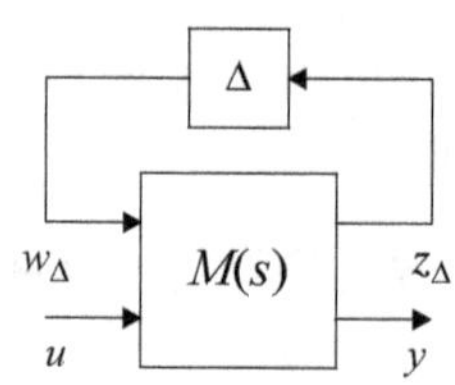

Figure 11.3 Structure of the open-loop LFR

are used to connect the aeroelastic controller as depicted in Figure 11.1, while the input w_a and the outputs $a_{cont}, a_{pil}, a_{cabin}, a_{rear}$ allow to check whether the objectives and constraints listed in Section 11.1.2 are satisfied.

11.1.4 Structure and notation

The chapter is organized as follows. An overview of the proposed low-order control-oriented parameter dependent model construction algorithm is presented in Section 11.2. Then, the two main steps, i.e., large-scale approximation and interpolation, are described in Sections 11.3 and 11.4, respectively. Concluding remarks are finally given in Section 11.5.

Mathematical notations are standard. The state vector of the initial (full-order) models and of the reduced-order models are denoted by $x(t) \in \mathbb{R}^n$ and $\hat{x}(t) \in \mathbb{R}^r$, respectively. The corresponding state-space representations and transfer matrices are denoted by $\mathbf{H} := (A, B, C, D)$ and $H(s) := C(sI_n - A)^{-1}B + D$, and $\hat{\mathbf{H}} := (\hat{A}, \hat{B}, \hat{C}, \hat{D})$ and $\hat{H}(s) := \hat{C}(sI_r - \hat{A})^{-1}\hat{B} + \hat{D}$, respectively. $\mathcal{V} = \mathbf{span}(V)$ means that the space $\mathcal{V}$ is spanned by the column vectors of the matrix V. $\lambda(.)$ holds for the eigenvalue operator, whereas the residual of the complex-valued function $f(z)$ at a given point λ is defined by $\mathbf{res}(f(z), \lambda) = \mathbf{lim}_{z \to \lambda}(z - \lambda)f(z)$, where z is a complex number. The variance of a vector v is denoted by $\mathbf{var}(v)$. The real part, the imaginary part, and the conjugate of a complex number z are denoted by $\Re(z)$, $\Im(z)$, and z^*, respectively. z^H denotes the hermitian of z i.e., conjugate transpose. Finally, the cardinality of a set $\mathcal{I}$ is denoted by $\#\mathcal{I}$.

11.2 Multi-LTI model approximation and interpolation algorithm overview

As illustrated in Figure 11.4, a systematic two-step approach, which can be applied to any set of purely numerical (and possibly unstructured) models, is proposed here. This structure has been initially proposed in [20] and is extended hereafter.

The first step aims at *generating a set of reduced-order models* (all with same dimension r). The main difference with classical LTI model approximation [2] is that, more than just preserving stability and input/output behavior, the proposed algorithm aims at enforcing the modal content of all models to be close to each other. This requirement is crucial for interpolating the models with simple polynomial or rational functions. Moreover, as the objective is to provide control-oriented models, a frequency-limited approximation is performed here in order to account for control objective and limitations while avoiding to accurately describe the system at control irrelevant frequencies.

The second step consists in *interpolating the reduced-order models*. As a necessary condition to obtain a suitable LPV/LFT model is to use the same state-space form for all of them, first, a basis change is performed, then an appropriate interpolation is computed.

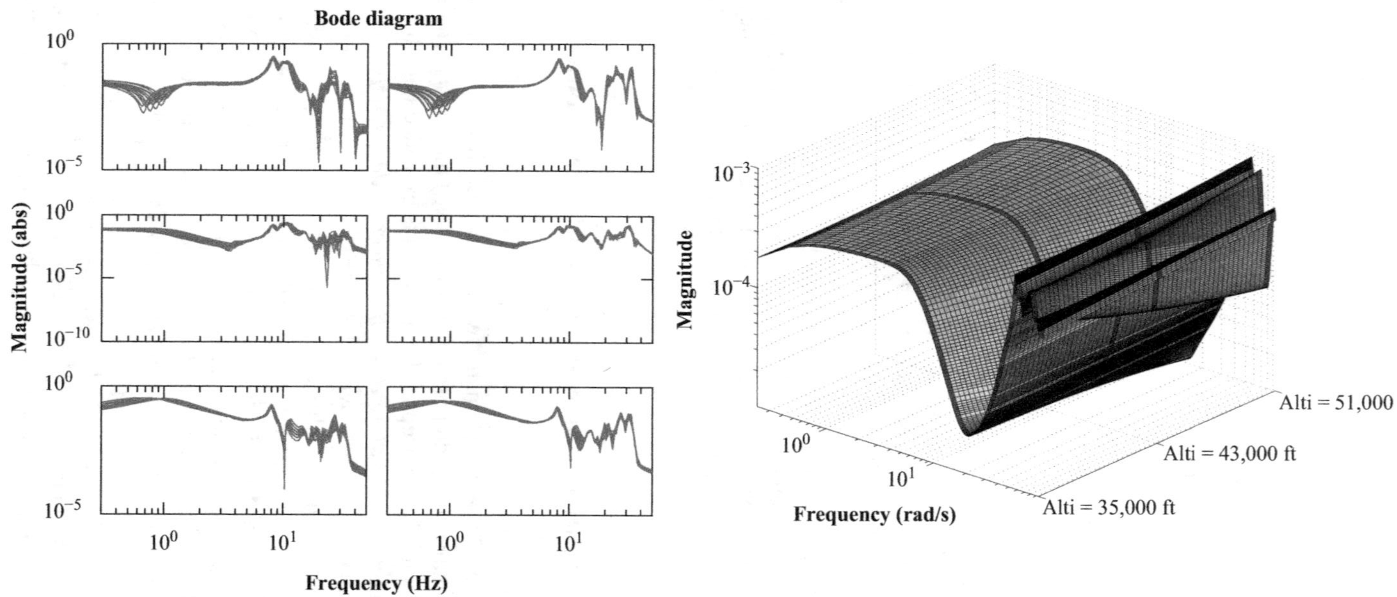

Figure 11.4 *Overview of the proposed method. From multiple large-scale MIMO initial models to a single MIMO LPV model (illustrated on the BizJet use-case)*

11.3 Frequency-limited large-scale MIMO multi-LTI models approximation

The first step of the reduced-order LPV model generation (see Figure 11.4), namely *the MIMO multi-LTI models approximation procedure*, is presented. Without loss of generality, the projection-based model reduction framework is used here (see [2] for details). Since the aim of model approximation is to capture the main system dynamics of interest and input/output behavior, while maintaining stability and achieving minimal model mismatch, the so-called $\mathcal{H}_2$ criterion is often considered.[3] This problem consists of seeking an approximation $\hat{H}(s)$ of $H(s)$, so that

$$\hat{H} := \underset{\substack{G \in \mathcal{H}_2^{n_y \times n_u} \\ \dim G = r}}{\textbf{argmin}} \quad ||H - G||_{\mathcal{H}_2} \tag{11.4}$$

Here, instead of using the standard $\mathcal{H}_2$-norm, its frequency limited version, denoted $\mathcal{H}_{2,\Omega}$-norm, is used to capture the model behavior over the support Ω (i.e., to focus on the most reliable frequency range). Indeed, in the case of generating control-oriented reduced-order models and in the view of designing an aeroelastic control law that rolls-off after a given frequency, a very appealing support Ω can be $\Omega := [0 \ \omega]$, where $\omega \in \mathbb{R}_+^*$ is the upper frequency the controller will operate. The optimization problem then becomes,

$$\hat{H} := \underset{\substack{G \in \mathcal{H}_\infty^{n_y \times n_u} \\ \dim G = r}}{\textbf{argmin}} \quad ||H - G||_{\mathcal{H}_{2,\Omega}} \tag{11.5}$$

In the case where Ω is simply $[0 \quad \omega]$, the $\mathcal{H}_{2,\Omega}$ is defined as

$$||H||^2_{\mathcal{H}_{2,\Omega}} := \text{trace}\left(\frac{1}{2\pi}\int_{-\omega}^{\omega} |H(j\nu)|^2 d\nu\right) \tag{11.6}$$

More specifically, in the aeroelastic controller case, $\omega = 15 \times 2\pi$ rad/s.

Remark 11.1 ($\mathcal{H}_{2,\Omega}$-norm implementation). More details on the $\mathcal{H}_{2,\Omega}$-norm and its spectral-based computation for large-scale systems, avoiding ill-conditioned Gramian use, is firstly described in [29]. From a practical point of view, the frequency limited $\mathcal{H}_2$-norm is also made available in the MORE toolbox [21, 29], running the following MATLAB® code:

```
opt.freqBand = [0 15]*2*pi
sysNorm = moreNorm(sys,2,opt)
```

where `sys` is the considered LTI dynamical model and `opt.freqBand` stands for the frequency band Ω over which the norm is computed.

[3] The $\mathcal{H}_2$-norm denotes the norm associated with $\mathcal{H}_2^{n_y \times n_u}$ space.

11.3.1 Preliminaries on projection-based LTI model approximation

Let us consider a stable, n_u inputs n_y outputs MIMO LTI continuous-time dynamical system of order n described by $H : \mathbb{C} \rightarrow \mathbb{C}^{n_y \times n_u}$, a $\mathscr{H}_\infty^{n_y \times n_u}$ matrix complex-valued function defined as

$$H(s) = C(sI_n - A)^{-1}B + D \in \mathbb{C}^{n_y \times n_u} \tag{11.7}$$

A realization $\mathbf{H} : (A, B, C, D)$ of $H(s)$ is defined as

$$\mathbf{H} : \begin{cases} \dot{x}(t) &= Ax(t) + Bu(t) \\ y(t) &= Cx(t) + Du(t) \end{cases} \tag{11.8}$$

where $A \in \mathbb{R}^{n \times n}$, $B \in \mathbb{R}^{n \times n_u}$, $C \in \mathbb{R}^{n_y \times n}$, and $D \in \mathbb{R}^{n_y \times n_u}$. The *projection-based approximation problem* consists of finding $\hat{\mathbf{H}} := (\hat{A}, \hat{B}, \hat{C}, \hat{D})$ (or $\hat{H}(s) := \hat{C}(sI_r - \hat{A})^{-1}\hat{B} + \hat{D}$), a reduced-order model of order $r \ll n$:

$$\hat{\mathbf{H}} : \begin{cases} \dot{\hat{x}}(t) &= \hat{A}\hat{x}(t) + \hat{B}u(t) \\ \hat{y}(t) &= \hat{C}\hat{x}(t) + \hat{D}u(t) \end{cases} \tag{11.9}$$

where $\hat{A} = W^T AV$, $\hat{B} = W^T B$, $\hat{C} = CV$, $\hat{D} = D$, and $W, V \in \mathbb{R}^{n \times r}$ are left and right projectors that satisfy $W^T V = I_r$, such that $\hat{\mathbf{H}}$ well approximates $\mathbf{H}$ in the sense of a given norm.

Problems (11.4) and (11.5) are non-convex (but smooth), so they are very difficult to solve in practice for large(medium)-scale models.[4] Nevertheless *first-order optimality conditions* can be obtained for problem (11.4), although they are non-convex (see [33, 35, 2, 10, 26, 27, 34]), and for problem (11.5) one (see [30, 28]).

Remark 11.2 (About frequency-limited model approximation). Some works have addressed Problem (11.5) using weighting functions (see e.g., [1]). The advantage of the approach of [30, 28] is that it alleviates the use of weight functions, and provides a direct and exact frequency-limited $\mathscr{H}_2$-norm (see [29] for the computation and [31] for an application in the model approximation LTI problem).

11.3.2 Large-scale single-LTI model approximation procedure

Methods allowing to reach first-order optimality conditions for the $\mathscr{H}_2$ problem (11.4) have been widely explored and are still of great interest in both the numerical and the control communities. Some very efficient iterative procedures allowing to converge toward a near optimal condition have notably been proposed. For more details, the reader is invited to refer to [11, 26, 3, 34], and references therein.

[4] Note also that the norms are also numerically complex to compute, especially when large-scale settings are considered.

Algorithm 11.1 Frequency-Limited Iterative SVD-Tangential Interpolation Algorithm (**FL-ISTIA**) [30]

Input: $A \in \mathbb{R}^{n \times n}, B \in \mathbb{R}^{n \times n_u}, C \in \mathbb{R}^{n_y \times n}, \sigma^{(0)} = \left[\sigma_1^{(0)} \ldots \sigma_r^{(0)}\right]^T \in \mathbb{C}^r,$
$\hat{b} = \left[\hat{b}_1 \ldots \hat{b}_r\right] \in \mathbb{C}^{n_u \times r}, \quad \varepsilon > 0, \quad \omega \in \mathbb{R}_+^*.$

1 Construct V such that $\mathbf{span}(V) = \left[(\sigma_1^{(0)} I_n - A)^{-1} B\hat{b}_1, \ldots, (\sigma_r^{(0)} I_n - A)^{-1} B\hat{b}_r\right].$
2 Compute the frequency-limited observability Gramian Q_ω by solving $Q_\omega A + A^T Q_\omega + W_o(\omega) = 0$, where $W_o(\omega) = S^H(\omega) C^T C + C^T C S(\omega)$.
3 Compute $W = QV(V^T QV)^{-1}$.
4 Let $k = 0$. **While** $|\sigma^{(k)} - \sigma^{(k-1)}| > \varepsilon$ **do**
5 Set $k \leftarrow k + 1$.
6 Compute $\hat{A} = W^T AV, \hat{B} = W^T B$.
7 Compute left eigenvectors L such that $L\hat{A} = \mathbf{diag}(\lambda(\hat{A}))L$.
8 Compute $\left[\hat{b}_1 \ldots \hat{b}_r\right] = \hat{B}^T L$.
9 Set $\sigma^{(k)} = -\lambda(\hat{A})$.
10 Construct V such that $\mathbf{span}(V) = \left[(\sigma_1^{(k)} I_n - A)^{-1} B\hat{b}_1, \ldots, (\sigma_r^{(k)} I_n - A)^{-1} B\hat{b}_r\right].$
11 Compute $W = QV(V^T QV)^{-1}$.

Output: $\hat{\mathbf{H}} : (W^T AV, W^T B, CV, D)$.
Ensure: $V, W \in \mathbb{R}^{n \times r}$, $W^T V = I_r$ and $\hat{\mathbf{H}}$ is Hurwitz.

Grounded on [8, 26] and [9] ideas, a *simple but very efficient hybrid algorithm* has been proposed by [18]. This algorithm, called **ISTIA**, for Iterative SVD-Tangential Interpolation Algorithm, allows (*i*) to preserve stability at each iteration (provided that the system is stable) thanks to the computation of a single Gramian, and (*ii*) to ensure good frequency response fitting through the use of the one-sided tangential interpolation (refer to [18] for additional details). In [30], this algorithm is extended to the frequency-limited case, and results in the following Algorithm 11.1.

With reference to Algorithm 11.1, the procedure constructs the right projector V through tangential interpolation subspace at step 1 and computes a single frequency-limited Gramian (here the observability one) at step 2. Then, to preserve Petrov-Galerkin constraints, the right projector W is obtained by enforcing *orthogonality* at step 3. From step 4 to step 11, this procedure is repeated by using the *mirror images* of the eigenvalues of the reduced-order model as interpolation points $\sigma^{(k)}$, and the right eigenvectors associated with these eigenvalues as interpolation directions $\hat{b}_i$ (steps 7–9). The process is repeated until the variation of the interpolation points becomes smaller than a given tolerance ε. The **FL-ISTIA** algorithm exhibits some nice properties when approximating large(medium)-scale LTI models, as illustrated in the aircraft application in Section 11.3.4. Nevertheless, it may lead to *inappropriate model spectra* (i.e., scattered eigenvalues) when applied to multi-LTI models. Consequently, this may lead to computational problems when model interpolation and LPV model generation are to be performed. This issue is addressed in the following section.

Remark 11.3 (Interest of using Krylov-like approaches). There are two main interests in using the Krylov-like approaches for model reduction purposes instead of the standard Balanced Truncation: first, it allows to obtain a minimal mismatch error (e.g., $\mathcal{H}_{2,\Omega}$), and second, it provides some additional degrees of freedom when selecting the interpolation points. In the BizJet context, this last point is crucial to facilitate the generation of an LPV model in Section 11.4, since it allows to generate reduced-order models with consistent spectra (i.e., eigenvalues that are close to each other).

Remark 11.4 (FL-ISTIA – Algorithm 11.1 – implementation). Reader should note that the above algorithm is embedded in the MORE toolbox.[5] It is then possible to run the above algorithm using the following MATLAB code:

```
opt.freqBand = [0 15]*2*pi
[sysHat,out] = moreLTI(sys,r,'ISTIA',opt)
```

where and `opt.freqBand` stands for the frequency band over which the model is reduced, `sys` is the LTI model realization, `r` is the approximation order and `'ISTIA'` the name of the approximation method.

11.3.3 Large-scale multi-LTI models approximation procedure

The previous statements are now extended to multi-LTI systems. Let us consider n_s stable, n_u inputs n_y outputs MIMO LTI continuous-time dynamical systems of order n_i $(i = 1 \dots n_s)$ described by $H_i : \mathbb{C} \rightarrow \mathbb{C}^{n_y \times n_u}$ corresponding to given parametric configurations $(\delta_i)_{i \in [1,n_s]}$, a $\mathcal{H}_\infty^{n_y \times n_u}$ matrix complex-valued function defined as

$$H_i(s) = C_i(sI_{n_i} - A_i)^{-1}B_i + D_i \in \mathbb{C}^{n_y \times n_u}, \quad i = 1, \dots, n_s \tag{11.10}$$

A realization $\mathbf{H}_i : (A_i, B_i, C_i, D_i)$ of $H_i(s)$ is defined as

$$\mathbf{H}_i : \left\{ \begin{array}{rcl} \dot{x}_i(t) &=& A_i x_i(t) + B_i u(t) \\ y(t) &=& C_i x_i(t) + D_i u(t) \end{array} \right., \quad i = 1, \dots, n_s, \tag{11.11}$$

where $A_i \in \mathbb{R}^{n_i \times n_i}$, $B_i \in \mathbb{R}^{n_i \times n_u}$, $C_i \in \mathbb{R}^{n_y \times n_i}$, and $D_i \in \mathbb{R}^{n_y \times n_u}$. Moreover, $x(t) \in \mathbb{R}^{n_i} := \mathcal{X}$, $u(t) \in \mathbb{R}^{n_u} := \mathcal{U}$, and $y(t) \in \mathbb{R}^{n_y} := \mathcal{Y}$. The objective consists of constructing n_s reduced-order stable MIMO LTI dynamical models $(\hat{\mathbf{H}}_i)_{i \in [1,n_s]}$ of order $r \ll n_i$:

$$\hat{\mathbf{H}}_i : \left\{ \begin{array}{rcl} \dot{\hat{x}}_i(t) &=& \hat{A}_i \hat{x}_i(t) + \hat{B}_i u(t) \\ \hat{y}(t) &=& \hat{C}_i \hat{x}_i(t) + \hat{D}_i u(t) \end{array} \right., \quad i = 1, \dots, n_s \tag{11.12}$$

[5] See webpage `http://w3.onera.fr/more`.

corresponding to the same parametric configurations $(\delta_i)_{i\in[1,n_s]}$, such that *the frequency limited mismatch error is as low as possible and modal consistency is guaranteed.* Note that this last statement is crucial in the perspective of interpolation (see Section 11.4) and is one of the justifications for using Krylov-like subspace approximation techniques.

In [6] and [14], a multi-model reduction technique is proposed in order to guarantee consistency of the reduced states in the perspective of interpolation: the reduced-order models are re-projected onto the dominant subspace of all projectors (V_i or W_i) obtained by SVD. While this approach leads to nice theoretical results, it fails on the considered application. The main reason is that the number of available full-order models is very small, i.e., the parametric grid $(\delta_i)_{i\in[1,n_s]}$ is very coarse. Thus, it requires the reduced-order models spectra (i.e., the eigenvalues of the $\hat{A}_i$ matrices) to be as close as possible from a parametric configuration δ_i to another to avoid unrealistic modal trajectories between the grid points after the state-space representations are interpolated. This is why a procedure is proposed in this section to force the eigenvalues of the reduced-order models to remain within the same areas of the complex plane. Based on the **FL-ISTIA** recalled above in Algorithm 11.1, the **mFL-ISTIA** – for *multi-models Frequency-Limited Iterative SVD-Tangential Interpolation Algorithm* – is presented in Algorithm 11.2.

The main innovation here consists in applying a three-step procedure. First, each initial full-order realization is approximated independently of the others (steps 1–4). Then, the modes of the resulting reduced-order models are sorted (step 5) and new starting interpolation points are selected (steps 7–9) as the mirror image of the mean value of each mode set $\tilde{\sigma}_j^{(1,\dots,n_s)}$ if the variance of the jth interpolation points is greater that the given user defined tolerance η, or simply as the mirror image of each modes otherwise (steps 7–9). The tangential directions are selected as the optimal ones if shift is in the given tolerance or as mean ones, if outside. Finally, a single iteration of Algorithm 11.1 is performed on each full-order model $\mathbf{H}_i$ using these new interpolation points and tangential direction as initialization. A set of n_s rth-order models is generated, whose modal content is now consistent (steps 10–12). The way of selecting the new interpolation points together with tangential directions at steps 7–9 has the advantage to keep the eigenvalues of all the reduced-order models in the same areas of the complex plane if it is already the case, i.e., if steps 1–4 already place them close enough, and to force them to belong to the same areas if it was not initially the case. The counterpart of this adjustment is that the optimality conditions may no longer be guaranteed. Indeed, as illustrated in the next section, the approximation error is higher with the **mFL-ISTIA** than with the **FL-ISTIA**. Nevertheless, this leads to reduced-order models, whose modal content is more appropriate in the perspective of interpolating their state-space representations. Therefore, the price paid by losing minimal mismatch optimality is balanced by a good modal dispersion, through the use of the η parameter.

Remark 11.5 (Comparison with Balanced truncation). Krylov-based approaches are especially interesting here, since they allow to choose some important frequencies to focus on, and to force eigenvalues to be close to each other, which is not possible with the Balanced Truncation (see also Section 11.3.4).

Algorithm 11.2 Multi-models Frequency-Limited Iterative SVD-Tangential Interpolation Algorithm (**mFL-ISTIA**)

Input: a set of LTI models $(\mathbf{H}_i)_{i\in[1,n_s]}$ corresponding to parametric configurations $(\delta_i)_{i\in[i,n_s]}$, $\sigma^{(0)} = \left[\sigma_1^{(0)} \ldots \sigma_r^{(0)}\right]^T \in \mathbb{C}^r$, $\hat{b} = \left[\hat{b}_1 \ldots \hat{b}_r\right] \in \mathbb{C}^{n_u \times r}$, $\varepsilon > 0$, and $\eta > 0$ the tolerated percentage of dispersion of each eigenvalues set.

For $i = 1, \ldots, n_s$ **do**

  Run Algorithm 11.1 on $\mathbf{H}_i = (A_i, B_i, C_i, D_i)$ using $\sigma^{(0)}$ and $\hat{b}$ as initialization shift points and tangential directions.

  Compute $\hat{\mathbf{H}}_i : (W_i^T A_i V_i, W_i^T B_i, C_i V_i)$ of order r.

  Compute the final interpolation points $\sigma^{[i]} = -\lambda(W_i^T A_i V_i) \in \mathbb{C}^r$ left eigenvectors L_i and tangential directions $\left[\hat{b}_1^{[i]} \ldots \hat{b}_r^{[i]}\right] = \hat{B}_i^T L_i$.

Sort $\sigma^{[1,\ldots,n_s]} = \left[\sigma^{[1]} \ldots \sigma^{[n_s]}\right] \in \mathbb{C}^{r\times n_s}$ by applying Algorithm 11.3 to $(\hat{\mathbf{H}}_i)_{i\in[1,n_s]}$, so that each row of $\sigma^{[1,\ldots,n_s]}$ corresponds to a consistent set of interpolation points.

For $j = 1, \ldots, r$ **do**

  Compute $\overline{\sigma}_j$ as the mean value of the jth row $\sigma_j^{[1,\ldots,n_s]}$ of $\sigma^{[1,\ldots,n_s]}$.

  Compute $\overline{L}_j$ as the mean value of the jth row $L_j^{[1,\ldots,n_s]}$ of $L^{[1,\ldots,n_s]}$.

  If $\mathbf{var}(\sigma_j^{[1,\ldots,n_s]}) > \eta\,\overline{\sigma}_j$ **then**

    Set $\tilde{\sigma}_j^{[i]} = \overline{\sigma}_j$ for all $i \in [1, n_s]$.

    Set $\left[\tilde{b}_1^{[i]} \ldots \tilde{b}_r^{[i]}\right] = \hat{B}_i \overline{L}_i$.

  Else

    Set $\left[\tilde{\sigma}_j^{[1]} \ldots \tilde{\sigma}_j^{[n_s]}\right] = \sigma_j^{[1,\ldots,n_s]}$.

    Set $\left[\tilde{b}_1^{[i]} \ldots \tilde{b}_r^{[i]}\right] = \hat{B}_i L_i$.

For $i = 1, \ldots, n_s$ **do**

  Construct V_i such that

  $\mathbf{span}(V_i) = \left[(\tilde{\sigma}_1^{[i]} I_{n_i} - A_i)^{-1} B_i \tilde{b}_1, \ldots, (\tilde{\sigma}_r^{[i]} I_{n_i} - A_i)^{-1} B_i \tilde{b}_r\right]$.

  Compute $W_i = Q_i V_i (V_i^T Q_i V_i)^{-1}$.

Output: a set of LTI models $(\hat{\mathbf{H}}_i = (W_i^T A_i V_i, W_i^T B_i, C_i V_i))_{i\in[1,n_s]}$ with consistent modal contents.

Ensure: $V_i, W_i \in \mathbb{R}^{n\times r}$, $W_i^T V_i = I_r$ and $\hat{\mathbf{H}}_i$ is Hurwitz for all $i \in [1, n_s]$.

Remark 11.6 (Eigenvalues pairing). Sorting the eigenvalues of $(\hat{\mathbf{H}}_i)_{i\in[1,n_s]}$ is not a trivial task, but Algorithm 11.3 is proposed below to address this issue. It is based on the natural pairing obtained with the companion form, which is illustrated in Figure 11.6 for the special case of the BizJet aircraft model considered in this chapter.

Algorithm 11.3 Sorting of the eigenvalues of a set of LTI models

Input: a set of LTI models $(\hat{\mathbf{H}}_i)_{i\in[1,n_s]}$ with (fairly) consistent modal content, but possibly with different state-space forms, associated to parametric configurations $(\delta_i)_{i\in[i,n_s]}$.

1 Let $\mathscr{I} = \{1\}$.

2 **While** $\#\mathscr{I} < n_s$ **do**

3 Determine the two models $\hat{\mathbf{H}}_j$ and $\hat{\mathbf{H}}_k$, where $j \in \mathscr{I}$ and $k \notin \mathscr{I}$, such that $|\delta_j - \delta_k|$ is minimum.

4 Write the state matrices $\hat{A}_j$ and $\hat{A}_k$ in companion form.

5 Generate a set of N matrices $\tilde{A}_i = \alpha_i\hat{A}_j + (1-\alpha_i)\hat{A}_k$ by linear interpolation of $\hat{A}_j$ and $\hat{A}_k$, where α_i takes N equally spaced values between 0 and 1.

6 Starting from the eigenvalues of $\hat{A}_j$, follow the trajectories formed by the eigenvalues of $(\tilde{A}_i)_{i\in[i,N]}$ until the eigenvalues of $\hat{A}_k$ are reached. This can be achieved first by computing for each $i \in [1, N-1]$ first- or second-order approximations of the eigenpairs of $\tilde{A}_{i+1}$ from the ones of $\tilde{A}_i$ (see e.g. [32]), and then by associating them to the true eigenpairs of $\tilde{A}_{i+1}$. The eigenvalues of $\hat{\mathbb{H}}_j$ and $\hat{\mathbb{H}}_k$ are now paired.

7 Set $\mathscr{I} = \mathscr{I} \cup \{k\}$.

Output: a matrix $\sigma^{[1,\ldots,n_s]} \in \mathbb{C}^{r\times n_s}$, where each row contains a consistent set of eigenvalues (represented by a given symbol / color in Figure 11.5 in the case of the BizJet aircraft model considered in this chapter).

11.3.4 Application to the BizJet model

The **FL-ISTIA** and the **mFL-ISTIA** are applied to the BizJet models introduced in Section 11.1 for different values of r and compared to the classical Balanced Truncation (**BT**) technique. For each method, the mean and worst case $\mathscr{H}_{2,\Omega}$ errors are computed as follows (results are gathered on Figure 11.6),

$$\mathbf{max}_{i\in[1,n_s]}\frac{\|H_i - \hat{H}_i\|_{\mathscr{H}_{2,\Omega}}}{\|H_i\|_{\mathscr{H}_{2,\Omega}}}$$

and

$$\mathbf{mean}_{i\in[1,n_s]}\frac{\|H_i - \hat{H}_i\|_{\mathscr{H}_{2,\Omega}}}{\|H_i\|_{\mathscr{H}_{2,\Omega}}}$$

With reference to Figure 11.6, it is clear that the **FL-ISTIA** provides the best results in terms of frequency-limited approximation error for all approximation orders, which is not that surprising (see [30]). It also appears that the approximation quality is not as good with the **mFL-ISTIA** as with the **FL-ISTIA** (still the difference is

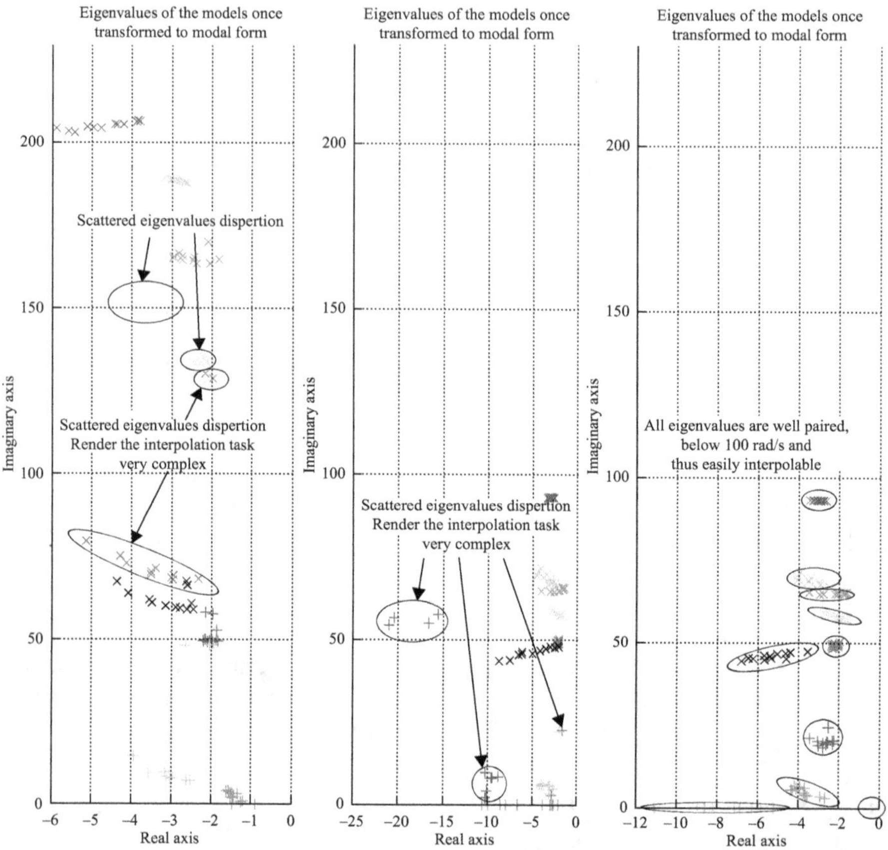

*Figure 11.5 Eigenvalues of the reduced-order models ($r = 18$) with the **BT** technique (left), the **FL-ISTIA** (middle) and the **mFL-ISTIA** (right). The **FL-ISTIA** (middle) and the **mFL-ISTIA** (right) provide eigenvalues below magnitude of 100 rad/s. Moreover, the **mFL-ISTIA** (right) exposes well grouped eigenvalues*

very small). This is due to the modification of the projectors V_i (steps 10–12 of Algorithm 11.2).

The counterpart of this loss of accuracy, illustrated on Figure 11.5, is the reduced-order models eigenvalues pairing. They are compared for the different reduction techniques (**BT**, **FL-ISTIA** and **mFL-ISTIA**) and for $r = 18$. With reference to the right frame of Figure 11.5, it is very interesting to notice that the spectra obtained with the **mFL-ISTIA** provides both nicely grouped eigenvalues (and thus easy to interpolate), with magnitude below 100 rad/s. Conversely, considering the first frame of Figure 11.5, the **BT** approach provides scattered eigenvalues around 50 rad/s and a large number of modal content at frequencies higher than 100 rad/s. The **FL-ISTIA**

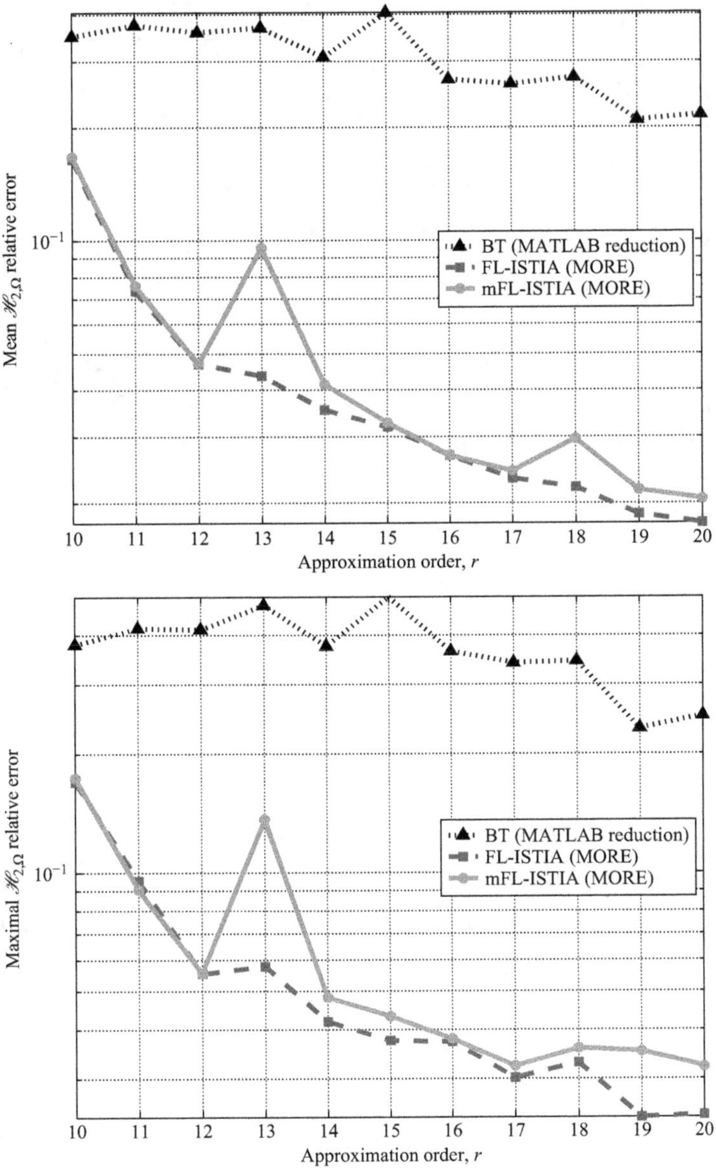

Figure 11.6 $\mathscr{H}_{2,\Omega}$ *(mismatch) mean (top) and maximal (bottom) errors for several reduction orders and methods*

(middle frame of Figure 11.5) guarantees poles below 100 rad/s but shows bad modal pairing.

Finally, Figure 11.7 illustrates the model frequency response associated to the highest mismatch error configuration (i.e., $\delta_{Mach} = 0.80$ and $\delta_{alti} = 51{,}000$ ft), for an

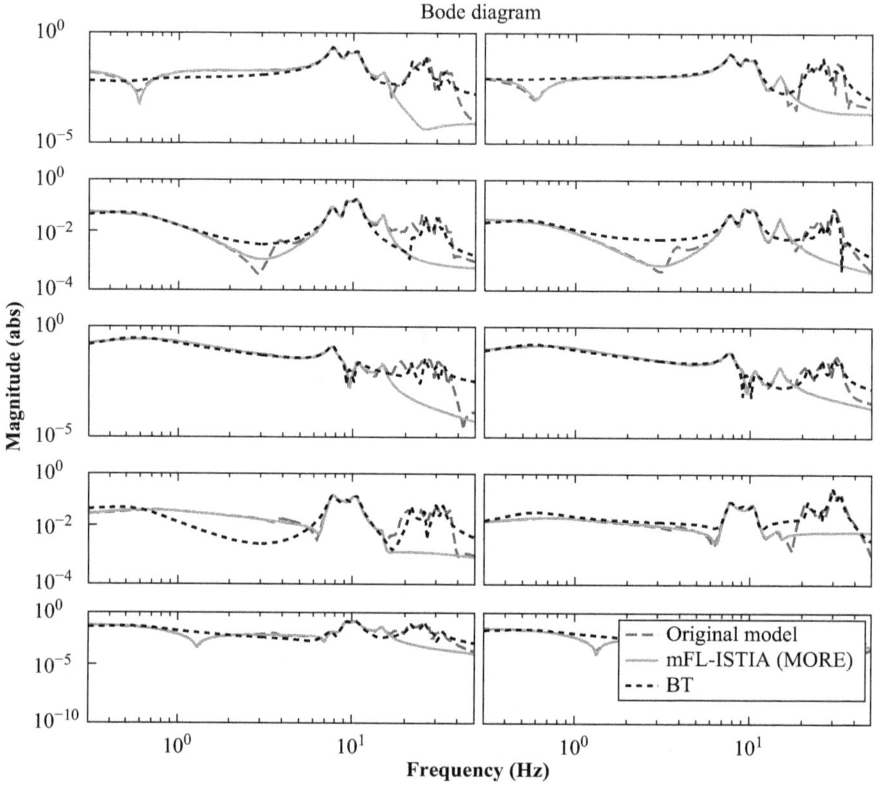

*Figure 11.7 Comparison for $\delta_{mach} = 0.80$ and $\delta_{alti} = 51{,}000$ ft (worst case approximation) of the frequency responses of the initial full-order model (dashed lines) and of the reduced-order models obtained with the **BT** technique (dotted lines) and the **mFL-ISTIA** (solid lines)*

approximation order $r = 16$. On Figure 11.7, the **mFL-ISTIA** and the **BT** are both compared with the original system for configuration $\delta_{Mach} = 0.80$ and $\delta_{alti} = 51{,}000$ ft.[6] This figure highlights the fact that the **mFL-ISTIA** approach well approximates the frequency response over the limited frequency support ($\Omega = [0\ 2\pi\,15]$) while the **BT** tends to approximate the model over all the spectrum and is thus inaccurate over the support of interest. Reader should note that implementing the **mFL-ISTIA** with $\Omega = [0 \quad \infty]$ still provides better results than the **BT** (see [20]).

[6] For sake of clarity, the **FL-ISTIA** is not plotted, but it is obvious that, as the mismatch error is better for this algorithm, better fitting is observed.

11.4 Interpolation of the reduced-order models

To convert a set of LTI models into a single LPV model, most local identification techniques require the LTI models to be represented in a consistent state-space form. The choice of a suitable representation is thus discussed in Section 11.4.1. An interpolation method is then introduced in Section 11.4.2 to build an LPV model, from which an LFR can be easily obtained. Several improvements are finally proposed in Section 11.4.3. They allow to avoid data overfitting by using as simple as possible approximation formulas, thus leading to an LFR of reasonable size.

11.4.1 Choice of a suitable state-space form

A preliminary task is to choose the same state-space form for all the reduced-order LTI models $(\hat{\mathbf{H}}_i)_{i\in[1,n_s]}$ obtained in Section 11.3. Several solutions can be considered:

- The balanced realization used in [15] has two drawbacks. First, the columns of the transformation matrix are uniquely determined to within sign. The elements of the state-space matrices can thus exhibit abrupt sign changes and a thorough validation must be carried out *a posteriori*. Moreover, most of the elements are usually nonzero, which makes it difficult to build a simple LPV model.
- The companion form does not suffer from these two drawbacks and appears to be a good candidate, provided the resulting state-space representations are scaled before interpolation to avoid bad conditioning [7]. Moreover, interpolating a companion representation is equivalent to interpolating the numerator and the denominator of the corresponding transfer function, which usually leads to an LPV model with good input/output behavior.
- The series interconnection of first- and second-order systems introduced in [4] is more difficult to obtain, since it requires to exactly pair the modes of the reduced-order models. Moreover, the number of non-constant elements in the resulting realizations is higher.
- The modal form used in [22] leads to representations with as few non-constant elements as in the companion form, but it also requires to pair the modes of the reduced-order models.

An intuitive way to compute a simple LPV/LFT model is to limit the number of varying elements in the state-space matrices of the LTI models used for interpolation [25], which speaks in favor of either the companion or the modal form. In this chapter, the emphasis is put on the companion form. Indeed, when applied to the BizJet aircraft model described in Section 11.1, the proposed interpolation methodology produces parameterized models with very smooth frequency responses (see Section 11.4.4). Moreover, the resulting LFR have smaller Δ matrices. But remember that a *scaled* companion form must be used to improve the conditioning of the state matrices [7].

11.4.2 Description of the interpolation method

The element-wise interpolation of the state-space matrices of $(\hat{\mathbf{H}}_i)_{i\in[1,n_s]}$ is described in this section. Let $(z_i^{h,v})_{i\in[1,n_s]}$, where $1 \leq h \leq r + n_y$ and $1 \leq v \leq r + n_u$, denote the values taken by the element located on the hth row and the vth column of $\left(\begin{bmatrix} \hat{A}_i & \hat{B}_i \\ \hat{C}_i & \hat{D}_i \end{bmatrix}\right)_{i\in[1,n_s]}$ for all the parametric configurations $(\delta_i)_{i\in[1,n_s]}$. In the perspective of building an LFR, either a polynomial or a rational interpolation must be performed for each of these $(r + n_u) \times (r + n_y)$ elements. The former prevents the appearance of discontinuities and is preferred here, although a higher degree is sometimes necessary to model accurately a large variation in the elements. The following expression is thus assumed:

$$z^{h,v}(\delta) = \sum_{k=1}^{n_p} \gamma_k p_k(\delta) \tag{11.13}$$

where $(p_k)_{k\in[1,n_p]}$ is a set of multivariate polynomials and $(\gamma_k)_{k\in[1,n_p]}$ are parameters to be determined. Let

$$P = \begin{bmatrix} p_1(\delta_1) & \dots & p_{n_p}(\delta_1) \\ \vdots & \ddots & \vdots \\ p_1(\delta_{n_s}) & \dots & p_{n_p}(\delta_{n_s}) \end{bmatrix} = \begin{bmatrix} P_1 & \dots & P_{n_p} \end{bmatrix} \tag{11.14}$$

$$\Gamma^T = \begin{bmatrix} \gamma_1 & \dots & \gamma_{n_p} \end{bmatrix}, \quad Z^T = \begin{bmatrix} z_1^{h,v} & \dots & z_{n_s}^{h,v} \end{bmatrix} \tag{11.15}$$

where $(P_k)_{k\in[1,n_p]}$ are called the modeling functions. The objective is to minimize the quadratic error between $z^{h,v}(\delta)$ and $(z_i^{h,v})_{i\in[1,n_s]}$, i.e., to compute:

$$\Gamma_{opt} = \underset{\Gamma\in\mathbb{R}^{n_p}}{\mathbf{argmin}} \; J(\Gamma) \tag{11.16}$$

where

$$J(\Gamma) = (Z - P\Gamma)^T(Z - P\Gamma) \tag{11.17}$$

An intuitive choice for $(p_k)_{k\in[1,n_p]}$ is

$$\left\{ \begin{array}{ccc} \mathbb{R}^l & \rightarrow & \mathbb{R} \\ (\delta(1), \dots, \delta(l)) & \rightarrow & \delta(1)^{i_1} \dots \delta(l)^{i_l} \end{array}, \quad i_1 \leq d_1, \dots, i_l \leq d_l \right\} \tag{11.18}$$

where $\delta(j)$ is the jth element of δ and $d_1, \dots, d_l$ are user-defined integers. In this case, the number of multivariate polynomials used for interpolation is $n_p = (d_1 + 1)\dots(d_l + 1)$, and the solution of (11.16) is

$$\Gamma_{opt} = (P^T P)^{-1} P^T Z \tag{11.19}$$

The interpolation strategy is then quite simple. Γ_{opt} is computed for each non-constant element of the state-space matrices of $(\hat{\mathbf{H}}_i)_{i\in[1,n_s]}$, and polynomial approximations $\hat{A}(\delta), \hat{B}(\delta), \hat{C}(\delta), \hat{D}(\delta)$ of $(\hat{A}_i, \hat{B}_i, \hat{C}_i, \hat{D}_i)_{i\in[1,n_s]}$ are thus obtained. The structured tree decomposition algorithm of [5] is finally applied to get an LFR. This algorithm is implemented in the function `symtreed` of the LFR Toolbox for MATLAB [16].

Remark 11.7. Linear least-squares are very popular, since they can be easily implemented and applied to high-dimensional systems. The considered interpolation problem can be solved element by element as detailed above, or in a single step using the Frobenius norm as proposed in [4] (although in this case, a preliminary rescaling might be necessary to ensure that all the elements of the state-space matrices have roughly the same magnitude). But it must be kept in mind that these methods suffer from two drawbacks. First, nothing can be assessed outside the parametric configurations used for interpolation and a thorough validation of the LPV model is thus necessary. A common strategy to address this issue is to evaluate the stability and the performance properties of the LPV model at each point of a very fine grid of the operating domain (see Section 11.4.4), as suggested in [25]. A more rigorous approach is to use μ-analysis based tools [23, 24]. Then, no input/output criterion is considered during interpolation. A solution to this problem is proposed in [17]. It consists in solving an SDP problem to minimize the $\mathscr{H}_2$ error between the LPV model and the local LTI models. Nevertheless, such an approach can only be applied to low-order systems due to a prohibitive computational cost.

11.4.3 Generation of a simplified LFR

It is generally desirable to have an LFR with as low complexity as possible, which can be achieved in two steps: the simplest possible LPV model is first generated and then converted into a linear fractional form. Several techniques such as the structured tree decomposition exist to perform the latter transformation. Although the minimality of the resulting LFR (in terms of the size m of the Δ matrix) cannot be guaranteed, symbolic preprocessing techniques, as well as numerical reduction, usually permit to overcome complexity. Moreover, efficient software tools such as the LFR Toolbox for MATLAB are available (see [16] and references therein for a comprehensive overview of LFR modeling). Thus, the issue of converting an LPV model into an LFR is not addressed in this chapter. On the other hand, the preliminary issue of converting tabulated data into simple yet accurate polynomial or rational expressions has been paid much less attention, although it can have a significant influence on the size of the resulting LFR (see for example [13, 12]). In this context, the following strategies can be applied to prevent data overfitting and to build a simplified LPV model, thus improving the interpolation method of Section 11.4.2.

- If $d_1, \ldots, d_l$ are chosen sufficiently large when generating the admissible set of monomials (11.18) used for interpolation, the error $J(\Gamma_{opt})$ is equal to zero, but the resulting LPV model is usually quite complicated. An intuitive and very simple solution thus consists of choosing lower values of $d_1, \ldots, d_l$.

- An efficient three-step procedure is described in [20], which generalizes the aforementioned approach:
  1. The couples (h, v) are identified, for which the values $(z_i^{h,v})_{i\in[1,n_s]}$ have no significant influence on the frequency responses of $(\hat{\mathbf{H}}_i)_{i\in[1,n_s]}$ and can be replaced by their mean value $\bar{z}^{h,v}$ over the whole set of models.
  2. The other couples are considered. For each of them, the values $(z_i^{h,v})_{i\in[1,n_s]}$ are interpolated using the method of Section 11.4.2. As small as possible values of $d_1, \ldots, d_l$ are first selected. An algorithm is then proposed, which identifies the most relevant monomials of (11.18) for which the interpolation error remains lower than a given threshold. All the other monomials are eliminated, thus leading to sparse polynomial expressions, which amounts to a certain extent to minimizing the complexity of the approximation formulas.
  3. The $\mathscr{H}_{2,\Omega}$ error between the resulting LFR and the models $(\hat{\mathbf{H}}_i)_{i\in[1,n_s]}$ is minimized by alternatively optimizing the matrices B, D and C, D of the LFR.
- In this chapter, the emphasis is put on polynomial approximation, but rational approximation can also be considered. Indeed, it offers additional degrees of freedom, which can be used to obtain simpler LPV models. Two algorithms are notably proposed in [13, 12] and could be implemented instead of the polynomial approximation technique of Section 11.4.2. The first one performs the data approximation by building a sparse modeling based on neural networks, before translating the result into a fractional form. A stepwise selection algorithm is used, combining the benefits of forward orthogonal least squares to estimate the regression parameters, and particle swarm optimization to determine the best location of the regressors. The second one looks for a rational approximation in a single step thanks to a symbolic regression technique. Genetic programming is implemented to select sparse monomials and coupled with a nonlinear iterative procedure to estimate the coefficients of the resulting rational functions.
- If the number n_s of models available for interpolation is sufficiently large, only some of them can be considered for interpolation, the other ones being used to evaluate the accuracy of the resulting LFR. Such an approach is very relevant, since it provides quantitative information to validate the behavior of the LFR between the parametric configurations used for interpolation.

The last two strategies can only by applied when the number of models available for interpolation is large enough, which is not the case here ($n_s = 12$). The second one is really efficient when the modal form is considered (see [20]) but improvements over the first one is usually marginal when the scaled companion form is used. This chapter thus focuses on the first strategy, which is quite easy to implement and yet efficient when applied to the considered BizJet aircraft model. It allows to significantly reduce the LFR complexity while maintaining a satisfactory accuracy, as shown in Section 11.4.4.

11.4.4 Application to the BizJet model

Three different strategies can be considered to build a closed-loop LFR including the aeroelastic controller, as depicted in Figure 11.1:

1. Build an open-loop (with respect to the aeroelastic feedback) LFR by interpolating the reduced-order models $(\hat{\mathbf{H}}_i)_{i\in[1,n_s]}$ obtained in Section 11.3.4, and connect it to the controller to get a closed-loop LFR.
2. Connect the controller to the reduced-order models $(\hat{\mathbf{H}}_i)_{i\in[1,n_s]}$ obtained in Section 11.3.4 to compute a set of closed-loop reduced-order LTI models, from which a closed-loop LFR can be obtained by interpolation.
3. Connect the controller to the initial full-order models $(\mathbf{H}_i)_{i\in[1,n_s]}$ to compute a set of closed-loop full-order LTI models, then apply the reduction and the interpolation algorithms described in Sections 11.3 and 11.4 to get a closed-loop LFR.

These strategies are listed by decreasing flexibility of implementation, and also by decreasing complexity of the resulting LFR (for a given level of accuracy). Indeed, if a new aeroelastic controller is designed, the first strategy requires neither reduction nor interpolation to get a new closed-loop LFR, while the third one requires both. Results obtained using the first two strategies are now presented. The third one is discarded because of its tedious implementation from a generic point of view.

11.4.4.1 First strategy

The interpolation method proposed in Section 11.4.2 is first applied with $d_1 = 2$ and $d_2 = 3$. With reference to Figure 11.3, an open-loop LFR with 2 inputs (u_a, w_a) and 5 outputs $(a_{cont}, q_{cont}, a_{pil}, a_{cabin}, a_{rear})$ is obtained, where $M(s)$ is a LTI model of order $r = 16$ and $\Delta = \text{diag}(\delta_{mach} I_{28}, \delta_{alti} I_{54})$ is a 82×82 real diagonal matrix. With these values of d_1 and d_2, the reduced-order models $(\hat{\mathbf{H}}_i)_{i\in[1,12]}$ are exactly interpolated, since a 3×4 parametric grid is considered. It means that the eigenvalues and the frequency responses obtained when the LFR is evaluated for each of the 12 considered parametric configurations precisely match those of $(\hat{\mathbf{H}}_i)_{i\in[1,12]}$. Moreover, the open-loop $\mathcal{H}_{2,\Omega}$-norm error between the initial full-order models $(\mathbf{H}_i)_{i\in[1,12]}$ and the LFR is equal to

$$\mathbf{max}_{i\in[1,12]} \frac{\|H_i - \hat{H}(\delta_i)\|_{\mathcal{H}_{2,\Omega}}}{\|H_i\|_{\mathcal{H}_{2,\Omega}}} = 3.47\%$$

where $\hat{H}(s, \delta)$ denotes the transfer matrix of the LFR. The 6th-order aeroelastic controller is then connected to the open-loop LFR. A closed-loop LFR with 22 states, 1 input (w_a), 4 outputs $(a_{cont}, a_{pil}, a_{cabin}, a_{rear})$, and a 82×82 Δ matrix is obtained. The closed-loop $\mathcal{H}_{2,\Omega}$-norm error is 4.47%.

The values of d_1 and d_2 are then progressively decreased as explained in Section 11.4.3, and results are summarized in Table 11.1. It appears that the size of the Δ matrix can be divided by more than 3 with no significant loss of accuracy.

Table 11.1 Size and accuracy of the LFR for different values of d_1 and d_2 (first strategy)

Values of d_1-d_2	Size of Δ	Open-loop $\mathcal{H}_{2,\Omega}$ error	Closed-loop $\mathcal{H}_{2,\Omega}$ error
2-3	$28+54=82$	3.47%	4.47%
1-3	$14+42=56$	4.20%	5.19%
1-2	$12+28=40$	4.29%	5.25%
1-1	$10+14=24$	6.92%	8.70%

Table 11.2 Size and accuracy of the LFR for different values of d_1 and d_2 (second strategy)

Values of d_1-d_2	Size of Δ	Closed-loop $\mathcal{H}_2$, Ω error
2-3	$10+45=55$	4.49%
1-3	$5+30=35$	4.91%
1-2	$5+20=25$	5.08%
1-1	$5+10=15$	6.43%

11.4.4.2 Second strategy

The 6th-order aeroelastic controller is now connected to each of the open-loop reduced-order models $(\hat{\mathbf{H}}_i)_{i\in[1,12]}$. A set of closed-loop models with 22 states, 1 input (w_a), and 4 outputs ($a_{cont}, a_{pil}, a_{cabin}, a_{rear}$) is obtained. A closed-loop LFR is then generated for different values of d_1 and d_2, and results are summarized in Table 11.2. It appears that the size of the Δ matrix can be divided by almost 4 with no significant loss of accuracy, which is confirmed by Figure 11.8. The LFR generated with $d_1 = d_2 = 1$ is thus considered in the sequel for further validation. Note also that the size of the Δ matrix, as well as the closed-loop $\mathcal{H}_{2,\Omega}$-norm error, is lower than for the first strategy.

Although no comparison can be made outside the grid points $(\delta_i)_{i\in[1,12]}$, a strong requirement is that the LFR behavior remains realistic, i.e., that its eigenvalues and frequency responses vary as smoothly as possible on the whole parametric domain. This is actually the case, as illustrated in Figures 11.9 and 11.10. Note that only partial results can be presented here due to space limitations, but similar ones are obtained when the other LFR presented in Tables 11.1 and 11.2, as well as the other outputs, are considered.

Both strategies thus prove very conclusive. The resulting LFRs are fully representative of the initial full-order BizJet aircraft model. Moreover, they are simple enough to allow the use of modern analysis tools such as μ-analysis [23, 24]. Finally, note that the second strategy leads to simpler and more accurate LFR, but the first one is easier to implement in case the aeroelastic controller is modified.

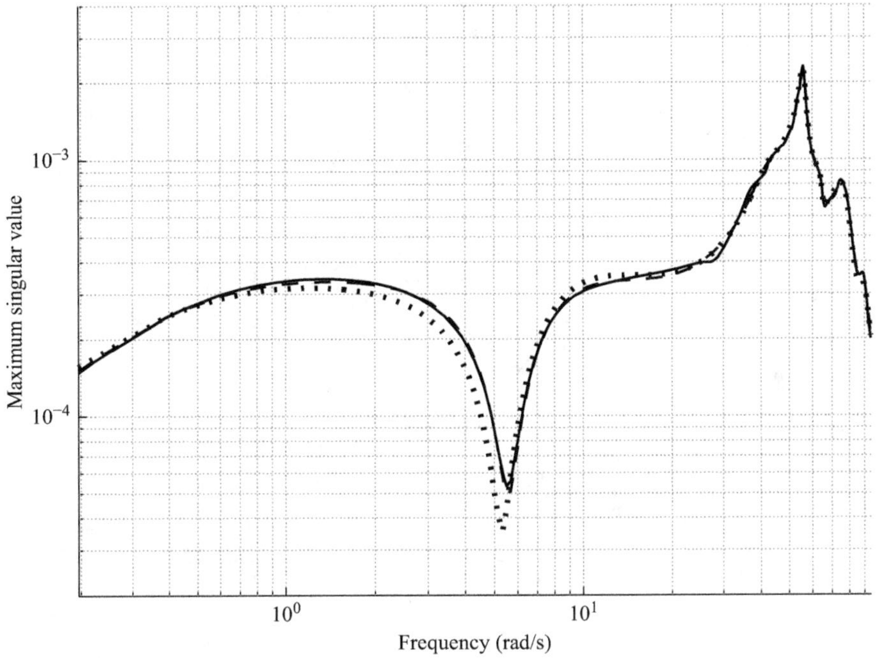

Figure 11.8 Comparison of the maximum singular values of $H_i(s)$ (straight line) and $\hat{H}(s, \delta_i)$ (dashed line $\leftrightarrow$ $d_1 = 2$ and $d_2 = 3$ – dotted line $\leftrightarrow$ $d_1 = 1$ and $d_2 = 1$) for the configuration δ_i with the largest $\mathcal{H}_{2,\Omega}$ error

11.5 Conclusion

Grounded on [20], an enhanced systematic methodology to convert a set of large-scale numerical models into an LFR of reasonable size is proposed here and applied on an accurate generic BizJet aircraft model set. This model generation allows then engineers to use modern robust analysis and synthesis techniques. Comparing to [20], the contributions are twofold.

(*i*) First, even if the problem is non-convex, a numerically stable and efficient Krylov-based procedure is proposed to approximate a set of reduced-order models with accurate frequency responses, *consistent modal content* and *over a bounded spectral support*. The main interest in this finite frequency approximation is that it is possible to accurately reproduce the system behavior over the frequency range where the aeroelastic control law will operate, and avoid over-description.

(*ii*) Then, starting from these consistent frequency-limited reduced-order models, an interpolation procedure allowing to avoid overfitting is presented, and a simple LPV/LFT model representing the system behavior on its whole operating domain is finally obtained, thanks to an *adequate interpolation basis selection*.

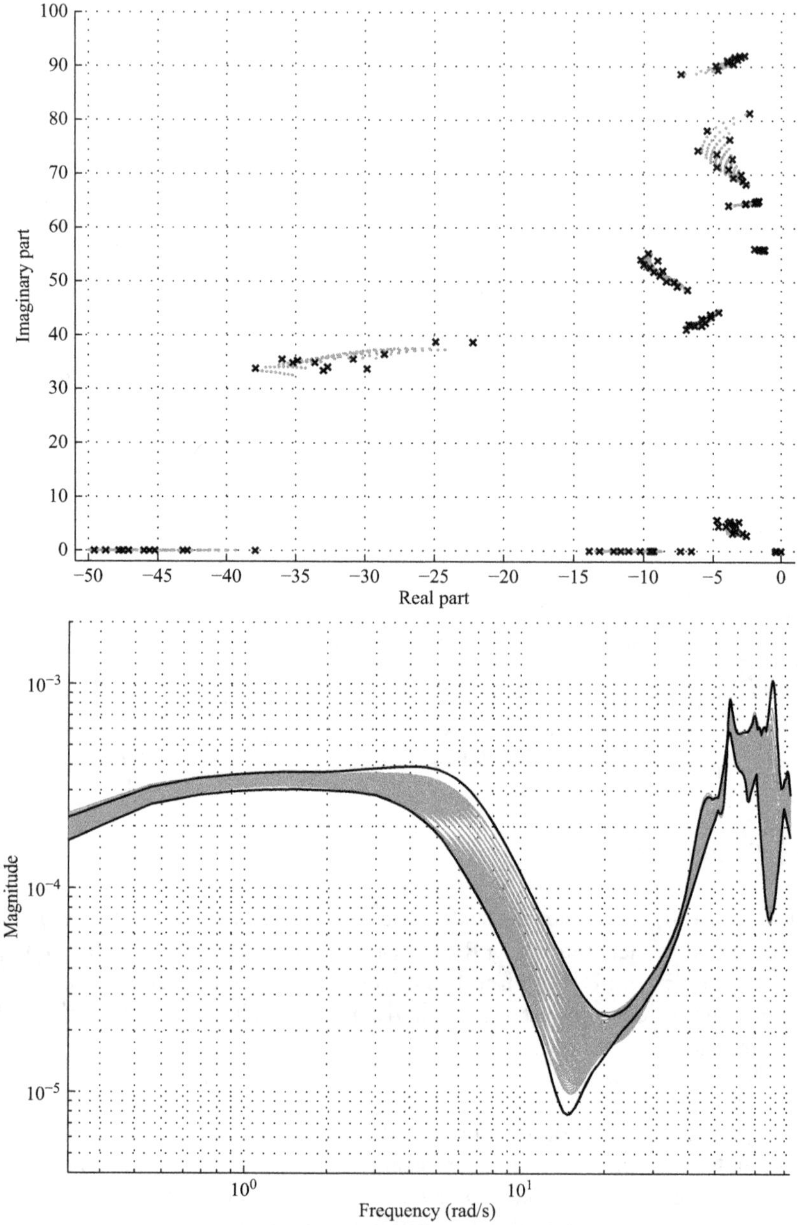

Figure 11.9 Eigenvalues (top) and frequency responses from w_a to a_{pil} (bottom) of the LFR computed on a 10 × 10 grid of the parametric domain (crosses on the top plot and dark lines on the bottom plot correspond to the reduced-order LTI models $(\hat{\mathbf{H}}_i)_{i\in[1,12]}$)

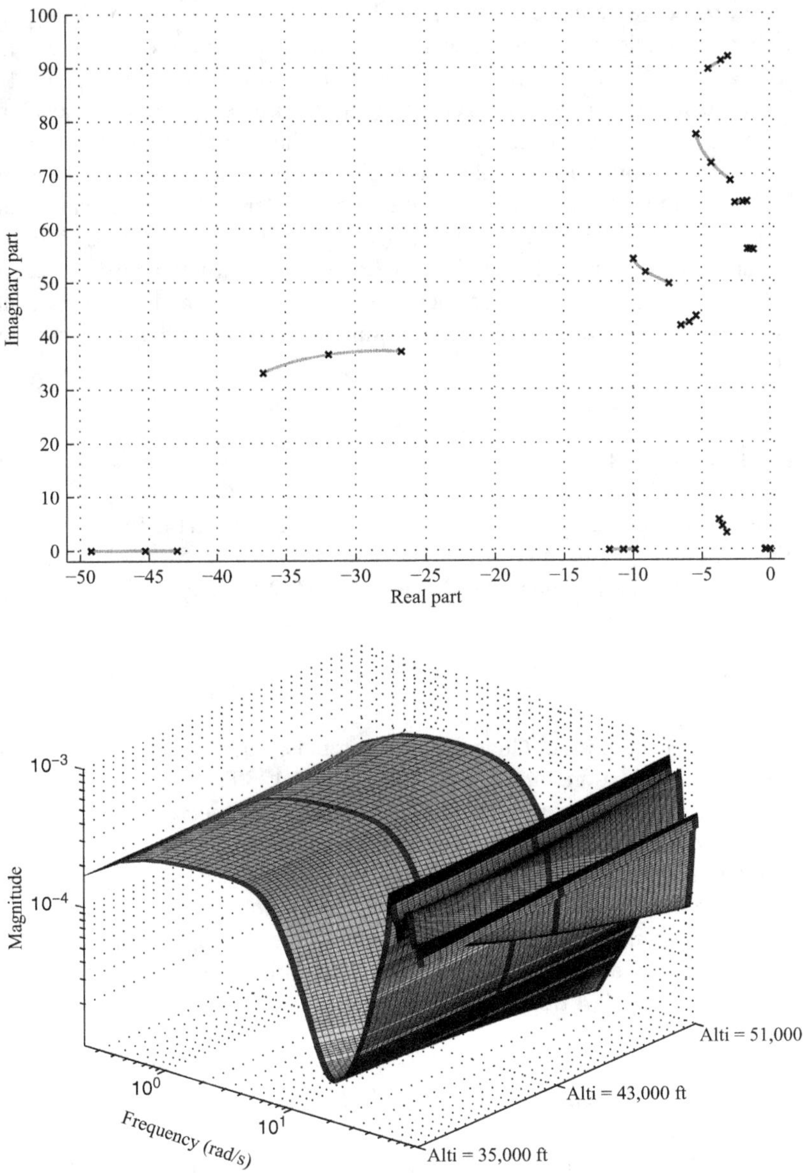

Figure 11.10 *Modal trajectories (top) and frequency responses from w_a to a_{pil} (bottom) of the LFR with respect to the altitude δ_{alti}, the Mach number δ_{mach} being fixed to its mean value 0.85 (crosses on the top plot and dark lines on the bottom plot correspond to $\delta_{alti} = 35,000$, 43,000, and 51,000 ft)*

The entire procedure is successfully applied to a set of generic MIMO large-scale BizJet, aircraft models corresponding to different flight configurations. Theses models are then used by Dassault-Aviation to adjust control laws.

A significant gain in computational cost can then be expected in the evaluation of the stability and the performance properties of the aircraft. Indeed, it takes several days if only Monte Carlo simulations are performed using the large-scale initial models. On the other hand, analyzing the reduced-order LPV model using modern techniques such as, delay, μ-analysis usually takes only a few seconds or minutes. In this context, a relevant strategy is first to determine quickly the most critical parametric configurations using the aforementioned LPV model and then to concentrate the Monte Carlo simulations on reduced parametric domains, thus minimizing the computational cost.

Acknowledgments

The research leading to these results has received funding from the European Union's Seventh Framework Program (FP7/2007-2013) for the Clean Sky Joint Technology Initiative under grant agreement CSJU-GAM-SFWA-2008-001.

References

[1] Anic, B., Beattie, C.A., Gugercin, S., Antoulas, A.C.: Interpolatory weighted-$\mathcal{H}_2$ model reduction. *Automatica* **49**(5), 1275–1280 (2013)

[2] Antoulas, A.C.: *Approximation of Large-Scale Dynamical Systems. Advanced Design and Control*. SIAM, Philadelphia (2005)

[3] Antoulas, A.C., Beattie, C.A., Gugercin, S.: *Efficient Modeling and Control of Large-Scale Systems*, chap. 1: *Interpolatory model reduction of large-scale dynamical systems*, pp. 3–58. Robotics. Springer-Verlag, London, UK (2010)

[4] de Caigny, J., Camino, J., Swevers, J.: Interpolation-based modeling of MIMO LPV systems. *IEEE Transactions on Control Systems Technology* **19**(1), 46–63 (2011)

[5] Cockburn, J., Morton, B.: Linear fractional representations of uncertain systems. *Automatica* **33**(7), 1263–1271 (1997)

[6] Eid, R., Lohmann, B., Panzer, H.: Interpolation-based Parametric Model Reduction using Krylov Subspaces. In: *Presented at the Autumn School on Future Developments in Model Order Reduction*. Terschelling, The Netherlands (2009)

[7] Ferreres, G.: Computation of a flexible aircraft LPV/LFT model using interpolation. *IEEE Transactions on Control Systems Technology* **19**(1), 132–139 (2011)

[8] Gallivan, K.A., Vanderope, A., Van Dooren, P.: Model reduction of MIMO systems via tangential interpolation. *SIAM Journal of Matrix Analysis and Application* **26**(2), 328–349 (2004)

[9] Gugercin, S.: An iterative SVD-Krylov based method for model reduction of large-scale dynamical systems. *Linear Algebra and its Applications* **428**(8–9), 1964–1986 (2008)

[10] Gugercin, S., Antoulas, A.C., Beattie, C.A.: A rational Krylov iteration for optimal $\mathcal{H}_2$ model reduction. In: *Proceedings of the 17th International Symposium on Mathematical Theory and Networks Systems*. Kyoto, Japan (2006)

[11] Gugercin, S., Antoulas, A.C., Beattie, C.A.: $\mathcal{H}_2$ Model reduction for large scale linear dynamical systems. *SIAM Journal on Matrix Analysis and Applications* **30**(2), 609–638 (2008)

[12] Hardier, G., Roos, C.: Creating sparse rational approximations for Linear Fractional Representations using genetic programming. In: *Submitted to the 3rd IFAC International Conference on Intelligent Control and Automation Science*. Chengdu, China (2013)

[13] Hardier, G., Roos, C.: Creating sparse rational approximations for Linear Fractional Representations using surrogate modeling. In: *Submitted to the 3rd IFAC International Conference on Intelligent Control and Automation Science*. Chengdu, China (2013)

[14] Lohmann, B., Eid, R.: Efficient Order Reduction of Parametric and Nonlinear Models by Superposition of Locally Reduced Models. In: *Methoden und Anwendungen der Regelungstechnik. Erlangen-Manchener Workshops*. Aachen, Germany (2009)

[15] Lovera, M., Mercère, G.: Identification for gain-scheduling: a balanced subspace approach. In: *Proceedings of the American Control Conference*, pp. 858–863. New York City, New York (2007)

[16] Magni, J.F.: User manual of the Linear Fractional Representation Toolbox (version 2.0). Tech. Rep. 5/10403.01F, ONERA (2006). Available at http://www.onera.fr/staff-en/jean-marc-biannic/

[17] Petersson, D., Löfberg, J.: Optimization based LPV-approximation of multimodel systems. In: *Proceedings of the 10th European Control Conference*. Budapest, Hungary (2009)

[18] Poussot-Vassal, C.: An iterative SVD-tangential interpolation method for medium-scale MIMO systems approximation with application on flexible aircraft. In: *Proceedings of the 50th IEEE Conference on Decision and Control – European Control Conference*, pp. 7117–7122. Orlando, Florida, USA (2011)

[19] Poussot-Vassal, C., Loquen, T., Vuillemin, P., Cantinaud, O., Lacoste, J.P.: Business Jet large-scale model approximation and vibration control. In: *Proceedings of the 11th IFAC ALCOSP*, pp. 199–204. Caen, France (2013)

[20] Poussot-Vassal, C., Roos, C.: Generation of a Reduced-Order LPV/LFT Model from a set of large-scale MIMO LTI flexible aircraft models. *Control Engineering Practice* **20**(9), 919–930 (2012)

[21] Poussot-Vassal, C., Vuillemin, P.: Introduction to MORE: a MOdel REduction toolbox. In: *Proceedings of the IEEE Multi-conference on Systems and Control*, pp. 776–781. Dubrovnik, Croatia (2012)

[22] Roos, C.: Generation of flexible aircraft LFT models for robustness analysis. In: *Proceedings of the 6th IFAC Symposium on Robust Control Design*. Haifa, Israel (2009)

[23] Roos, C.: Systems Modeling, Analysis and Control (SMAC) Toolbox: an insight into the robustness analysis library. In: *Proceedings of the IEEE Multiconference on Systems and Control*. Hyderabad, India (2013)

[24] Roos, C., Lescher, F., Biannic, J.M., Döll, C., Ferreres, G.: A set of μ-analysis based tools to evaluate the robustness properties of high-dimensional uncertain systems. In: *Proceedings of the IEEE Multiconference on Systems and Control*, pp. 644–649. Denver, Colorado (2011)

[25] Steinbuch, M., van de Molengraft, R., van der Voort, A.: Experimental modelling and LPV control of a motion system. In: *Proceedings of the American Control Conference*, pp. 1374–1379. Denver, Colorado (2003)

[26] Van Dooren, P., Gallivan, K.A., Absil, P.A.: $\mathcal{H}_2$-optimal model reduction of MIMO systems. *Applied Mathematics Letters* **21**(12), 53–62 (2008)

[27] Van Dooren, P., Gallivan, K.A., Absil, P.A.: $\mathcal{H}_2$-optimal model reduction with higher order poles. *SIAM Journal on Matrix Analysis and Applications* **31**(5), 2738–2753 (2010)

[28] Vuillemin, P., Poussot-Vassal, C., Alazard, D.: A frequency-limited iterative SVD-tangential medium-scale model approximation method. In: *Proceedings of the 2nd Euro GNC*. Deflt, The Netherlands (2013)

[29] Vuillemin, P., Poussot-Vassal, C., Alazard, D.: A Spectral Expression for the Frequency-Limited $\mathcal{H}_2$-norm. arXiv:1211.1858 (2013)

[30] Vuillemin, P., Poussot-Vassal, C., Alazard, D.: $\mathcal{H}_2$ optimal and frequency limited approximation methods for large-scale LTI dynamical systems. In: *Proceedings of the 6th IFAC Symposium on Systems Structure and Control*, pp. 719–724. Grenoble, France (2013)

[31] Vuillemin, P., Poussot-Vassal, C., Alazard, D.: Poles residues descent algorithm for optimal frequency-limited $\mathcal{H}_2$ model approximation. In: *Submitted to CDC*. Florence, Italy (2013)

[32] Wilkinson, J.: *The Algebraic Eigenvalue Problem*. Chap. 2: Perturbation theory. Oxford University Press, Clarendon Press, Oxford, UK (1965)

[33] Wilson, D.A.: Model reduction for multivariable systems. *International Journal of Control* **20**(1), 57–64 (1974)

[34] Xu, Y., Zeng, T.: Optimal $\mathcal{H}_2$ model reduction for large scale MIMO systems via tangential interpolation. *International Journal of Numerical Analysis* **8**(1), 174–188 (2010)

[35] Zhou, K., Doyle, J.C.: *Essentials of Robust Control*. Prentice Hall, Upper Saddle River, New Jersey, USA (1997)

Chapter 12

Active vibration control using subspace predictive control

Gijs van der Veen[1], *Jan-Willem van Wingerden*[2] *and Michel Verhaegen*[3]

Abstract

In this chapter we apply the subspace predictive control algorithm to the problem of active vibration control of a flexible beam. The flexible beam is equipped with piezoelectric transducers. This example application demonstrates that computations can be performed in real time for a realistic system and it shows how the scheme rapidly adapts when a sudden significant change in structural dynamics is introduced by changing one of the structural parameters.

12.1 Introduction

In this chapter we consider vibration control of a flexible beam. Vibration control of flexible structures continues to be an area of active research. In particular, there is interest in 'smart' structures equipped with electrically deformable materials. Recent overviews of control design techniques for such structures are given in, e.g., [8] and [6]. It is frequently the case in practice that the characteristics of flexible structures change, e.g., due to variable loading or changing boundary conditions. Adaptive control methodologies for resonant structures have been proposed in [7]. Alternatively, robust control techniques can be applied to synthesise controllers that take uncertainty into account, but dealing with shifting resonance frequencies in such a framework is challenging and may require resorting to linear, parameter-varying (LPV) or gain-scheduling control techniques. Recently, the feasibility of model predictive control (MPC) for vibration control has been demonstrated, e.g., in [9].

The subspace predictive control (SPC) algorithm presented in Chapter 7 has been derived before in the literature [2, 4], but it has rarely been applied to real systems

[1] Delft University of Technology, Delft, The Netherlands, e-mail: G.J.vanderVeen@tudelft.nl
[2] Delft University of Technology, Delft, The Netherlands, e-mail: J.W.vanWingerden@tudelft.nl
[3] Delft University of Technology, Delft, The Netherlands, e-mail: M.Verhaegen@tudelft.nl

[10, 2]. Hence, in this chapter we consider an experiment in which we test the SPC algorithm on a real system.

12.2 Experimental set-up

The subspace predictive control formulation derived in Chapter 7 is applied to a vibration control problem on a beam equipped with piezoelectric transducers; see Figure 12.1. Two of these transducers are configured as actuators and two are configured as sensors. A schematic representation of the beam is provided in Figure 12.2. The beam is an aluminium strip of 95 cm length. At locations 1 and 2, two (almost) collocated sensor–actuator pairs are situated. The actuators are flexible Macro Fiber Composite (MFC) devices, type M8528, from Smart Material Corp. The sensors are connected to high-impedance buffers and the actuators are driven by high-voltage amplifiers. As shown in Figure 12.1, the beam is clamped at the near end. At the far end either of two boundary conditions can be activated by means of an electromagnetic holding device. If the holding device is inactive, the beam is free at the far end, resulting in clamped-free boundary conditions; see Figure 12.2(a). If the holding device is active, the beam is effectively pinned at the far end, resulting in clamped-pinned boundary conditions; see Figure 12.2(b). These two different configurations result in two radically different dynamic behaviours, as can be seen from the expected vibration modes in Figure 12.2. The beam has a potentially infinite number of vibration modes, the lowest of which have very low intrinsic damping. For the given beam, the first two modes of Figure 12.2(a) are located at $f_1^{(a)} = 3.7$ Hz

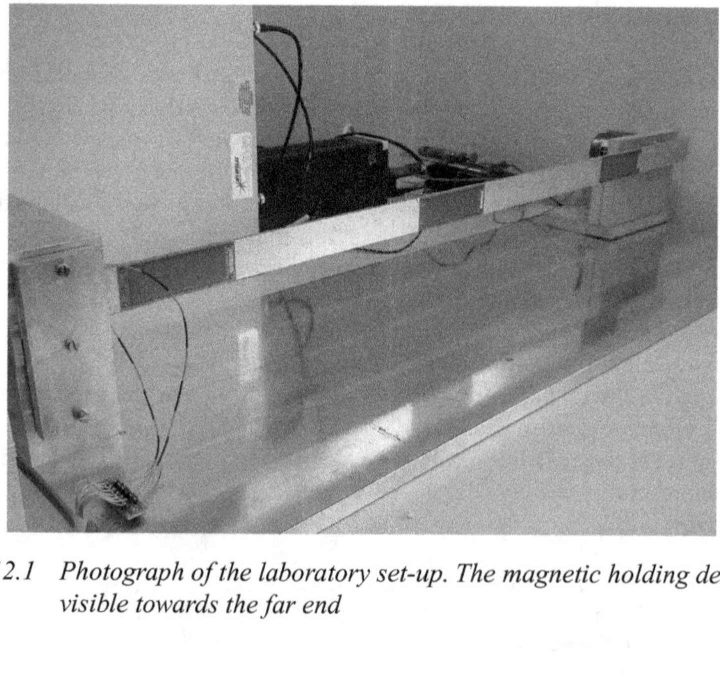

Figure 12.1 Photograph of the laboratory set-up. The magnetic holding device is visible towards the far end

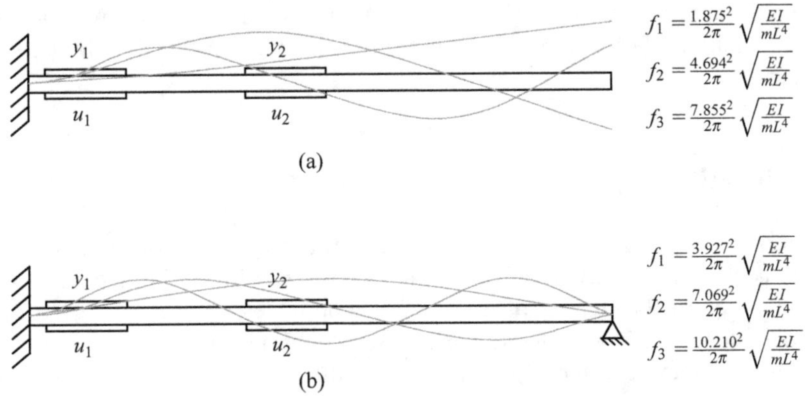

Figure 12.2 Schematic view of the two beam configurations and the associated natural vibration modes [5]: (a) clamped, (b) clamped-pinned

and $f_2^{(a)} = 23.7$ Hz, respectively. The first mode of Figure 12.2(b) is located at $f_1^{(b)} = 18.6$ Hz.

12.2.1 Control design

The objective of the control algorithm will be to augment damping of the first two vibration modes of Figure 12.2(a) and the first mode of Figure 12.2(b). During operation, the configuration can switch and the algorithm should track the associated changes in the dynamic behaviour.

To allow the digital control algorithm to operate at a rate of 200 Hz, data was sampled at 2 kHz, then filtered using a second-order Butterworth anti-aliasing filter with a corner frequency at 60 Hz and then downsampled to 200 Hz. The control signal was upsampled to 2 kHz and then low-pass filtered at 250 Hz to eliminate the high-frequency content introduced by the zero-order hold reconstruction.

In the current experiment the following parameter values were chosen: a past window dimension of $p = 25$; a forgetting factor of $\lambda = 0.99995$, a prediction horizon of $N_p = 50$ ($= 0.25$ s) and a control horizon of $N_c = 10$. Furthermore, we have applied the recursive least-squares (RLS) scheme with directional forgetting as mentioned in Chapter 7.

12.2.2 Notes on the implementation

The algorithm described in Chapter 7 has been implemented using a tailored C++ implementation. For matrix manipulations, use was made of the Eigen template library [3]. The QP is solved using the online active set strategy implemented in the qpOASES package [1]. Although we are dealing with a time-varying QP, the use of an active set strategy is still beneficial since the Hessian in the quadratic program is expected to vary slowly most of the time. All code was accessed through Simulink® S-functions and executed in real time using the xPC Target™ real time kernel.

12.3 Results

An experiment was performed in which the beam was subjected to random excitations on both actuators. These random excitations consisted of band-limited white noise within the control bandwidth of 60 Hz. The amplitudes were chosen such that – in an open-loop situation – the sensor signals would not saturate. The experiment was initiated in an open-loop fashion to allow the identification algorithm to start converging to an estimate of the observer Markov parameters. At $t = 10$ s the controller is switched on. Then, at $t = 50$ s, an electromagnetic actuator clamps the tip of the beam. This momentarily introduces an impulse disturbance (shock) and results in permanently changed characteristics. During this switching, the controller is kept switched on. From Figure 12.3 one can see that almost immediately after switching

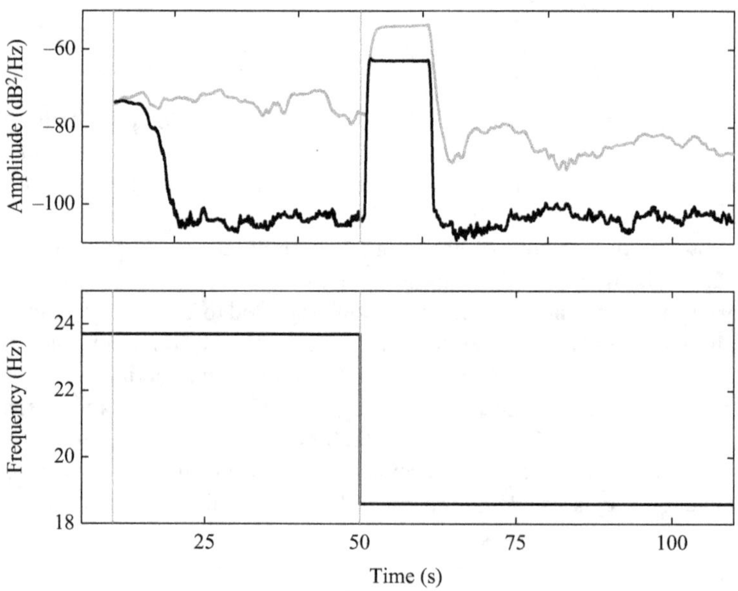

Figure 12.3 Plot showing the spectral amplitude (estimated) and frequency of the controlled mode. This is the second bending mode of Figure 12.2(a) and the first bending mode of Figure 12.2(b), measured at sensor 1 (y_1) as a function of time: open-loop (grey), closed-loop (black). The vertical lines indicate switching on of the controller ($t = 10$ s) and the instant at which the beam is clamped ($t = 50$ s)

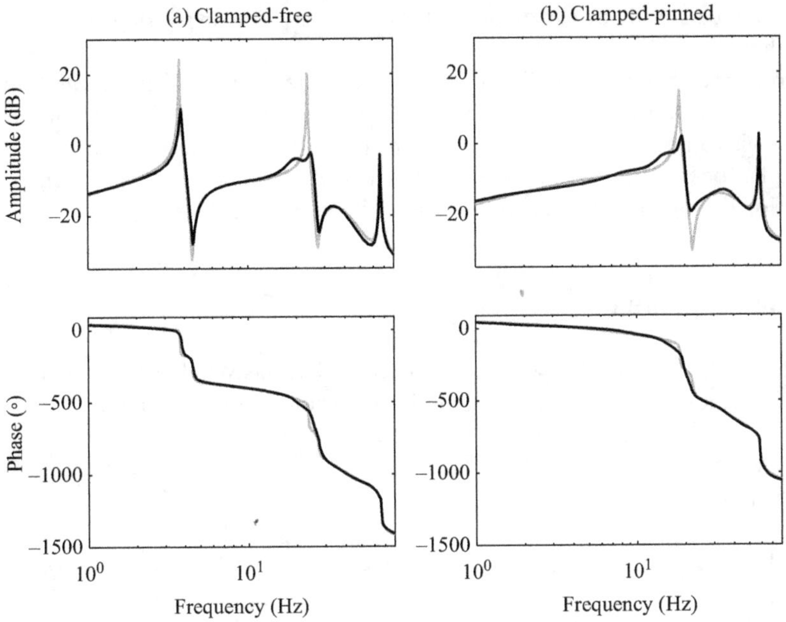

Figure 12.4 Identified frequency responses of the open-loop (grey) and closed-loop (black) system under SPC control from disturbance u_1 to output y_1 for the clamped-free configuration (a) and the clamped-pinned configuration (b)

on the controller the second mode is almost completely eliminated. Furthermore, after the change in dynamics, the controller adapts to the new configuration in seconds, after which the first mode of Figure 12.2(b) is also almost eliminated.

In Figure 12.4 we compare the results of system identifcation performed for the open and closed-loop systems for the first input–output pair and for both configurations. These responses clearly show that the modes at $f_2^{(a)} = 23.7$ Hz and $f_1^{(b)} = 18.6$ Hz, respectively, for the two configurations are attenuated by approximately 20 dB. The first mode at $f_1^{(a)} = 3.7$ Hz in Figure 12.4(a) is attenuated by about 10 dB.

12.4 Conclusions

In this chapter subspace predictive control was applied to an illustrative example. It was demonstrated how subspace predictive control can be applied to vibration control problems in flexible structures and that damping can be achieved when the dynamics change considerably. Advantages are the adaptation to changes in dynamics and the limited set of tuning parameters. Although the scheme may be computationally complex, it can easily be run on present-day processors by exploiting structure in the algorithm.

To maintain proper conditioning of the covariance matrix in a forgetting RLS scheme it is necessary that conventional persistence of excitation conditions are satisfied. This can be achieved by applying an appropriate reference perturbation that is as small as possible (to avoid loss of performance). Guaranteeing a sufficient level of excitation during operation is an active area of further research.

Acknowledgements

The authors gratefully acknowledge the support of Vestas Wind Systems A/S. We also thank Ivo Houtzager and Rufus Fraanje for the valuable discussions.

References

[1] Ferreau, H.J., Kirches, C., Potschka, A., Bock, H.G., Diehl, M.: *qpOASES: a parametric active-set algorithm for quadratic programming.* Mathematical Programming Computation 1–37 (2014). URL: http://link.springer.com/article/10.1007/s12532-014-0071-1

[2] Dong, J., Verhaegen, M., Holweg, E.: Closed-loop subspace predictive control for fault tolerant MPC design. In: *Proceedings of the 17th IFAC World Congress*, vol. 17, Seoul, South Korea (2008)

[3] Guennebaud, G., Jacob, B., et al.: Eigen v3. URL: http://eigen.tuxfamily.org/index.php?title=BibTeX (2010)

[4] Hallouzi, R.: *Multiple-Model Based Diagnosis for Adaptive Fault-Tolerant Control.* Ph.D. thesis, Delft University of Technology (2008).

[5] Inman, D.J.: *Engineering Vibration*, 2nd edn. Prentice-Hall, Upper Saddle River, New Jersey (2001)

[6] Moheimani, S.O.R., Fleming, A.J.: *Piezoelectric Transducers for Vibration Control and Damping.* Springer, Berlin, Germany (2006)

[7] Niederberger, D., Fleming, A., Moheimani, S.O.R., Morari, M.: Adaptive multi-mode resonant piezoelectric shunt damping. *Smart Materials and Structures* **13**(5), 1025 (2004)

[8] Preumont, A., Seto, K.: *Active Control of Structures.* Wiley, New York City, New York (2008)

[9] Wills, A., Bates, D., Fleming, A., Ninness, B., Moheimani, S.: Model predictive control applied to constraint handling in active noise and vibration control. *IEEE Transactions on Control Systems Technology*, **16**(1), 3–12 (2008). doi 10.1109/TCST.2007.903062

[10] Woodley, B.R.: *Model Free Subspace Based H_∞ Control.* Ph.D. thesis, Stanford University (2001)

Chapter 13

Rotorcraft system identification: an integrated time–frequency-domain approach

Marco Bergamasco[1] *and Marco Lovera*[2]

Abstract

The problem of rotorcraft system identification is considered and a novel, two-step technique is proposed, which combines the advantages of time-domain and frequency-domain methods. In the first step, the identification of a black-box model using a subspace model identification method is carried out, using a technique which can deal with data generated under feedback; subsequently, in the second step, *a-priori* information on the model structure is enforced in the identified model using an $\mathscr{H}_\infty$ model matching method. A bootstrap-based approach is used to estimate model uncertainty for the identified models. A simulation study is used to illustrate the proposed approach.

13.1 Introduction

The problem of system identification of helicopter aeromechanics has been studied extensively in the last few decades, as identification has been known for a long time as a viable approach to the derivation of control-oriented dynamic models in the rotorcraft field (see for example the recent books [26, 14] and the references therein). Model accuracy is becoming more and more important, as progressively stringent requirements are being imposed on rotorcraft control systems: as the required control bandwidth increases, accurate models become a vital part of the design problem (see, e.g., [22] and the references therein).

In the system identification literature, on the other hand, one of the main novelties of the last two decades has been the development of the so-called subspace model identification (SMI) methods (see for example the books [28, 31]), which have proven extremely successful in dealing with the estimation of state-space models for multiple-inputs multiple-outputs (MIMO) systems. Surprisingly enough, even though SMI can

[1] Politecnico di Milano, Milano, Italy, e-mail: bergamasco@elet.polimi.it
[2] Politecnico di Milano, Milano, Italy, e-mail: lovera@elet.polimi.it

be effectively exploited in dealing with MIMO modeling problems, until recently these methods have received limited attention from the rotorcraft community, with the partial exception of some contributions such as [30, 8, 18]. SMI methods are particularly well suited for rotorcraft problems, for a number of reasons. First of all, the subspace approach can deal in a very natural way with MIMO problems; in addition, all the operations performed by subspace algorithms can be implemented with numerically stable and efficient tools from numerical linear algebra. Finally, information from separate data sets (such as generated during different experiments on the system, i.e., different test flights) can be merged in a very simple way into a single state-space model. Recently, see [17], the interest in SMI for helicopter model identification has been somewhat revived and the performance of subspace methods has been demonstrated on flight test data. However, so far only methods and tools that go back 10 to 15 years in the SMI literature (such as the MOESP algorithm of [29] and the bootstrap-based method for uncertainty analysis of [9]) have been considered. Therefore, the further potential benefits offered by the latest developments in the field have not been fully exploited. Among other things, present-day approaches can provide:

- unbiased model estimates from data generated during closed-loop operation, as is frequently the case in experiments for rotorcraft identification (see, e.g., [10, 13]);
- the possibility to quantify model uncertainty using analytical expressions for the variance of the estimates instead of relying on computational statistics (see [10]);
- the direct estimation of continuous-time models from (possibly nonuniformly) sampled input-output data (see [5] and the references therein).

Some preliminary results in the application of continuous-time SMI to the rotorcraft problem have been presented in [4, 6].

The only, well known, downside of the SMI approach to state-space model identification, on the other hand, is the impossibility to impose a fixed basis to the state-space representation. This, in turn, implies that it is hard to impose a parameterization to the state-space matrices in this framework, and therefore recovering physically motivated models is a challenging problem. This, to date, prevents the successful application of SMI methods to the problem of initialising iterative methods for the identification of structured state-space representations and constitutes a major stumbling block for the application of such methods in communities in which physically motivated models represent the current practice.

In this chapter the problem of bridging the gap between "unstructured" models obtained using SMI and structured ones deriving from flight mechanics is addressed as an input-output model matching one, in terms of the $\mathcal{H}_\infty$ norm of the difference between the two models (see also [7]). The solution of the problem is then computed using recent results in non-smooth optimization techniques, see [1], which yield effective computational tools (see [12]). As a final step, the bootstrap is proposed as a simple method to quantify uncertainty associated with the identified models.

In view of the above discussion, this chapter has the following objectives. First, a set of methods suitable for time-domain, continuous-time identification of rotorcraft dynamics using SMI is presented. The proposed technique can deal with data generated in closed-loop operation as it does not require restrictive assumptions in this sense. Subsequently, a frequency-domain $\mathscr{H}_\infty$ approach to the problem of deriving a structured model from the unstructured one is proposed. Finally, the achievable model accuracy is illustrated by means of simulation results for a full-scale helicopter.

The chapter is organized as follows. In Section 13.2 the problem statement is given and some definitions are provided. Section 13.3 provides a summary of the proposed two-step approach, while the main idea about the bootstrap are summarized in Section 13.4. Finally, some simulation results are presented in Section 13.5 to illustrate the performance of the proposed method.

13.2 Problem statement and preliminaries

Consider the linear, time-invariant continuous-time system

$$\mathscr{M}_s(\lambda) : \begin{cases} \dot{x}(t) = A(\lambda)x(t) + B(\lambda)u(t) + w(t), \; x(0) = x_0 \\ y(t) = C(\lambda)x(t) + D(\lambda)u(t) + v(t) \end{cases} \tag{13.1}$$

where $x \in \mathbb{R}^n$, $u \in \mathbb{R}^m$, and $y \in \mathbb{R}^p$ are, respectively, the state, input and output vectors and $w \in \mathbb{R}^n$ and $v \in \mathbb{R}^p$ are the process and the measurement noise, respectively, with covariance given by

$$E\left\{\begin{bmatrix} w(t_1) \\ v(t_1) \end{bmatrix} \begin{bmatrix} w(t_2) \\ v(t_2) \end{bmatrix}^T\right\} = \begin{bmatrix} Q & S \\ S^T & R \end{bmatrix} \delta(t_2 - t_1)$$

The system matrices $A(\lambda)$, $B(\lambda)$, $C(\lambda)$, and $D(\lambda)$ are dependent on the constant parameter vector $\lambda \in \mathbb{R}^{n_\lambda}$ such that $(A(\lambda), C(\lambda))$ is observable and $(A(\lambda), [B(\lambda), Q^{1/2}])$ is controllable.

Assume now that a dataset $\{u(t_i), y(t_i)\}$, $i \in [1, N]$ of sampled input/output data (possibly associated with a non equidistant sequence of sampling instants) obtained from system (13.1) is available. Then, the problem is to provide an estimate of the parameter λ on the basis of the available data. Note that unlike most identification techniques, in this setting incorrelation between u and w, v is not required, so that this approach is viable also for systems operating under feedback.

In the following sections a number of definitions will be used, which are summarized hereafter for the sake of clarity (see, e.g., [32, 15, 19, 2] for further details).

Definition 13.1. (Laguerre basis) Let $\mathscr{L}_2(0, \infty)$ denote the space of square integrable and Lebesgue measurable functions of time $0 < t < \infty$. Consider the first order all-pass (inner) transfer function

$$w(s) = \frac{s - a}{s + a} \tag{13.2}$$

$a > 0$. $w(s)$ generates the family of Laguerre filters, defined as

$$\mathscr{L}_i(s) = w^i(s)\mathscr{L}_0(s) = \sqrt{2a}\frac{(s-a)^i}{(s+a)^{i+1}} \tag{13.3}$$

Denote with $\ell_i(t)$ the impulse response of the i-th Laguerre filter. Then, it can be shown that the set

$$\{\ell_0, \ell_1, \ldots, \ell_i, \ldots\} \tag{13.4}$$

is an orthonormal basis of $\mathscr{L}_2(0, \infty)$, i.e., all signals in $\mathscr{L}_2(0, \infty)$ can be represented by means of the set of their projections on the Laguerre basis.

Definition 13.2. (**$\mathscr{H}_\infty$ norm**) Consider an asymptotically stable, linear time-invariant system with transfer function $G(s)$. Then the $\mathscr{H}_\infty$ norm of the system is defined as

$$\|G\|_\infty = \sup_{\alpha>0}\left\{\sup_{\omega}\overline{\sigma}\left(G(\alpha + j\omega)\right)\right\} = \sup_{\omega}\overline{\sigma}\left(G(j\omega)\right) \tag{13.5}$$

where $\overline{\sigma}$ is the maximum singular value.

Identifiability is an important issue in system identification problems; for the purpose of this study we adopt the following definitions:

Definition 13.3. (**Local identifiability**) Let $\lambda^o \in \Lambda \subset \mathbb{R}^{n_\lambda}$, the model structure is said to be locally identifiable in λ^o if $\forall\lambda_1, \lambda_2$ in the neighborhood of λ^o it holds that

$$\mathscr{M}_s(\lambda_1) = \mathscr{M}_s(\lambda_2) \Rightarrow \lambda_1 = \lambda_2$$

Definition 13.4. (**Global identifiability**) The model structure $\mathscr{M}_s(\lambda)$ is said to be globally identifiable if it is locally identifiable $\forall\lambda \in \Lambda$, i.e., over the entire parameter space.

In the following the model structure $\mathscr{M}_s(\lambda)$ is considered globally identifiable.

13.3 An integrated time–frequency-domain approach

The problem formulated in the previous section can be faced using a two-step approach: in the first step a black-box model is identified using a continuous-time SMI method, which can deal with data generated under feedback but generates an "unstructured" model; in the subsequent step *a-priori* information on the model structure is enforced in the model using an $\mathscr{H}_\infty$ model matching method.

In Section 13.2 the gray-box model $\mathscr{M}_s(\lambda)$ was introduced, while a generic "unstructured" black-box model $\mathscr{M}_{ns}$ can be described as the linear time-invariant system

$$\mathscr{M}_{ns}: \begin{cases} \dot{x}(t) = \hat{A}x(t) + \hat{B}u(t) + w(t), \ x(0) = x_0 \\ y(t) = \hat{C}x(t) + \hat{D}u(t) + v(t) \end{cases} \tag{13.6}$$

where x, u, y, w, and v are defined as in Section 13.2. The system matrices $\hat{A}$, $\hat{B}$, $\hat{C}$, and $\hat{D}$ have been estimated from a dataset $\{u(t_i), y(t_i)\}$, $i \in [1, N]$ of sampled input/output data using the continuous-time predictor-based subspace model identification algorithm introduced in the Section 13.3.1. Suppose $\mathscr{M}_{ns}$ belonging to the same model structure of $\mathscr{M}_s(\lambda)$, and that (13.1) and (13.6) describe the same system with different state-space basis. Therefore the problem becomes to provide estimates of λ such that the input-output behaviors of $\mathscr{M}_{ns}$ and $\mathscr{M}_s(\lambda)$ are equivalent under some criterion, and it is faced using an $\mathscr{H}_\infty$ approach described in Section 13.3.2.

13.3.1 Continuous-time predictor-based subspace model identification

13.3.1.1 From continuous-time to discrete-time using Laguerre projections

The main issue in the application of subspace model identification methods to continuous-time model identification is the need of computing the high order derivatives of input-output measurements arising from the continuous-time data equation. This problem can be faced using a method, based on the results first presented in [21, 19], and further expanded in [16, 20], that transforms a continuous-time system and signals to their discrete-time representations. First note that under the assumptions stated in the previous section, (13.6) can be written in innovation form as

$$\begin{aligned} \dot{x}(t) &= Ax(t) + Bu(t) + Ke(t) \\ y(t) &= Cx(t) + Du(t) + e(t) \end{aligned} \tag{13.7}$$

and it is possible to apply the results of [21] to derive a discrete-time equivalent model, as follows. Note that the notation $\hat{(\cdot)}$ has been dropped for clarity. Consider the first order inner function $w(s)$ defined in (13.2) and apply to the input u, the output y and the innovation e of (13.7) the transformations

$$\begin{aligned} \tilde{u}(k) &= \int_0^\infty \ell_k(t)u(t)dt \\ \tilde{y}(k) &= \int_0^\infty \ell_k(t)y(t)dt \\ \tilde{e}(k) &= \int_0^\infty \ell_k(t)e(t)dt \end{aligned} \tag{13.8}$$

where $\tilde{u}(k) \in \mathbb{R}^m$, $\tilde{e}(k) \in \mathbb{R}^p$ and $\tilde{y}(k) \in \mathbb{R}^p$. Then (see [21] for details) the transformed system has the state-space representation

$$\begin{aligned} \xi(k+1) &= A_o\xi(k) + B_o\tilde{u}(k) + K_o\tilde{e}(k), \ \xi(0) = 0 \\ \tilde{y}(k) &= C_o\xi(k) + D_o\tilde{u}(k) + \tilde{e}(k) \end{aligned} \tag{13.9}$$

where the state-space matrices are given by

$$\begin{aligned} A_o &= (A - aI)^{-1}(A + aI) \\ B_o &= \sqrt{2a}(A - aI)^{-1}B \\ K_o &= \sqrt{2a}(I - C(A - aI)^{-1}K)^{-1}(A - aI)^{-1}K \\ C_o &= -\sqrt{2a}C(A - aI)^{-1} \\ D_o &= D - C(A - aI)^{-1}B \end{aligned} \tag{13.10}$$

It is worth to underline that in this context k is not a time index, but refers to the projection of the signals onto the k-th basis function.

13.3.1.2 Predictor-based subspace model identification

In this section a summary of the continuous-time PBSID algorithm proposed in [3, 5], called CT-PBSID$_o$, is provided, and its implemention is discussed. More precisely, starting from system (13.7), a sketch of the derivation of a PBSID-like approach to the estimation of the state-space matrices A_o, B_o, C_o, D_o, K_o is presented. Considering the sequence of sampling instants t_i, $i = 1, \ldots, N$, the input u, the output y, and the innovation e of (13.7) are subjected to the transformations

$$\begin{aligned} \tilde{u}_i(k) &= \int_0^\infty \ell_k(\tau)u(t_i + \tau)d\tau \\ \tilde{y}_i(k) &= \int_0^\infty \ell_k(\tau)y(t_i + \tau)d\tau \\ \tilde{e}_i(k) &= \int_0^\infty \ell_k(\tau)e(t_i + \tau)d\tau \end{aligned} \tag{13.11}$$

(or to the equivalent ones derived from (13.8)), where $\tilde{u}_i(k) \in \mathbb{R}^m$, $\tilde{e}_i(k) \in \mathbb{R}^p$ and $\tilde{y}_i(k) \in \mathbb{R}^p$. Then (see [21] for details) the transformed system has the state-space representation

$$\begin{aligned} \xi_i(k+1) &= A_o\xi_i(k) + B_o\tilde{u}_i(k) + K_o\tilde{e}_i(k), \ \xi_i(0) = x(t_i) \\ \tilde{y}_i(k) &= C_o\xi_i(k) + D_o\tilde{u}_i(k) + \tilde{e}_i(k) \end{aligned} \tag{13.12}$$

where the state-space matrices are given by (13.10).

Letting now

$$\tilde{z}_i(k) = \begin{bmatrix}\tilde{u}_i^T(k) & \tilde{y}_i^T(k)\end{bmatrix}^T$$

and

$$\begin{aligned}\overline{A}_o &= A_o - K_o C_o \\ \overline{B}_o &= B_o - K_o D_o \\ \widetilde{B}_o &= \begin{bmatrix} \overline{B}_o & K_o \end{bmatrix}\end{aligned}$$

system (13.12) can be written in predictor form as

$$\begin{aligned}\xi_i(k+1) &= \overline{A}_o \xi_i(k) + \widetilde{B}_o \tilde{z}_i(k), \ \xi_i(0) = x(t_i) \\ \tilde{y}_i(k) &= C_o \xi_i(k) + D_o \tilde{u}_i(k) + \tilde{e}_i(k)\end{aligned} \tag{13.13}$$

to which the PBSID$_{opt}$ algorithm, summarized hereafter, can be applied to compute estimates of the state-space matrices A_o, B_o, C_o, D_o, K_o. To this purpose note that iterating $p-1$ times the projection operation (i.e., propagating $p-1$ forward in the index k the first of equations (13.13), where p is the so-called past window length) one gets

$$\begin{aligned}\xi_i(k+2) &= \overline{A}_o^2 \xi_i(k) + \begin{bmatrix} \overline{A}_o \widetilde{B}_o & \widetilde{B}_o \end{bmatrix} \begin{bmatrix} \tilde{z}_i(k) \\ \tilde{z}_i(k+1) \end{bmatrix} \\ &\ \ \vdots \\ \xi_i(k+p) &= \overline{A}_o^p \xi_i(k) + \mathcal{K}^p Z_i^{0,p-1}\end{aligned} \tag{13.14}$$

where

$$\mathcal{K}^p = \begin{bmatrix} \overline{A}_o^{p-1} \widetilde{B}_0 & \dots & \widetilde{B}_o \end{bmatrix} \tag{13.15}$$

is the extended controllability matrix of the system in the transformed domain and

$$Z_i^{0,p-1} = \begin{bmatrix} \tilde{z}_i(k) \\ \vdots \\ \tilde{z}_i(k+p-1) \end{bmatrix}$$

Under the considered assumptions, $\overline{A}_o$ has all the eigenvalues inside the open unit circle, so the term $\overline{A}_o^p \xi_i(k)$ is negligible for sufficiently large values of p and we have that

$$\xi_i(k+p) \simeq \mathcal{K}^p Z_i^{0,p-1}$$

As a consequence, the input-output behavior of the system is approximately given by

$$\begin{aligned} \tilde{y}_i(k+p) &\simeq C_o\mathcal{K}^p Z_i^{0,p-1} + D_o\tilde{u}_i(k+p) + \tilde{e}_i(k+p) \\ &\vdots \\ \tilde{y}_i(k+p+f) &\simeq C_o\mathcal{K}^p Z_i^{f,p+f-1} + D_o\tilde{u}_i(k+p+f) \\ &\quad + \tilde{e}_i(k+p+f) \end{aligned} \tag{13.16}$$

so that introducing the vector notation

$$\begin{aligned} Y_i^{p,f} &= \begin{bmatrix} \tilde{y}_i(k+p) & \tilde{y}_i(k+p+1) & \dots & \tilde{y}_i(k+p+f) \end{bmatrix} \\ U_i^{p,f} &= \begin{bmatrix} \tilde{u}_i(k+p) & \tilde{u}_i(k+p+1) & \dots & \tilde{u}_i(k+p+f) \end{bmatrix} \\ E_i^{p,f} &= \begin{bmatrix} \tilde{e}_i(k+p) & \tilde{e}_i(k+p+1) & \dots & \tilde{e}_i(k+p+f) \end{bmatrix} \\ \Xi_i^{p,f} &= \begin{bmatrix} \xi_i(k+p) & \xi_i(k+p+1) & \dots & \xi_i(k+p+f) \end{bmatrix} \\ \overline{Z}_i^{p,f} &= \begin{bmatrix} Z_i^{0,p-1} & Z_i^{1,p} & \dots & Z_i^{f,p+f-1} \end{bmatrix} \end{aligned} \tag{13.17}$$

(13.14) and (13.16) can be rewritten as

$$\begin{aligned} \Xi_i^{p,f} &\simeq \mathcal{K}^p \overline{Z}_i^{p,f} \\ Y_i^{p,f} &\simeq C_o\mathcal{K}^p \overline{Z}_i^{p,f} + D_o U_i^{p,f} + E_i^{p,f}. \end{aligned} \tag{13.18}$$

Considering now the entire dataset for $i = 1, \dots, N$, the data matrices become

$$\begin{aligned} Y^{p,f} &= [\tilde{y}_1(k+p)\dots\tilde{y}_N(k+p)\dots \\ &\qquad \tilde{y}_1(k+p+f)\dots\tilde{y}_N(k+p+f)] \end{aligned} \tag{13.19}$$

and similarly for $U_i^{p,f}$, $E_i^{p,f}$, $\Xi_i^{p,f}$ and $\overline{Z}_i^{p,f}$. The data equations (13.18), in turn, are given by

$$\begin{aligned} \Xi^{p,f} &\simeq \mathcal{K}^p \overline{Z}^{p,f} \\ Y^{p,f} &\simeq C_o\mathcal{K}^p \overline{Z}^{p,f} + D_o U^{p,f} + E^{p,f} \end{aligned} \tag{13.20}$$

From this point on, the algorithm can be developed along the lines of the discrete-time PBSID$_{opt}$ method, i.e., by carrying out the following steps. Considering $p = f$, estimates for the matrices $C_o\mathcal{K}^p$ and D_o are first computed by solving the least-squares problem

$$\min_{C_o\mathcal{K}^p, D_o} \|Y^{p,p} - C_o\mathcal{K}^p \overline{Z}^{p,p} - D_o U^{p,p}\|_F \tag{13.21}$$

where by $\|\cdot\|_F$ we denote the Frobenius norm of a matrix. Defining now the extended observability matrix Γ^p as

$$\Gamma^p = \begin{bmatrix} C_o \\ C_o\overline{A}_o \\ \vdots \\ C_o\overline{A}_o^{p-1} \end{bmatrix} \tag{13.22}$$

and noting that the product of Γ^p and $\mathscr{K}^p$ can be written as

$$\Gamma^p \mathscr{K}^p \simeq \begin{bmatrix} C_o\overline{A}^{p-1}\widetilde{B}_o & \dots & C_o\widetilde{B}_o \\ 0 & \dots & C_o\overline{A}\widetilde{B}_o \\ \vdots & & \\ 0 & \dots & C_o\overline{A}^{p-1}\widetilde{B}_o \end{bmatrix} \tag{13.23}$$

such product can be computed using the estimate $\widehat{C_o\mathscr{K}^p}$ of $C_o\mathscr{K}^p$ obtained by solving the least squares problem (13.21).

Recalling now that

$$\Xi^{p,p} \simeq \mathscr{K}^p \overline{Z}^{p,p} \tag{13.24}$$

it also holds that

$$\Gamma^p \Xi^{p,p} \simeq \Gamma^p \mathscr{K}^p \overline{Z}^{p,p} \tag{13.25}$$

Therefore, computing the singular value decomposition

$$\Gamma^p \mathscr{K}^p \overline{Z}^{p,p} = U \Sigma V^T \tag{13.26}$$

an estimate of the state sequence can be obtained as

$$\widehat{\Xi}^{p,p} = \Sigma_n^{1/2} V_n^T = \Sigma_n^{-1/2} U_n^T \Gamma^p \mathscr{K}^p \overline{Z}^{p,p} \tag{13.27}$$

from which, in turn, an estimate of C_o can be computed by solving the least squares problem

$$\min_{C_o} \| Y^{p,p} - \widehat{D}_o U^{p,p} - C_o \widehat{\Xi}^{p,p} \|_F \tag{13.28}$$

The final steps consist of the estimation of the innovation data matrix $E^{p,p}$

$$E^{p,p} = Y^{p,p} - \widehat{C}_o \widehat{\Xi}^{p,p} - \widehat{D}_o U^{p,p} \tag{13.29}$$

and of the entire set of the state-space matrices for the system in the transformed domain, which can be obtained by solving the least squares problem

$$\min_{A_o,B_o,K_o} \|\widehat{\Xi}^{p+1,p} - A_o\widehat{\Xi}^{p,p-1} - B_oU^{p,p-1} - K_oE^{p,p-1}\|_F \tag{13.30}$$

The state-space matrices of the original continuous-time system can then be retrieved by inverting the (bilinear) transformations (13.10).

13.3.2 From unstructured to structured models with an $\mathscr{H}_\infty$ approach

Suppose that the linear continuous-time time-invariant system $\mathscr{M}_{ns}$ has been estimated from a dataset of sampled input/output data using the CT-PBSID$_o$ algorithm presented in the previous section. Consider now the model class $\mathscr{M}_s(\lambda)$ introduced in Section 13.1. $\mathscr{M}_{ns}$ and $\mathscr{M}_s(\lambda)$ should have the same input-output behavior. This problem can be faced in a computationally effective way by defining the input-output operators associated with $\mathscr{M}_{ns}$ and $\mathscr{M}_s(\lambda)$ and seeking the values of the parameters corresponding to the solution of the optimisation problem

$$\lambda^\star = \arg\min_{\lambda} \|\mathscr{M}_{ns} - \mathscr{M}_s(\lambda)\| \tag{13.31}$$

for a suitably chosen norm. In the linear time-invariant case, the input-output operators can be represented as the transfer functions $\hat{G}_{ns}(s)$ and $G_s(s;\lambda)$ and the $\mathscr{H}_\infty$ norm is considered, so that the model matching problem can be recast as

$$\lambda^\star = \arg\min_{\lambda} \|\hat{G}_{ns}(s) - G_s(s;\lambda)\|_\infty \tag{13.32}$$

Note that the open-loop dynamics of a helicopter is unstable in most flight conditions and so the $\mathscr{H}_\infty$ norm is undefined. In this case the eigenvalues of $\mathscr{M}_s(\lambda)$ and $\mathscr{M}_{ns}$ are shifted on the real axis by a suitable value μ as follows

$$\tilde{G}_s(s;\lambda) = C(\lambda)((s-\mu)I - A(\lambda))^{-1}B(\lambda) + D(\lambda) \tag{13.33}$$

$$\tilde{G}_{ns}(s) = \hat{C}((s-\mu)I - \hat{A})^{-1}\hat{B} + \hat{D} \tag{13.34}$$

where μ is chosen such that all eigenvalues of $\mathscr{M}_{ns}$ have negative real part. Then the model matching problem is reformulated as

$$\lambda^\star = \arg\min_{\lambda} \|\tilde{G}_{ns}(s) - \tilde{G}_s(s;\lambda)\|_\infty \tag{13.35}$$

As mentioned in the Introduction, this is a non-convex, non-smooth optimization problem, which has been studied extensively in recent years in the framework of the fixed-structured controller design problem and for which reliable computational tools (see [12]) are presently available.

13.4 Bootstrap uncertainty estimation in subspace identification methods

In this section a procedure for the evaluation of the uncertainty associated with invariants such as the frequency response of the estimated models is proposed. The statistical tool we will resort to is the bootstrap method [11, 24], along the lines of the results in [9]. The bootstrap is a computational statistical method which was originally introduced in order to solve the following problem ([11, 24]): given a random, independent, identically distributed (i.i.d.) sample $x = (x_1, x_2, \dots, x_n)$ drawn from an unknown distribution F, one computes an estimate $\hat{\theta}$ of the parameter $\theta = t(F) = t[x]$ on the basis of the available data, and would like to assess the accuracy of the obtained estimate, in terms of its standard deviation or its variance.

Various approaches have been proposed in order to apply the bootstrap for variance estimation in time series analysis ([24]), signal processing ([23]), and system identification ([9, 27]).

For the present purposes and with specific reference to the problem of evaluating the standard deviation for the frequency response of the estimated model, the method of bootstrapping residuals can be synthesized as follows:

1. Estimate the linear model $[\hat{A}, \hat{B}, \hat{C}, \hat{D}, \hat{K}]$ from the available input/output data (u, y) and compute the estimate for the points of interest of its frequency response $\hat{G}(j\omega_k)$, $k = 1, \dots, N$.
2. Compute $e(t) = y(t) - \hat{y}(t)$.
3. Obtain an estimate $\hat{F}_e$ for the distribution F_e of the prediction error. In this work a parametric estimate will be considered and the normality assumption for the distribution of $e(t)$ will be made.
4. Generate B replications $(u^{*(i)}, y^{*(i)})$ of the original data set (u, y), with $u^{*(i)} = u$ and $y^{*(i)}$ obtained by feeding the identified model $[\hat{A}, \hat{B}, \hat{C}, \hat{D}, \hat{K}]$ with the deterministic input $u^{*(i)} = u$ and the stochastic input $e^{*(i)}$, $i = 1, \dots, B$ where $e^{*(i)}$ is constructed by resampling (with replacement) from the distribution $\hat{F}_e$.
5. Estimate B replications of the identified model and of the points of interest for the frequency response $\hat{G}^{*(i)}(j\omega_k)$, $k = 1, \dots, N$.
6. The estimate of the standard error for the frequency response of the model is finally given by

$$\hat{\sigma}^*_{\hat{G}(j\omega_k)} = \frac{1}{\sqrt{B-1}} \left(\sum_{i=1}^{B} (\hat{G}^{*(i)}(j\omega_k) - \overline{\hat{G}}^*(j\omega_k))^2 \right)^{\frac{1}{2}} \tag{13.36}$$

where

$$\overline{\hat{G}}^*(j\omega_k) = \frac{1}{B} \sum_{i=1}^{B} \hat{G}^{*(i)}(j\omega_k) \tag{13.37}$$

In a similar way one can obtain estimates of the standard deviation for the poles and zeros of the estimated model.

13.5 Simulation study: model identification for the BO-105 helicopter

The simulation example considered in this chapter is based on a model of the dynamics of the BO-105 helicopter (see [25]). The BO-105 is a light, twin-engine, multi-purpose utility helicopter, which is possibly the most studied helicopter in the rotorcraft system identification literature. It is here considered in forward flight at 80 knots, a flight condition which corresponds to unstable dynamics, with the aim of demonstrating the identification of a nine-DOF state-space model with test data extracted from a simulator based on the nine-DOF model from [25]. As described in the cited reference, the model includes the classical six-DOF and some additional states to account for some additional effects, namely:

- The BO-105 exhibits highly coupled body-roll and rotor-flapping responses; their interaction is represented in the model with a dynamic equation that describes the flapping dynamics using the cyclic controls.
- A second order dipole is appended to the model of roll rate response to lateral stick in order to account for the effect of lead-lag rotor dynamics.

Therefore, the simulator includes a nine-DOF linear model including the six-DOF quasi steady dynamics, the flapping equations and the lead-lag dynamics modelled with a complex dipole. Delays at the input of the model are also taken into account in the simulation, though they are not estimated. The state vector and the trim values are

$$x = \begin{bmatrix} u & v & w & p & q & r & \phi & \theta & a_{1s} & b_{1s} & x_1 & x_2 \end{bmatrix}$$

and, respectively,

$$u_0 = 40 \text{ m/s}, \quad v_0 = 3 \text{ m/s}, \quad w_0 = -5 \text{ m/s}, \quad \phi_0 = 0 \text{ rad}, \quad \theta_0 = 0 \text{ rad} \tag{13.38}$$

In details, the state vector includes the longitudinal flapping a_{1s}, the lateral flapping b_{1s} and two state variables x_1 and x_2, coming from the lead-lag dynamics complex dipole. The corresponding equations of motion are

$$\begin{aligned}
\dot{u} &= X_u u + X_w w + X_p p + (X_q - w_0)q + v_0 r - g\theta + X_{\delta_{lon}}\delta_{lon} + X_{\delta_{col}}\delta_{col} \\
\dot{v} &= Y_v v + Y_w w + (Y_p + w_0)p + Y_q q + (Y_r - u_0)r - g\phi + Y_{\delta_{lat}}\delta_{lat} + Y_{\delta_{col}}\delta_{col} \\
\dot{w} &= Z_u u + Z_w w + (Z_p - v_0)p + u_0 q + Z_{\delta_{col}}\delta_{col} \\
\dot{p} &= L_u u + L_v v + L_w w + L_{\overline{q}} q + L_{\delta_{b_{1s}}} b_{1s} + L_{\delta_{lon}}\delta_{lon} + L_{\delta_{ped}}\delta_{ped} + L_{\delta_{col}}\delta_{col}
\end{aligned} \tag{13.39}$$

$$\dot{q} = M_v v + M_w w + M_p p + M_r r + M_{\delta_{a_{1s}}} a_{1s} + M_{\delta_{ped}} \delta_{ped} + M_{\delta_{col}} \delta_{col}$$

$$\dot{r} = N_v v + N_w w + N_q q + N_r r + N_{\delta_{lon}} \delta_{lon} + N_{\delta_{lat}} \delta_{lat} + N_{\delta_{ped}} \delta_{ped} + N_{\delta_{col}} \delta_{col}$$

$$\dot{\phi} = p$$

$$\dot{\theta} = q$$

$$\dot{a}_{1s} = -q - \frac{1}{\tau_f} a_{1s} + \frac{K_{a_{1s}}}{\tau_f} \delta_{lon},$$

$$\dot{b}_{1s} = -p - \frac{1}{\tau_f} b_{1s} + \frac{K_{b_{1s}}}{\tau_f} \delta_{lat} + K_{x_1} x_1 + K_{x_2} x_2$$

$$\dot{x}_1 = x_2$$

$$\dot{x}_2 = C_1 x_1 + C_2 x_2 + \delta_{lat}$$

Finally, the output vector is

$$y = \begin{bmatrix} u & v & w & p & q & r & a_x & a_y & a_z & \phi & \theta \end{bmatrix}$$

where

$$\begin{aligned} a_x &= \dot{u} + w_0 q - v_0 r + g\theta \\ a_y &= \dot{v} - w_0 p + u_0 r - g\phi \\ a_z &= \dot{w} + v_0 p - u_0 q \end{aligned}$$

i.e., the state variables related to quasi-steady dynamics and the linear accelerations are measured. Considering (13.39), λ contains the stability derivatives, the control derivatives, the flapping and lead-lag rotor dynamics parameters, for a total of 47 parameters.

The identification experiment has been carried out under closed-loop conditions, because of the instability of the model, with the helicopter operating under feedback from an LQG controller tuned in order to maintain the helicopter close enough to trim to justify the identification of a linear model. In the identification experiment, additive perturbations have been applied to the input variables (δ_{lat}, δ_{lon}, δ_{ped}, δ_{col}) computed by the controller; in particular, all the channels have been excited in the same experiment with pseudorandom binary signals with a duration of 60 s and a dwell time of 0.8 s. The perturbation of the control inputs has a 1% amplitude and the sampling time is 0.008 s. For the purpose of the present preliminary study, measurement noise has not been included in the simulated data. The parameters of the SMI algorithm presented in the previous section have been chosen as $p = 40$ and $a = 45$. The obtained results, in terms of a comparison between the eigenvalues of the simulation models and the ones of the identified black-box model, are illustrated in Table 13.1.

Table 13.1 Comparison between simulator and black-box identified eigenvalues

	Simulator				Identified model (CT-PBSID$_o$)			
	Real	**Imag**	**Omega**	**Zeta**	**Real**	**Imag**	**Omega**	**Zeta**
Pitch phugoid	0.119	0.278	0.302	−0.394	0.119	0.278	0.302	−0.394
Dutch roll	−0.571	2.546	2.609	0.219	−0.571	2.546	2.609	0.219
Roll/flapping	−9.904	7.740	12.569	0.788	−9.901	7.7399	12.568	0.788
Lead-Lag	−0.868	15.567	15.592	0.0557	−0.867	15.566	15.590	0.0556
Spiral	−0.0510				−0.0507			
Pitch$_1$	−0.448				−0.448			
Pitch$_2$	−5.843				−5.844			
Long. flapping	−15.930				−15.901			

Table 13.2 Comparison between simulator and gray-box identified eigenvalues

	Simulator				Identified model (Gray-box)			
	Real	**Imag**	**Omega**	**Zeta**	**Real**	**Imag**	**Omega**	**Zeta**
Pitch phugoid	0.119	0.278	0.302	−0.394	0.119	0.278	0.302	−0.394
Dutch roll	−0.571	2.546	2.609	0.219	−0.571	2.546	2.609	0.219
Roll/flapping	−9.904	7.740	12.569	0.788	−9.903	7.740	12.568	0.788
Lead-Lag	−0.868	15.567	15.592	0.0557	−0.868	15.566	15.590	0.557
Spiral	−0.0510				−0.0507			
Pitch$_1$	−0.448				−0.448			
Pitch$_2$	−5.843				−5.843			
Long. flapping	−15.930				−15.929			

As can be seen from the table, the CT-PBSID$_o$ algorithm is able to identify the dynamics of the system very accurately, with just a slight loss of accuracy at high frequency.

The second step of the proposed identification approach requires to fit a structured model to the black-box one obtained by means of the SMI approach. The study in the reconstruction of the structured state-space representation has been carried out by applying the approach presented in Section 13.3 to estimate the relevant parameters. In order to evaluate the performance of the proposed method the relative estimation error is defined as

$$\lambda_{err} = \frac{\lambda^0 - \bar{\lambda}}{\lambda^0} \tag{13.40}$$

where $\hat{\lambda}$ and λ^0 are, respectively, the estimated and the actual value of the parameter λ. In this example the relative errors of the estimated physical parameters in (13.39) are below 0.03%. It is clear from Table 13.2, where the eigenvalues of the real system and the identified gray-box model are shown, that using *a-priori* information, i.e., exploting the model structure, the estimation accuracy increases.

A time-domain validation of the identified models has been also carried out, by measuring the simulation accuracy of the models in response to a doublet input signal on each input channel. The input sequence used in the validation experiment is illustrated in Figure 13.1, while the time history for two of the outputs (u and w) is presented in Figure 13.2. Again, even though the open-loop system is unstable, the simulated outputs obtained from the identified models (dashed lines: black-box; cross: gray-box) match very well the ones computed from the nine-DOF model (solid lines).

In quantitative terms, considering the root mean square error, defined as

$$RMS = \sqrt{\frac{1}{N}\sum_{i=1}^{N}\left(y(i) - \hat{y}(i)\right)^2} \tag{13.41}$$

where y is the real output and $\hat{y}$ is the estimated one, its value is below 0.01 on all the considered output variables as shown in Table 13.3. Note that most of the error is due the unestimated input delays, as can be seen in Figure 13.2.

Finally, a frequency-domain analysis of model accuracy has been carried out, by comparing the results of a Monte Carlo study and the corresponding ones obtained by applying the bootstrap method described in the previous section. The results are

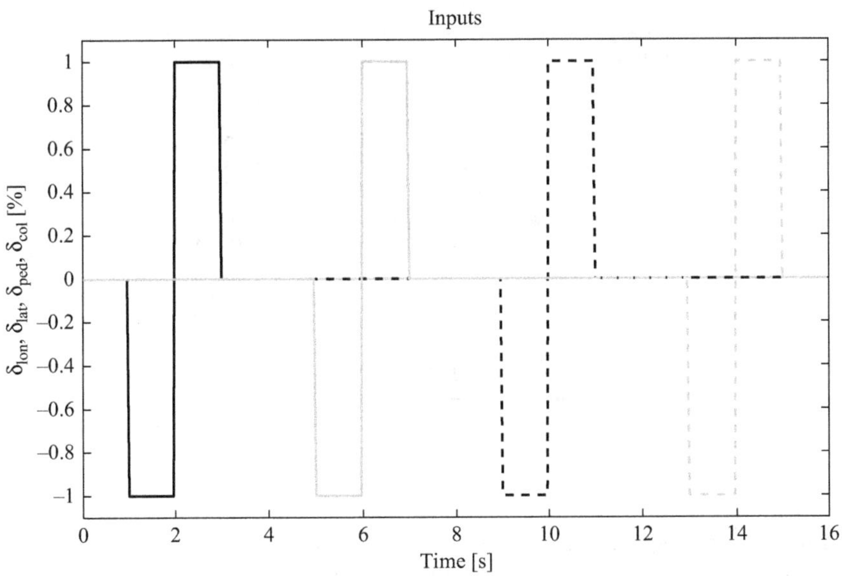

Figure 13.1 Doublet input signal used for model validation

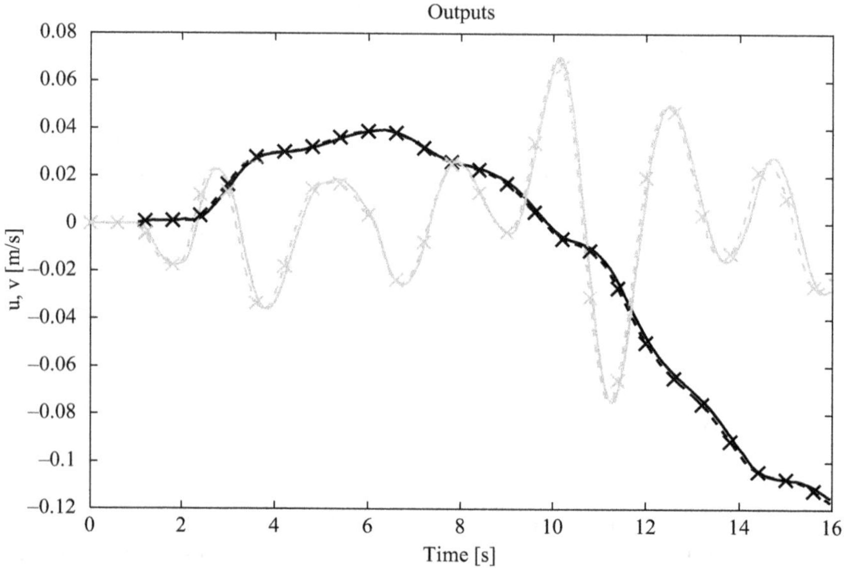

Figure 13.2 Doublet output signals (real: solid line; black-box: dashed line; gray-box: cross)

Table 13.3 Relative errors norm

Output	$RMS_{CT\text{-}PBSID_o}$	$RMS_{Gray\text{-}Box}$
u	0.0013	0.0013
v	0.0044	0.0044
w	0.0026	0.0026
p	0.0002	0.0002
q	0.0003	0.0003
r	0.0003	0.0003
a_x	0.0013	0.0013
a_y	0.0017	0.0017
a_z	0.0077	0.0077
ϕ	0.0001	0.0001
θ	0.0001	0.0001

presented in Figures 13.3–13.10, in which the magnitude of the frequency response of the identified model is depicted, together with the uncertainty regions computed by means of a Monte Carlo approach and of the bootstrap method described in the previous section. As can be seen from the figures, the agreement between the bootstrap results and the "true" Monte Carlo ones is very satisfactory.

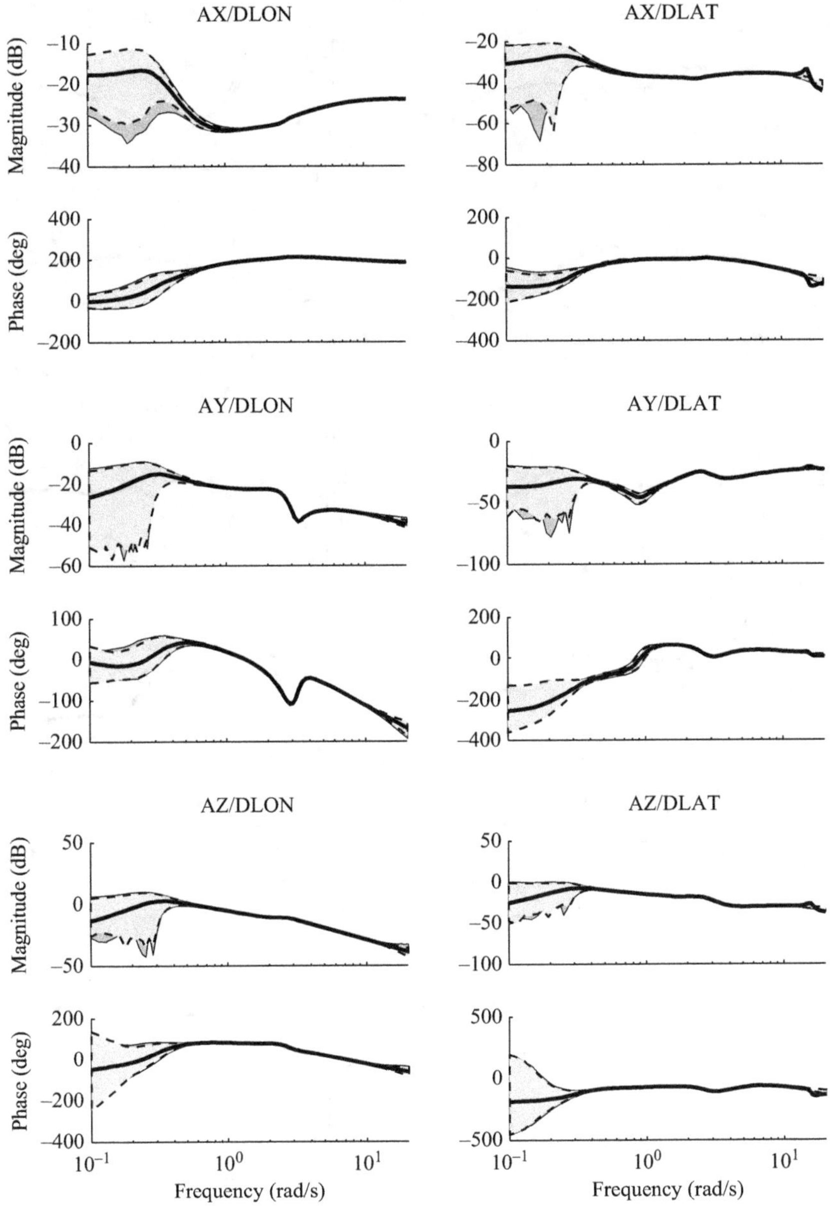

Figure 13.3 Frequency responses from longitudinal and lateral inputs to linear accelerations. (bootstrap method (light-gray) and Monte Carlo method (gray))

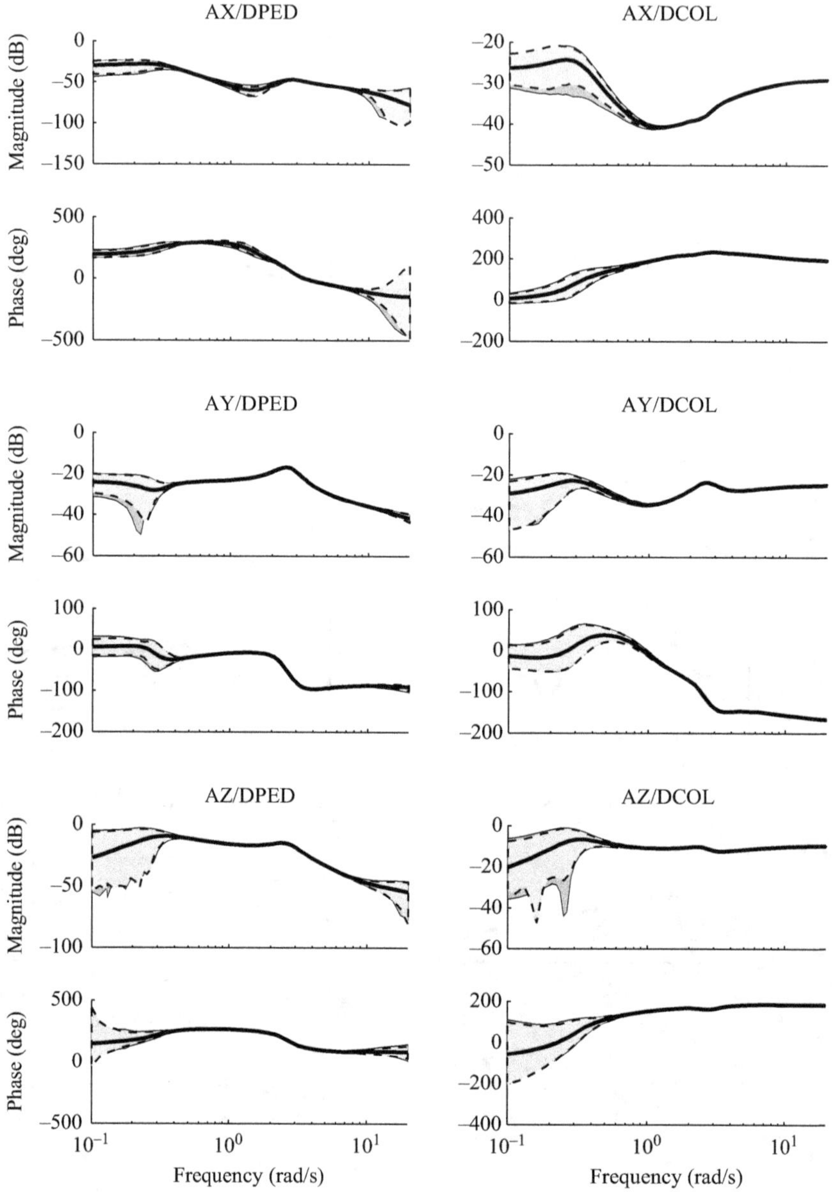

Figure 13.4 Frequency responses from pedal and collective inputs to linear accelerations. (bootstrap method (light-gray) and Monte Carlo method (gray))

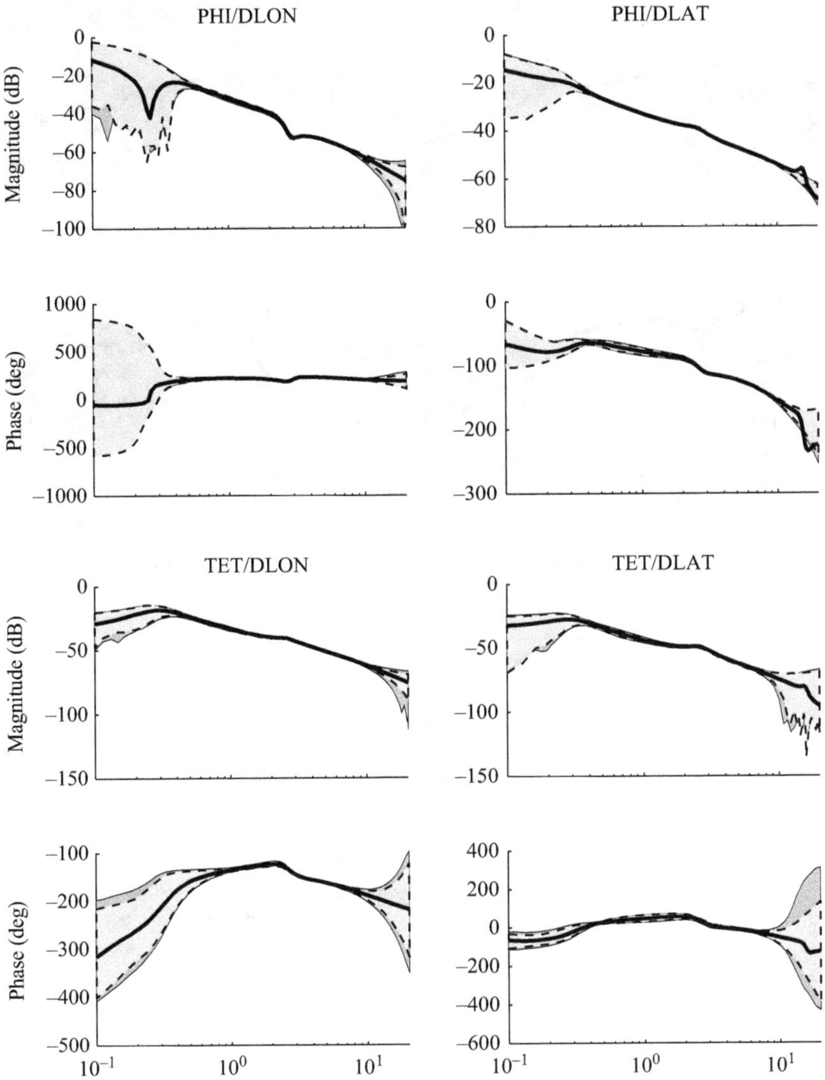

Figure 13.5 *Frequency responses from longitudinal and lateral inputs to roll and pitch angles. (bootstrap method (light-gray) and Monte Carlo method (gray))*

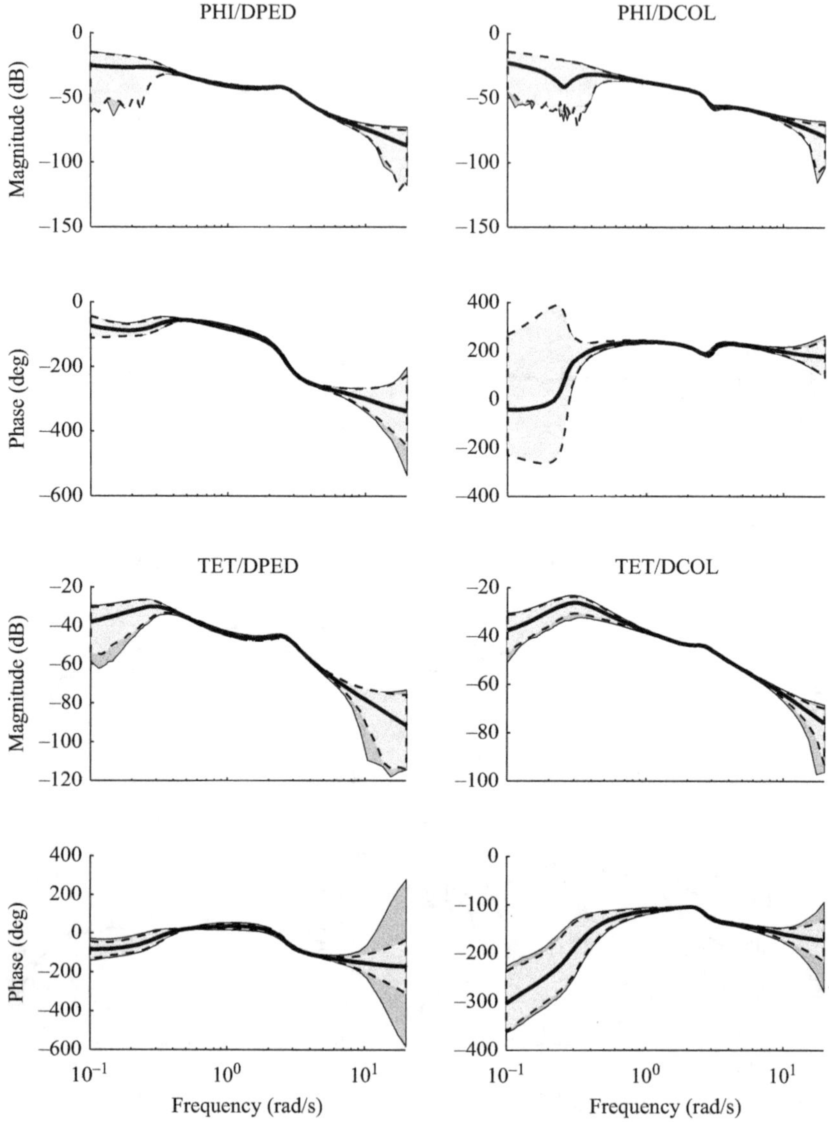

Figure 13.6 Frequency response from pedal and collective inputs to roll and pitch angles. (bootstrap method (light-gray) and Monte Carlo method (gray))

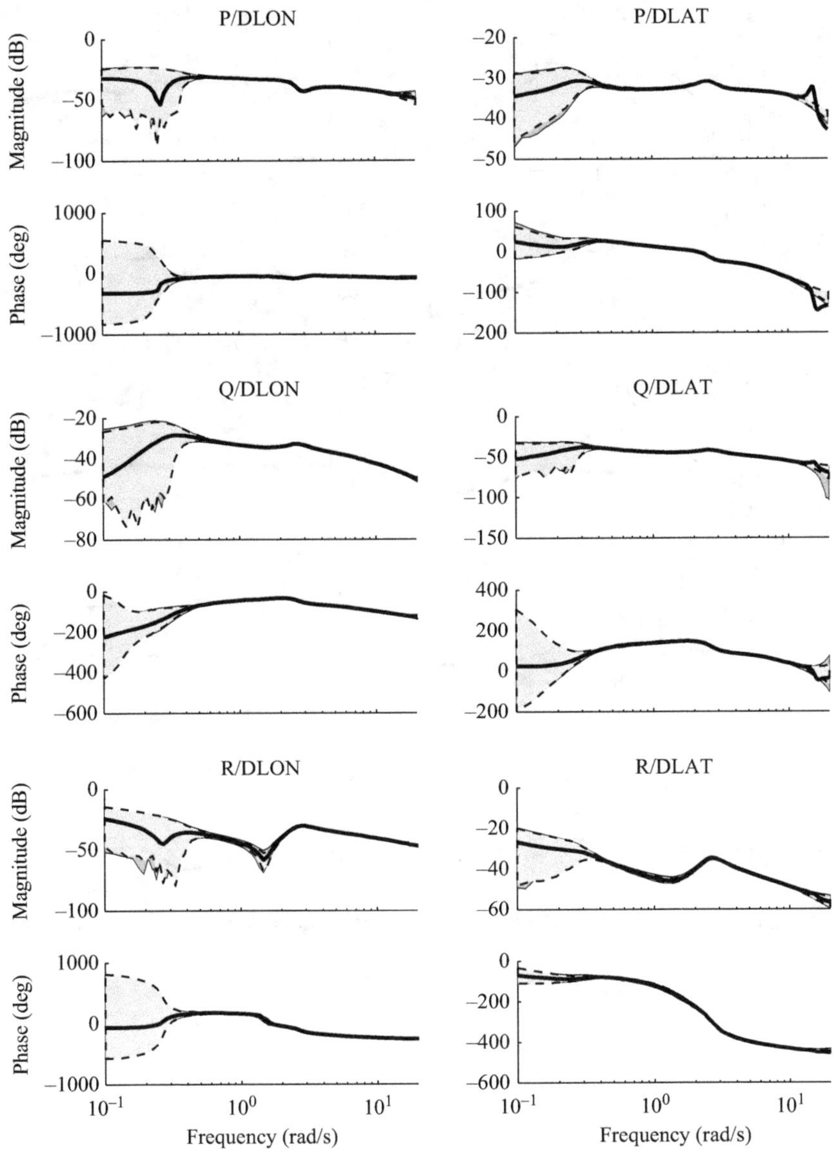

Figure 13.7 Frequency responses from longitudinal and lateral inputs to body angular rates. (bootstrap method (light-gray) and Monte Carlo method (gray))

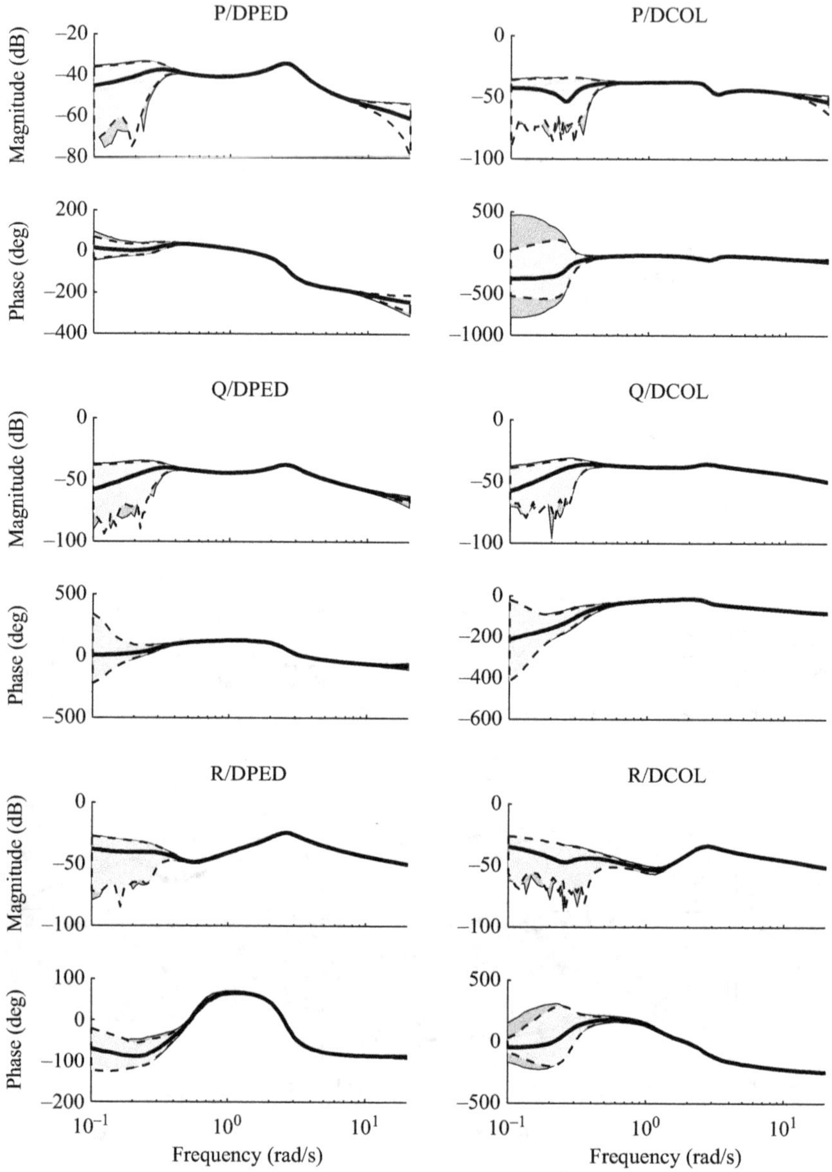

Figure 13.8 Frequency response from pedal and collective inputs to body angular rates. (bootstrap method (light-gray) and Monte Carlo method (gray))

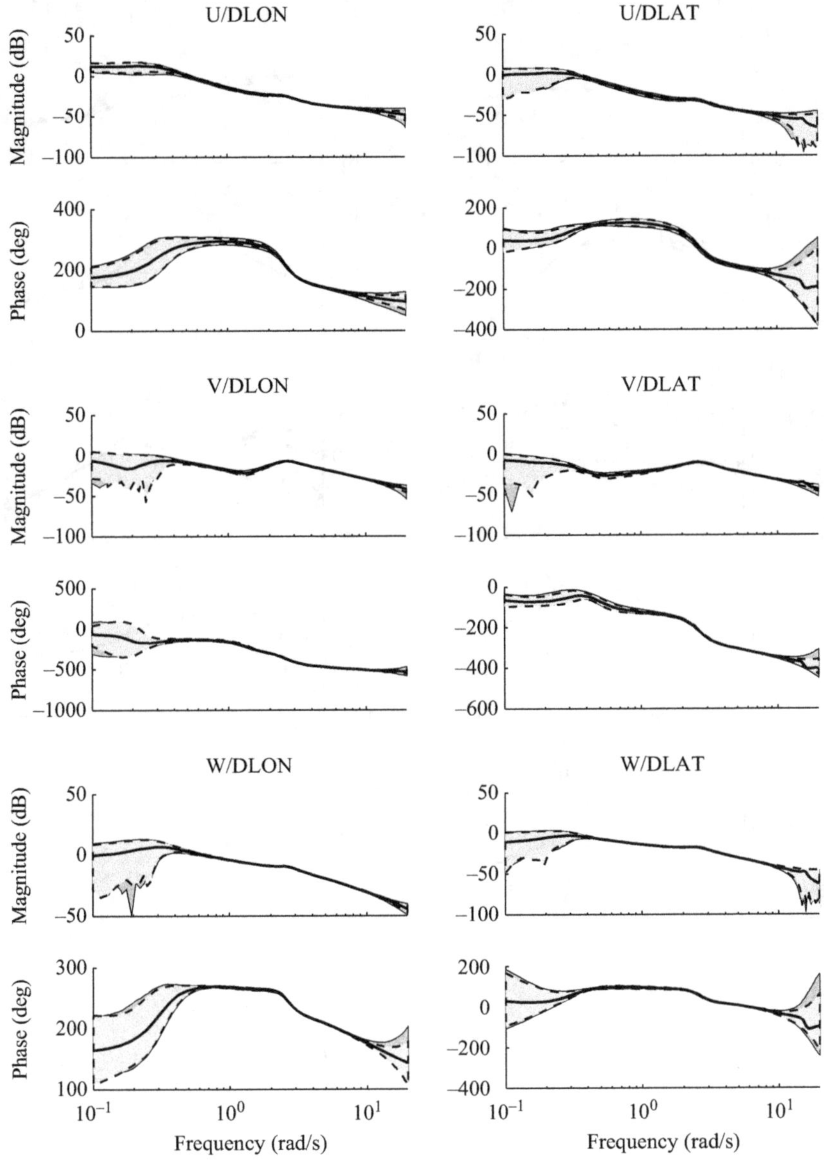

Figure 13.9 Frequency responses from longitudinal and lateral inputs to linear body rates. (bootstrap method (light-gray) and Monte Carlo method (gray))

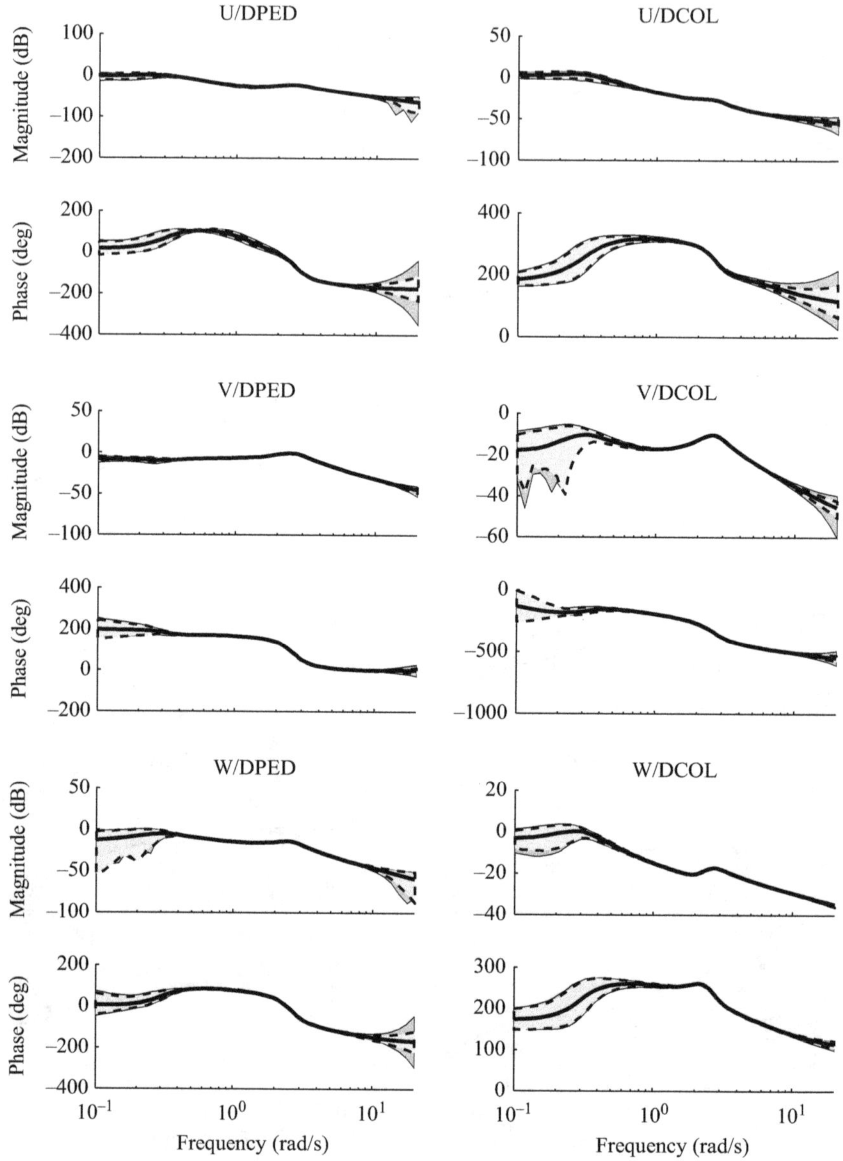

Figure 13.10 Frequency response from pedal and collective inputs to linear body rates. (bootstrap method (light-gray) and Monte Carlo method (gray))

13.6 Concluding remarks

The problem of rotorcraft system identification has been considered and a two-step technique combining time-domain and frequency-domain methods has been proposed. A simulation study based on a model of the BO-105 helicopter has been used to illustrate the proposed approach. Simulation results show that the proposed schemes are viable for rotorcraft applications and provide accurate models on the basis of data generated during closed-loop experiments. The bootstrap approach for uncertainty analysis in the identified models also provides realiable estimates of the frequency distribution of modeling error for the identified model.

References

[1] Apkarian, P., Noll, D.: Nonsmooth H_∞ synthesis. *IEEE Transactions on Automatic Control* **51**(1), 71–86 (1996)

[2] Bergamasco, M.: Continuous-time model identification with applications to rotorcraft dynamics. Ph.D. thesis, Politecnico di Milano (2012)

[3] Bergamasco, M., Lovera, M.: Continuous-time subspace identification in closed-loop. In: 19th International Symposium on Mathematical Theory of Networks and Systems, Budapest, Hungary (2010)

[4] Bergamasco, M., Lovera, M.: Continuous-time predictor-based subspace identification for helicopter dynamics. In: 37th European Rotorcraft Forum, Gallarate, Italy (2011)

[5] Bergamasco, M., Lovera, M.: Continuous-time predictor-based subspace identification using Laguerre filters. *IET Control Theory and Applications* **5**(7), 856–867 (2011). Special issue on Continuous-time Model Identification

[6] Bergamasco, M., Lovera, M.: Rotorcraft system identification: an integrated time-frequency domain approach. In: 2nd CEAS Specialist Conference on Guidance, Navigation & Control, Delft, The Nederlands (2013)

[7] Bergamasco, M., Lovera, M.: State space model identification: from unstructured to structured models with an H_∞ approach. In: 5th IFAC Symposium on System Structure and Control, Grenoble, France (2013)

[8] Bittanti, S., Lovera, M.: Identification of linear models for a hovering helicopter rotor. In: Proceedings of the 11th IFAC Symposium on system identification, Fukuoka, Japan (1997)

[9] Bittanti, S., Lovera, M.: Bootstrap-based estimates of uncertainty in subspace identification methods. *Automatica* **36**(11), 1605–1615 (2000)

[10] Chiuso, A., Picci, G.: Consistency analysis of certain closed-loop subspace identification methods. *Automatica* **41**(3), 377–391 (2005)

[11] Efron, B., Tibshirani, R.: *An introduction to the bootstrap*. Chapman and Hall/CRC Press, Boca Raton, Florida, USA (1994)

[12] Gahinet, P., Apkarian, P.: Decentralized and fixed-structure H_∞ control in MATLAB. In: 50th IEEE Conference on Decision and Control and European Control Conference, Orlando, USA (2011)

[13] Huang, B., Ding, S., Qin, S.: Closed-loop subspace identification: an orthogonal projection approach. *Journal of Process Control* **15**(1), 53–66 (2005)

[14] Jategaonkar, R.: Flight Vehicle System Identification. AIAA, New York, USA (2006)

[15] Johansson, R., Verhaegen, M., Chou, C.: Stochastic theory of continuous-time state-space identification. *IEEE Transactions on Signal Processing* **47**(1), 41–51 (1999)

[16] Kinoshita, Y., Ohta, Y.: Continuous-time system identification using compactly supported filter kernels generated from Laguerre basis functions. In: 49th IEEE Conference on Decision and Control, Atlanta, USA (2010)

[17] Li, P., I., P.: Subspace and bootstrap-based techniques for helicopter model identification. *Journal of the American Helicopter Society* **56**(1), 012,002 (2011)

[18] Lovera, M.: Identification of MIMO state space models for helicopter dynamics. In: 13th IFAC Symposium on System Identification, Rotterdam, The Nederlands (2003)

[19] Ohta, Y.: Realization of input-output maps using generalized orthonormal basis functions. *Systems & Control Letters* **22**(6), 437–444 (2005)

[20] Ohta, Y.: System transformation of unstable systems induced by a shift-invariant subspace. In: 50th IEEE Conference on Systems and Control (2011). Submitted

[21] Ohta, Y., Kawai, T.: Continuous-time subspace system identification using generalized orthonormal basis functions. In: 16th International Symposium on Mathematical Theory of Networks and Systems, Leuven, Belgium (2004)

[22] Panza, S., Lovera, M.: Rotor state feedback in helicopter flight control: robustness and fault tolerance. In: IEEE Multi-Conference on Systems and Control, Antibes, France (2014)

[23] Shamsunder, S.: Signal processing applications of the bootstrap. *IEEE Signal Processing Magazine* **15**, 38 (1998)

[24] Shao, J., Tu, D.: *The jackknife and bootstrap*. Springer (1995)

[25] Tischler, M., Cauffman, M.: Frequency-response method for rotorcraft system identification: Flight applications to BO-105 coupled rotor/fuselage dynamics. *Journal of the American Helicopter Society* **37**(3), 3–17 (1992)

[26] Tischler, M., Remple, R.: *Aircraft and Rotorcraft System Identification: Engineering Methods with Flight-test Examples*. AIAA (2006)

[27] Tjarnstrom, F., Ljung, L.: Using the bootstrap to estimate the variance in the case of undermodeling. *IEEE Transactions on Automatic Control* **47**(2), 395–398 (2002)

[28] Van Overschee, P., De Moor, B.: Subspace identification: theory, implementation, application. Kluwer Academic Publishers, Dordrecht, The Netherlands (1996)

[29] Verhaegen, M.: Identification of the deterministic part of MIMO state space models given in innovations form from input-output data. *Automatica* **30**(1), 61–74 (1994)

[30] Verhaegen, M., Varga, A.: Some experience with the MOESP class of subspace model identification methods in identifying the BO105 helicopter. Tech. Rep. TR R165-94, DLR (1994)

[31] Verhaegen, M., Verdult, V.: Filtering and System Identification: A Least Squares Approach. Cambridge University Press, Cambridge, UK (2007)

[32] Zhou, K., Doyle, J., Glover, K.: Robust and optimal control. Prentice Hall, Upper Saddle River, New Jersey, USA (1996)

Chapter 14
Parameter identification of a reduced order LFT model of anaerobic digestion

Alessandro Della Bona[1], *Gianni Ferretti*[2], *Elena Ficara*[3] *and Francesca Malpei*[4]

Abstract

Owing to their complexity, accurate and detailed models of anaerobic digestion cannot be used for online monitoring and control. To this aim reduced order models have to be considered. In this chapter, a modification of the well-known AMOCO model is first proposed in order to widen its field of applicability. Then, to perform parameter identification, a linear fractional transformation (LFT) formulation is derived, thanks to the use of a symbolic manipulation tool applied to an object-oriented model formulation. The approach has been applied to two case tests: in the first test, the data used for identification have been generated by a simulation of the fully detailed Anaerobic Digestion Model no. 1 (ADM1) model, assuming waste activated sludge as influent substrate, and in the second, the data have been collected on a real plant, used for anaerobic digestion of agricultural wastes.

14.1 Introduction

Modelling of anaerobic digestion processes is of fundamental importance not only in order to design wastewater treatment and biogas power plants but also to study the sensitivity of the plant behaviour to operational parameters, to monitor and control the plant performance and to assess to feasibility of the use of new substrates with varying characteristics, biodegradability and operational conditions.

In the literature, many anaerobic digestion models have been proposed for specific applications or fermenters, fed with a very specific substrate, which can be roughly grouped into three main categories. The simplest ones [14] are single-step

[1] Politecnico di Milano, Milano, Italy, e-mail: alessandro.dellabona@polimi.it
[2] Politecnico di Milano, Milano, Italy, e-mail: gianni.ferretti@polimi.it
[3] Politecnico di Milano, Milano, Italy, e-mail: elena.ficara@polimi.it
[4] Politecnico di Milano, Milano, Italy, e-mail: francesca.malpei@polimi.it

models involving a single bacterial population, endowed with a limited description of inhibition effects. Models of intermediate complexity [3] consider a higher number of processes and bacterial populations, as well as a more accurate description and implementation of inhibition factors. Finally, complex models [2, 24] take into account a large number of processes and specific bacterial populations, along with an in-depth description of the inhibition effects and of relevant chemical equilibria.

The best known and the most sophisticated model, able to describe the anaerobic degradation of various substrates (even if designed considering activated sludge as substrate), is the Anaerobic Digestion Model no. 1 (ADM1) [2], developed by the IWA Task Group for Mathematical Modelling and later modified by several authors [4, 13] to improve accuracy and robustness and to fit the model to other specific applications. While being very detailed in the description of the anaerobic digestion process, it can be hardly used for design and control purposes. In fact, a large number of parameters (about a hundred), depending on the specific substrate, need to be estimated, which is particularly difficult in complex plant operations and also because of the scarce data available in the literature.

Moreover, many parameters of the ADM1 model cannot be identified on the basis of field data typically included in conventional protocols applied in digesters operation and monitoring. Therefore, dedicated measurement campaigns are to be planned in order to get the required characterization of the chemical nature of the substrate to be degraded and its specific degradation pathways. The campaigns are complex and time consuming, although valuable when seeking a deeper process understanding.

This fact has motivated the research of simpler models, focused for example on a few number of processes or specifically designed for particular substrates. Among them, the most important is the AMOCO model [3], which reaches a good compromise between simplicity and accuracy. The AMOCO model has been developed mainly as a tool to monitor and control the anaerobic digestion process, rather than as a tool for accurate numerical simulation. Nonetheless, this model has been conceived to describe the degradation of soluble organic matter that is readily acidified under anaerobic conditions, and is not adequate to be applied to the degradation of particulate matter that needs an initial disintegration and hydrolysis prior to be acidified. Therefore, in order to widen its field of applicability, a modified AMOCO model has been proposed including a first-order hydrolysis step, during which alkalinity is produced because of the release of ammonium from protein hydrolysis [1].

While being of reduced order, the modified AMOCO model is however still non-linear; therefore, the application of a maximum likelihood approach for the identification of its parameters, considered in this work, results in a non-linear least-squares minimization problem, which can be solved through non-linear programming. In the framework of non-linear programming methods a main problem must be anyway faced, namely the need of gradient and/or Hessian computations [7], performed in general via computationally inefficient finite-difference approximations. On the other hand, if an equivalent linear fractional transformation (LFT) formulation of the original non-linear model is considered, the said quantities can be very efficiently computed by simply simulating index-1, semi-explicit DAE systems. However, manually deriving an LFT model formulation is a non-trivial task in general, so symbolic computing tools are needed. In this work, one of these tools, discussed in detail in Chapter 4 of this book, has been fully implemented for application to the general non-linear case and, in particular, to the modified AMOCO model.

Two different test cases have been considered. In a first case, the data used for parameter identification have been generated by simulating the ADM1 model, considering waste-activated sludge as substrate. A-first set of (simulated) data has been considered for the identification of six parameters: five stoichiometric coefficients and the specific hydrolysis rate, while a second set of data has been used for model validation, obtained by simulating a nominal loading operation. In the second test case, an experimental data set collected on a real plant, used for anaerobic digestion of agricultural wastes, has been considered for the identification of two parameters: the half-saturation constant relevant to the volatile fatty acids and again the specific hydrolysis rate.

The chapter is organized as follows. In section 14.2 the ADM1 model is fully described, while the modified AMOCO model proposed is described in section 14.3. In section 14.4 the identification of a general LFT model is outlined, while briefly mentioning the tool used for the automatic generation of an LFT formulation from an object-oriented (O-O) model description. In section 14.5 the identification of the modified AMOCO model based on data obtained by simulation of the ADM1 model is described, while in section 14.6 the results of the identification performed on experimental data collected on a real plant is discussed. Section 14.7 concludes together with some hints for future research.

14.2 ADM1 model

In this work, the ADM1 model version proposed in [4] and [23] is considered, in the case of a single stage CSTR with constant liquid volume V_l (m^3) and temperature T (K) and with no biomass retention.

The model consists of a DAE system of 35 differential and 1 algebraic equation; 29 state variables are given by the concentrations in the liquid outflow and in the gas outflow and are listed in Table 14.1, all expressed in (kgCOD m^{-3}), except for concentrations of total inorganic carbon S_{IC}, total inorganic nitrogen S_{IN}, cations S_{cat}, anions S_{an}. Concentrations in Table 14.2 are expressed in (kmol m^{-3}). The other six variables are given by the concentrations of ionized volatile fatty acids (VFAs) (kgCOD m^{-3}), bicarbonate (kmolC m^{-3}) and free ammonia (kmolN m^{-3}) and are reported in Table 14.3. The algebraic variable is the hydrogen ion concentration S_H (kmol m^{-3}).

The differential equations are given by the mass balances of the dynamic (state) variables, and involve 19 biochemical processes, listed in Table 14.4, 3 gas–liquid transfer processes and 6 additional acid–base dissociation processes.

Starting from the soluble and particulate matter we have (the time unit is day):

$$\frac{\mathrm{d}C_i^l}{\mathrm{d}t} = \frac{1}{t_{HR}}\left(C_{i,in}^l - C_i^l\right) + \sum_{j=1}^{19} \rho_j \upsilon_{i,j}\,, \qquad \begin{array}{l} i = 1,\ldots,24 \\ i \neq 10, 11 \end{array} \tag{14.1}$$

Table 14.1 ADM1 state variables relevant to the liquid phase

$C_1^l = S_{su}$	Sugars	$C_2^l = S_{aa}$	Amino acids
$C_3^l = S_{fa}$	Long chain fatty acids	$C_4^l = S_{va}$	Valeric acid
$C_5^l = S_{bu}$	Butyric acid	$C_6^l = S_{pro}$	Propionic acid
$C_7^l = S_{ac}$	Acetic acid	$C_8^l = S_{H_2}$	Dissolved hydrogen
$C_9^l = S_{CH_4}$	Dissolved methane	$C_{10}^l = S_{IC}$	Total inorganic carbon
$C_{11}^l = S_{IN}$	Total inorganic nitrogen	$C_{12}^l = S_i$	Soluble inerts
$C_{13}^l = X_c$	Composite	$C_{14}^l = X_{ch}$	Carbohydrates
$C_{15}^l = X_{pr}$	Proteins	$C_{16}^l = X_{li}$	Lipids
$C_{17}^l = X_{su}$	Sugars degraders	$C_{18}^l = X_{aa}$	Aminoacid degraders
$C_{19}^l = X_{fa}$	Long chain fatty acid degraders	$C_{20}^l = X_{C_4}$	Valerate and butyrate degraders
$C_{21}^l = X_{pro}$	Propionate degraders	$C_{22}^l = X_{ac}$	Acetate degraders
$C_{23}^l = X_{H_2}$	Hydrogen degraders	$C_{24}^l = X_i$	Particulate inerts
$C_{25}^l = S_{cat}$	Cations	$C_{26}^l = S_{an}$	Anions

Table 14.2 ADM1 state variables relevant to the gas phase

$C_1^g = S_{CO_2}$	Carbon dioxide	$C_2^g = S_{CH_4}$	Methane	$C_3^g = S_{H_2}$	Hydrogen

Table 14.3 Concentrations of the dissociated acids and free ammonia

$C_{27}^l = S_{vam}$	Valerate	$C_{28}^l = S_{bum}$	Butyrate	$C_{29}^l = S_{prom}$	Propionate
$C_{30}^l = S_{acm}$	Acetate	$C_{31}^l = S_{HCO_3}$	Bicarbonate	$C_{32}^l = S_{NH_3}$	Free ammonia

Table 14.4 Biochemical processes

1	Disintegration	2	Hydrolysis of carbohydrates	3	Hydrolysis of proteins
4	Hydrolysis of lipids	5	Uptake of sugars	6	Uptake of aminoacids
7	Uptake of LCFA	8	Uptake of valerate	9	Uptake of butyrate
10	Uptake of propionate	11	Uptake of acetate	12	Uptake of hydrogen
13	Decay of X_{su}	14	Decay of X_{aa}	15	Decay of X_{fa}
16	Decay of X_{C_4}	17	Decay of X_{pro}	18	Decay of X_{ac}
19	Decay of X_{H_2}				

where t_{HR} is the hydraulic retention time (days), ρ_j (kgCOD m^{-3} d^{-1} or kmol m^{-3} d^{-1}) is the reaction rate for process j and $\upsilon_{i,j}$ is the stoichiometric coefficient of the component i involved in the process j. In turn, the latter quantities are usually grouped in a Petersen matrix, where columns are related to the state variables and rows are related to the processes, while an extra column contains the kinetic equation corresponding to each process.

In order to close the inorganic carbon and nitrogen balances, additional terms were introduced in the original Petersen matrix [4] and then developed in the literature by several authors. These terms explicit the uptakes and releases of N and C in every process and affect the differential equations relevant to S_{IC} and S_{IN} (C^l_{10} and C^l_{11}):

$$\frac{dC^l_{10}}{dt} = \frac{1}{t_{HR}}\left(C^l_{10,in} - C^l_{10}\right) + \sum_{j=1}^{19} \rho_j \upsilon'_{10,j} - \rho_{T,10} \tag{14.2}$$

$$\frac{dC^l_{11}}{dt} = \frac{1}{t_{HR}}\left(C^l_{11,in} - C^l_{11}\right) + \sum_{j=1}^{19} \rho_j \upsilon'_{11,j} \tag{14.3}$$

where $\upsilon'_{10,j}$ and $\upsilon'_{11,j}$ account, respectively, for the carbon and nitrogen content of the components and $\rho_{T,10}$ is the gas transfer rate[1] relevant to component C^g_{10}. For the sake of brevity the Petersen matrix is not reported here; the reader is referred to [4] or [1].

Anions and cations dynamics is simply defined by the dilution effect of the reactor, since they are non-reactive species:

$$\frac{dC^l_i}{dt} = \frac{1}{t_{HR}}\left(C^l_{i,in} - C^l_i\right), \qquad i = 25, 26 \tag{14.4}$$

while the dynamics of the dissociated acid and free ammonia are given by

$$\frac{dC^l_i}{dt} = -\rho_{A,i}\,, \qquad i = 27, \ldots, 32 \tag{14.5}$$

where $\rho_{A,i}$ (kgCOD m^{-3} d^{-1}) is the acid–base reaction rate relevant to component C^l_i [23].

The dynamic equations for the gas phase are given by

$$\frac{dC^g_i}{dt} = -\frac{q_g}{V_g} C^g_i + \rho_{T,i}\frac{V_l}{V_g}\,, \qquad i = 1, 2, 3 \tag{14.6}$$

where V_g (m^3) is the total gas volume inside the reactor and $\rho_{T,i}$ (kmol m^{-3} d^{-1}) is the gas transfer rate relevant to component C^g_i [23].

The only algebraic equation defining the charge balance is

$$S_{cat} + S_{NH4} - S_{HCO3} - \frac{S_{acm}}{64} - \frac{S_{prom}}{112} - \frac{S_{bum}}{160} - \frac{S_{vam}}{208} - S_{an} + S_H - \frac{K_w}{S_H} = 0 \tag{14.7}$$

where $S_{NH_4} = S_{IN} - S_{NH_3}$ is the concentration of ammonium and K_W (kmol^2 L^{-2}) is the dissociation constant of water. This equation determines the molar concentration of hydrogen ion S_H, which in turn determines the $pH = -\log_{10} S_H$, which affects the dissociation of VFAs and ammonia.

[1] No nitrogen goes in the gas phase.

Finally, the biogas output flow rate q_g ($m^3\ d^{-1}$) can be computed as

$$q_g = k_p(p_g - p_{atm})\frac{p_g}{p_{atm}} \tag{14.8}$$

where k_p ($m^3\ bar^{-1}\ d^{-1}$) is the outflow coefficient, p_g (bar) is the biogas pressure and p_{atm} (bar) is the atmospheric pressure. In turn, the biogas pressure is given by the sum of the partial pressures of hydrogen, methane, carbon dioxide and water:

$$p_g = p_{H_2} + p_{CH_4} + p_{CO_2} + p_{H_2O} = S_{H_2}\frac{RT_g}{16} + S_{CH_4}\frac{RT_g}{64} + S_{CO_2}RT_g + p_{H_2O} \tag{14.9}$$

with R being the ideal gas constant and T_g (K) the gas temperature.

As it is apparent, the ADM1 model is quite complex, but a number of simplifications can be then considered, outlined in the following paragraphs.

First of all, the organic matter produced by bacterial decay has been neglected with respect to the one contained in the inflow substrate; as such, the corresponding terms of the Petersen matrix can be cancelled.

Other possible simplifications concern the inhibition terms in the kinetic rates. As an example, consider the rate ρ_8, relevant to the uptake of valerate:

$$\rho_8 = k_{m,C_4}\frac{S_{va}}{S_{va} + K_{S,C_4}}X_{C_4}\frac{S_{va}}{S_{va} + S_{bu}}I_{pH,aa}I_{IN,lim}I_{H_2,C_4} \tag{14.10}$$

In nominal operating conditions, where the pH ranges around 7, the inhibition terms, reported in Table 14.5, are not active (equal to 1) but, in order to extend the validity range of the model, the inhibition terms $I_{pH,ac}$ and I_{NH_3}, both relevant to the methanogenic process, have been considered:

$$I_{pH,aa} = \begin{cases} \exp\left[-3\left(\dfrac{pH - pH_{\max,ac}}{pH_{\max,ac} - pH_{\min,ac}}\right)\right], & pH < pH_{\max,ac} \\ 1, & pH > pH_{\max,ac} \end{cases} \tag{14.11}$$

$$I_{NH_3} = \frac{1}{1 + S_{NH_3}/K_{I,NH_3}} \tag{14.12}$$

Table 14.5 ADM1 inhibition and limiting factors

$I_{pH,aa}$	inhib. fact. for pH of amino acids degr.	$I_{H_2,fa}$	inhib. fact. for H_2 of fatty acid uptake
$I_{pH,ac}$	inhib. fact. for pH of acetate degr.	I_{H_2,C_4}	inhib. fact. for H_2 of valerate uptake
I_{pH,H_2}	inhib. fact. for pH of hydrogen degr.	$I_{H_2,pro}$	inhib. fact. for H_2 of propionic acid uptake
$I_{IN,lim}$	N-dependent limiting factor	I_{NH_3}	ammonia-dependent inhib. factor

Always taking into account the nominal operating conditions, the fractions of undissociated VFAs are negligible and their dissociation dynamics can be neglected, so that the following algebraic relations can be considered:

$$S_{\{vam,bum,prom,acm\}} = S_{\{va,bu,pro,ac\}} \tag{14.13}$$

Some other simplifications are relevant to the contribution of hydrogen: the contribution of the rate $\rho_{T,3}$ can be neglected ($\rho_{T,3} = 0$) as well as the contribution of hydrogen to the total pressure; this means that one of the equations (14.6), namely for $i = 3$, can be dropped.

Finally, the dynamics of inerts has no influence on biogas production and can also be dropped ($i = 12, 24$ in (14.1)).

14.3 Modified AMOCO model

Even introducing the above-mentioned simplifications, the ADM1 model is not suitable for identification and control. Apart from its complexity, it must be considered that it depends on more than a hundred of unknown parameters, which cannot be identified from the usually available measurements. It is therefore necessary to consider simpler models, characterized by a much smaller number of variables and parameters, such as the AMOCO model [3].

The original AMOCO model was then developed to support monitoring and control system design, rather than as a tool for numerical simulation of the process behaviour, and was mainly focused on the description of the anaerobic digestion of soluble substrates or with a negligible particulate content. Only two bacterial populations were considered, in particular acidogenic and methanogenic, while the hydrolysis and acetogenic phases were no longer considered explicitly. In the first step, the acidogenic bacteria X_1 (gVS L^{-1}) consume the organic substrate S_1 (gCOD L^{-1}) and produce CO_2 (mmol L^{-1}) and volatile fatty acids S_2 (mmol L^{-1}). The population of methanogenic bacteria X_2 (gVS L^{-1}) uses, in the second step, the VFAs as substrate for growth and produces CO_2 and methane.

In this work, three modifications are introduced with respect to the original model. At first, the hydrolysis step is taken into account, where the particulate organic matter fed to the digester (X_0) is solubilized into a soluble and degradable organic compound (S_1).

Second, a decay term in the growth rates of the biomasses (which can be estimated as 10% of the maximum bacterial growth rates) has been considered, becoming increasingly important at high hydraulic retention times t_{HR}, defined by adimensional kinetic constants k_{d1} and k_{d2}.

Finally, the contribution of inorganic nitrogen to alkalinity Z (mM) is taken into account [11]. In fact, the AMOCO model considers alkalinity as non-reactive and, consequently, its dynamics is just described by the dilution effect of the reactor. On the contrary, the dynamics of alkalinity is determined by its constituents: bicarbonates,

VFAs, hydroxide ions and, above all, free ammonia. The alkalinity state equation has been therefore modified with respect to the original AMOCO model to account for the inorganic nitrogen, introducing as parameters the nitrogen content of the substrate S_1 dependent on its protein content, N_{S1} (gN/gCOD), which is released into the reactor liquor during the acidogenic process and the nitrogen content in the biomass N_{bac} (gN/gVS) respectively uptaken from or released into the reactor liquor during biomass growth or decay.

The modified AMOCO model is thus defined by seven differential equations: one for the hydrolysis dynamics, two for the mass balances of the bacterial populations X_1 and X_2, two for the organic substrate S_1 and the VFAs S_2 and, finally, two for alkalinity Z and inorganic carbon C (mM):

$$\frac{\mathrm{d}X_0}{\mathrm{d}t} = \frac{1}{t_{HR}}(X_{0,in} - X_0) - \mu_0 X_0 \tag{14.14}$$

$$\frac{\mathrm{d}S_1}{\mathrm{d}t} = \frac{1}{t_{HR}}(S_{1,in} - S_1) + k_0\mu_0 X_0 - k_1\mu_{1,\max}\frac{S_1}{S_1 + K_{S1}}X_1 \tag{14.15}$$

$$\frac{\mathrm{d}S_2}{\mathrm{d}t} = \frac{1}{t_{HR}}(S_{2,in} - S_2) + k_2\mu_{1,\max}\frac{S_1}{S_1 + K_{S1}}X_1 - k_3\mu_{2,\max}\frac{S_2}{S_2 + K_{S2} + S_2^2/K_{I2}}X_2 \tag{14.16}$$

$$\frac{\mathrm{d}X_1}{\mathrm{d}t} = -\frac{1}{t_{HR}}X_1 + \mu_{1,\max}\left(\frac{S_1}{S_1 + K_{S1}} - k_{d1}\right)X_1 \tag{14.17}$$

$$\frac{\mathrm{d}X_2}{\mathrm{d}t} = -\frac{1}{t_{HR}}X_2 + \mu_{2,\max}\left(\frac{S_2}{S_2 + K_{S2} + S_2^2/K_{I2}} - k_{d2}\right)X_2 \tag{14.18}$$

$$\frac{\mathrm{d}Z}{\mathrm{d}t} = \frac{1}{t_{HR}}(Z_{in} - Z) + (k_1 N_{S1} - N_{bac})\,\mu_{1,\max}\frac{S_1}{S_1 + K_{S1}}X_1 - N_{bac}\mu_{2,\max}\frac{S_2}{S_2 + K_{S2} + S_2^2/K_{I2}}X_2 + k_{d1}N_{bac}\mu_{1,\max}X_1 + k_{d2}N_{bac}\mu_{2,\max}X_2 \tag{14.19}$$

$$\frac{\mathrm{d}C}{\mathrm{d}t} = \frac{1}{t_{HR}}(C_{in} - C) + k_4\mu_{1,\max}\frac{S_1}{S_1 + K_{S1}}X_1 - k_5\mu_{2,\max}\frac{S_2}{S_2 + K_{S2} + S_2^2/K_{I2}}X_2 - r_C \tag{14.20}$$

$$\varphi = C + S_2 - Z + K_H P_T + \frac{k_6}{k_{La}} \mu_{2,\max} \frac{S_2}{S_2 + K_{S2} + S_2^2/K_{I2}} X_2 \tag{14.21}$$

$$r_C = k_{La} \left[(C + S_2 - Z) - \frac{\varphi - \sqrt{\varphi^2 - 4K_H P_T (C + S_2 - Z)}}{2} \right] \tag{14.22}$$

$$r_{CH_4} = k_6 \mu_{2,\max} \frac{S_2}{S_2 + K_{S2} + S_2^2/K_{I2}} X_2 \tag{14.23}$$

where μ_0 (d^{-1}) is the specific hydrolysis rate, k_i, ($i = 0, \ldots, 6$) are stoichiometric coefficients, $\mu_{i,\max}$, ($i = 1, 2$) (d^{-1}) are the maximum specific growth rates, K_{S1} (gCOD L^{-1}) and K_{S2} (mmol L^{-1}) are the half-saturation constants, K_{I2} (mmol L^{-1}) is the inibition constant, K_H is Henry's constant for CO_2 (mmol L^{-1} atm^{-1}), $P_T = 1$ (atm) is the atmospheric pressure, k_{La} (d^{-1}) is the liquid–gas transfer coefficient, r_C (mmol L^{-1} d^{-1}) is the carbon dioxide production rate and r_{CH_4} (mmol L^{-1} d^{-1}) is the methane production rate. The quantities $X_{0,in}$, $S_{1,in}$, $S_{2,in}$, Z_{in} and C_{in} are model inputs.

Note that the inorganic carbon is assumed mainly composed of dissolved carbon dioxide CO_2 (mmol L^{-1}) and bicarbonate B (mmol L^{-1}), neglecting the amount of carbonate in the normal operating conditions,[2] while the total alkalinity Z is defined as the sum of dissociated acids in the liquid phase (bicarbonate and VFAs, assumed as completely dissociated in the pH range of interest).

14.4 LFT modelling and identification

While being of reduced order, the modified AMOCO model is however still non-linear, and the identification if its parameters cannot be performed through classical methods, generally based on a linear time-invariant (LTI) model formulation [20]. Therefore, in this work, a different model formulation is investigated, namely a linear fractional transformation (LFT) model formulation, a widely used formalism, both in modern control [25, 15] and identification [19, 17, 5, 16].

It must be pointed out, however, that deriving such model formulation directly from eqs. (14.14) to (14.23) is non-trivial if carried out manually. To this aim, the automated approach discussed in Chapter 4 of this book is adopted to bridge the gap between physical and control/estimation-oriented system modelling. In this approach, an LFT formulation can be automatically derived starting from O-O models of non-linear plants.

At first, the model is formulated in the Modelica language [21, 12], an O-O modelling language for general physical models, described by generic differential-algebraic equations (DAEs) following an acausal, declarative approach. To be further

[2] Values of pH range between 6 and 8, temperature between 35 and 38°C.

processed, the model has to be time-invariant (i.e. the `time` variable does not appear explicitly within any model equation), and all variables and equations are continuous-time[3] (i.e. there are no `when` statements and discrete variables in the model). Furthermore, the model is supplemented by a list of the uncertain parameters, with their minimum and maximum values (this information can be either supplied separately or embedded into the Modelica model by means of the `min`/`max`/`nominal` attributes, as well as of custom annotations to flag the uncertain parameters).

The subsequent steps must be performed within the Modelica compiler, and are similar to the steps routinely performed to produce simulation code:

- The O-O representation of the system is first *flattened*, i.e. reduced to a single DAE model without any hierarchical structure left.
- If the obtained DAE system has a structural index greater than one (in our case the index is 0) it is transformed into an equivalent index-1 formulation by using Pantelides' algorithm and the dummy derivative algorithm [22].
- The equations and the variables, given by the state derivatives and the algebraic variables of the DAE system, are re-ordered, so that the incidence matrix (equations on the rows, unknowns on the columns) is brought in block-lower-triangular (BLT) form. This task is accomplished by using Tarjan's algorithm [10].

It must be remarked that the O-O model might contain non-linear equations involving uncertain parameters, and this would cause the basic algorithm to fail. However, this case can be easily dealt with by defining new auxiliary variables in the O-O model,[4] as shown in an example in Appendix A.

After BLT re-ordering the model can be actually considered as defined by the cascade of r submodels, corresponding to the r blocks of the BLT partition and, in this form, exported from the Modelica compiler and imported into a suitable symbolic manipulation environment. This intermediate environment has the main tasks of defining the dependence of unknown from known parameters and of casting the model in a cascade connection of LFTs, to be finally exported in the LFR Toolbox [15].

Finally, after a suitable order reduction step performed in the Toolbox environment the overall LFT model is given by

$$\dot{x} = Ax + B_1 w + B_2 \zeta + B_3 u \tag{14.24}$$

$$z = C_1 x + D_{11} w + D_{12} \zeta + D_{13} u \tag{14.25}$$

$$\omega = C_2 x + D_{21} w + D_{22} \zeta + D_{23} u \tag{14.26}$$

$$y = C_3 x + D_{31} w + D_{32} \zeta + D_{33} u \tag{14.27}$$

[3] Discontinuous expressions like (14.11) can be smoothed through continuous ones.

[4] An automatic processing of these cases is planned in future implementation of the symbolic manipulation environment.

$$w = \Delta z = \text{diag}\{\delta_1 I_{r_1}, \ldots, \delta_q I_{r_q}\} z \tag{14.28}$$

$$\zeta = \Theta(\omega) \tag{14.29}$$

where x, y, u are the state, output and input vectors respectively, z, ω, w, ζ and z are auxiliary variables, δ_i ($i = 1, \ldots, q$) are the unknown parameters, A, B_i, C_i, D_{ij} are known constant matrices, r_i are the sizes of the corresponding identity matrices I_{r_i} in the Δ block and $\Theta(\omega)$ is a known non-linear vector function.

It is also worth mentioning that, in order to deal with parameter estimation, it is essential to rewrite model (14.24)–(14.29) by introducing normalized unknown parameters $\bar{\delta}_i$, varying between ± 1 as parameter δ_i varies between a maximum $\delta_{i,\max}$ and a minimum $\delta_{i,\min}$.[5]

The problem of model identification formulated over LFT model structures has been a subject of active research for more than 10 years (see, e.g. [19, 6, 17]). In particular, the parameter estimation method proposed in [17] is here extended to account for the non-linear functions $\Theta(\omega)$.

First, with a slight abuse of notation, denote with $y(k)$ the value of $y(t)$ at the sampling instant $t = kT$, with T being the sampling interval, then the identification problem can be formulated as given model (14.24)–(14.29) and given input and output measurements $\{u(k), y(k)\}$, $k = 1, \ldots, N$ generated by the plant, find the values of the unknown parameters $\tilde{\delta}$ minimizing the cost function

$$J(\delta) = \frac{1}{2}\langle e(\delta), e(\delta)\rangle_N = \frac{1}{2N}\sum_{k=1}^{N} e^T(k,\delta)e(k,\delta) \tag{14.30}$$

where $e(k,\delta) = y(k) - \hat{y}(k,\delta)$ is the prediction error between the measured output $y(k)$ and the output $\hat{y}(k,\delta)$ predicted by the model, computed by simulation of model (14.24)–(14.29).

As it is well known, $\tilde{\delta}$ is a *maximum-likelihood* estimate of the model parameters δ for output-error plants, under the assumption of Gaussian white noise [20], and can be obtained through well-known iterative optimization procedures such as the Gauss–Newton algorithm:

$$\hat{\delta}(\nu + 1) = \hat{\delta}(\nu) - \alpha(\nu)\hat{H}^{-1}(\hat{\delta}(\nu))g(\hat{\delta}(\nu)) \tag{14.31}$$

where ν is the iteration number, $\alpha(\nu)$ is the step size, $g(\delta) : \mathbb{R}^q \rightarrow \mathbb{R}^q$ and $\hat{H}(\delta) : \mathbb{R}^q \rightarrow \mathbb{R}^{q\times q}$ are respectively the gradient vector and a positive semi-definite approximation of the Hessian of the cost function with respect to the unknown

[5] $\delta_i = (\delta_{i,\max} + \delta_{i,\min})/2 + \bar{\delta}_i(\delta_{i,\max} - \delta_{i,\min})/2$

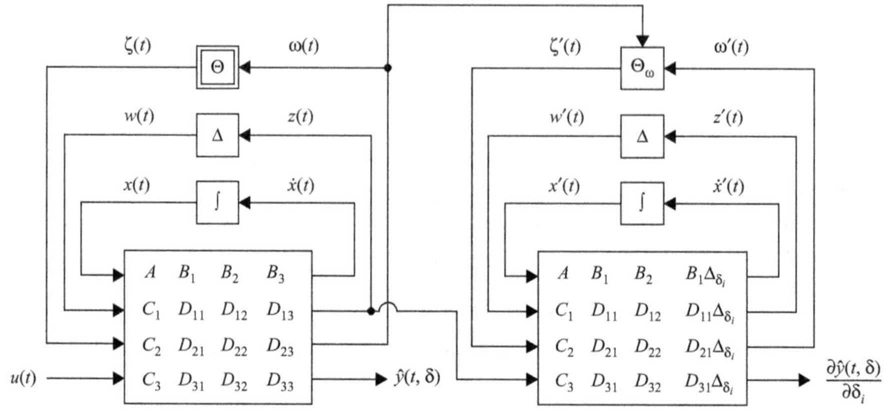

Figure 14.1 Simulation scheme for the computation of the sensitivity functions

parameters:

$$g(\delta) = \frac{1}{N}\sum_{k=1}^{N} E^T(k,\delta)e(k,\delta) \tag{14.32}$$

$$\hat{H}(\delta) = \frac{1}{N}\sum_{k=1}^{N} E^T(k,\delta)E(k,\delta) \tag{14.33}$$

where $E(k,\delta) \in \mathbb{R}^{n_y \times q}$ is the Jacobian of $e(k,\delta)$ and is given by

$$E(k,\delta) = \left[\begin{array}{cccc} \frac{\partial e(k,\delta)}{\partial \delta_1} & \frac{\partial e(k,\delta)}{\partial \delta_2} & \cdots & \frac{\partial e(k,\delta)}{\partial \delta_q} \end{array}\right] \tag{14.34}$$

In turn, each sensitivity function $\partial e(k,\delta)/\partial\delta_i$ can be computed by sampling the simulation of the cascade of two LFT systems shown in Figure 14.1, where

$$\Delta_{\delta_i} = \frac{\partial \Delta}{\partial \delta_i} = \text{diag}\{0_{r_1\times r_1}, \ldots, I_{r_i}, \ldots, 0_{r_q\times r_q}\} \tag{14.35}$$

$$\Theta_\omega = \frac{\partial \Theta(\omega)}{\partial \omega} \tag{14.36}$$

It must be pointed out that when parameter identification is dealt with based on an LFT model structure, the resulting computations of gradients and (approximated) Hessians simply consist in the simulation of index-1, semi-explicit DAE systems, without the need of numerical differentiation.

An important issue is the notion of identifiability itself, particularly in the case of non-linearly parameterized models [8, 9]. Local identifiability in $\tilde{\delta}$, minimizing

(14.30), can be verified by testing if the Hessian $\hat{H}$ is positive definite in $\tilde{\delta}$, i.e. if the rank of $\partial^2 J(\delta)/\partial\delta^2$ at $\tilde{\delta}$ is equal to q. If the Hessian $\hat{H}$ in $\tilde{\delta}$ is not full rank then the estimated parameters are not unique and the system is not identifiable. In this case, to select the identifiable parameter space, we can consider the singular value decomposition (SVD) of the Hessian:

$$\hat{H} = \begin{bmatrix} U_1 & U_2 \end{bmatrix} \begin{bmatrix} \Sigma_1 & 0 \\ 0 & \Sigma_2 \end{bmatrix} \begin{bmatrix} V_1^T \\ V_2^T \end{bmatrix} \tag{14.37}$$

where the separation between Σ_1 and Σ_2 is chosen in such a way that the singular values in Σ_2 are considerably smaller than those in Σ_1. Accordingly, the column space of U_1 represents the subspace of the original parameters δ_i that will be identifiable from the measurements.

14.5 Parameter identification based on ADM1 model simulation data

The data used for parameter identification have been first generated by simulating the ADM1 model, which has been therefore considered as the 'plant'. In particular, the case study is the waste sludge digestion as described in [23]. According to the authors, the input values may not be completely realistic for all variables but they have been chosen such that every input is active (i.e. non-zero) and able to excite all internal modes of the ADM1 model.

Since model (14.14)–(14.23) has a cascade structure, where the first block is made up by the first five equations and its outputs, namely X_0, S_1, S_2, X_1, X_2, appear as inputs for the remaining equations, the identification can be focused on the parameters governing the dynamics of substrates and biomasses only. Then, the following reduced model can be considered ($d = 1/t_{HR}$):

$$\dot{x}_1 = -dx_1 - \delta_6 x_1 + du_1 \tag{14.38}$$

$$\dot{x}_2 = -dx_2 + \delta_1\delta_6 x_1 - \delta_2\mu_{1,\max}\frac{x_2}{x_2 + K_{S1}}x_4 + du_2 \tag{14.39}$$

$$\begin{aligned}\dot{x}_3 = {} & -dx_3 + du_3 + \delta_3\mu_{1,\max}\frac{x_2}{x_2 + K_{S1}}x_4 \\ & - \delta_4\mu_{2,\max}\frac{x_3}{x_3 + K_{S2} + x_3^2/K_{I2}}x_5\end{aligned} \tag{14.40}$$

$$\dot{x}_4 = -dx_4 + \mu_{1,\max}\left(\frac{x_2}{x_2 + K_{S1}} - k_{d1}\right)x_4 \tag{14.41}$$

Table 14.6 Fixed parameters for identification based on ADM1 model simulations

d (d^{-1}) 0.05	$\mu_{1,\max}$ (d^{-1}) 0.206	$\mu_{2,\max}$ (d^{-1}) 0.2
K_{S1} (kgCOD m^{-3}) 1.096	K_{S2} (mmol L^{-1}) 6.86	K_{I2} (mmol L^{-1}) 433968
k_{d1} 0.1	k_{d2} 0.1	c (gCOD $(gVS)^{-1}$) 1.55

$$\dot{x}_5 = -dx_5 + \mu_{2,\max}\left(\frac{x_3}{x_3 + K_{S2} + x_3^2/K_{I2}} - k_{d2}\right)x_5 \tag{14.42}$$

$$y_1 = x_3 \tag{14.43}$$

$$y_2 = x_1 + x_2 + c(x_4 + x_5) \tag{14.44}$$

$$y_3 = \delta_5 \mu_{2,\max} \frac{x_3}{x_3 + K_{S2} + x_3^2/K_{I2}} x_5 \tag{14.45}$$

where

$$x = [\,x_1 \quad x_2 \quad x_3 \quad x_4 \quad x_5\,]^T = [\,X_0 \quad S_1 \quad S_2 \quad X_1 \quad X_2\,]^T \tag{14.46}$$

$$u = [\,u_1 \quad u_2 \quad u_3\,]^T = [\,X_{0,in} \quad S_{1,in} \quad S_{2,in}\,]^T \tag{14.47}$$

$$\delta = [\,\delta_1 \quad \delta_2 \quad \delta_3 \quad \delta_4 \quad \delta_5 \quad \delta_6\,]^T = [\,k_0 \quad k_1 \quad k_2 \quad k_3 \quad k_6 \quad \mu_0\,]^T \tag{14.48}$$

Of course, it is assumed that all inputs and outputs are measurable, while the parameters reported in Table 14.6 are assumed fixed and known. In particular, the measurable outputs are respectively given by the volatile fatty acid concentration $y_1 = x_3 = S_2$ (mmol L^{-1}), the sum of the concentration of other substances $y_2 = x_1 + x_2 + c(x_4 + x_5) \triangleq S_3$ (gCOD L^{-1}), where c is a scaling factor from (gVS L^{-1}) to (gCOD L^{-1}), and the methane flow rate $y_3 = r_{CH_4}$ (mmol L^{-1} d^{-1}). Accordingly, the relevant LFT model is defined in Appendix A.

A first main issue in the comparison among the outputs of the two models is related to the need of lumping several variables of the ADM1 model into single variables of the modified AMOCO model. In this respect, it must be recalled that all substrates and biomasses concentrations in the ADM1 model are expressed in (gCOD L^{-1}).

First of all, having extended the model to account for the hydrolysis step, the particulate substrate, defined by the ADM1 variables $X_c, X_{ch}, X_{pr}, X_{li}$, corresponding respectively to composite, carbohydrates, proteins and lipids concentrations, has been aggregated in the variable X_0.

The soluble organic matter in the input substrate, sugar, amino acids and fatty acids concentrations, corresponding to the ADM1 variables S_{su}, S_{aa}, S_{fa}, has been aggregated in the variable S_1.

The variable S_2 has been considered accounting for the total concentration of VFAs, given by the soluble compounds valeric, butyric, propionic and acetic acids, corresponding to the ADM1 variables $S_{va}, S_{bu}, S_{pro}, S_{ac}$.

The ADM1 bacterial populations in charge of the degradation of sugars, amino acids, fatty acids and volatile acids (with exception of acetic acid), described by ADM1 variables $X_{su}, X_{aa}, X_{fa}, X_{C_4}, X_{pro}$, can be grouped into the variable X_1.

Finally, the concentration of biomasses converting acetic acid and hydrogen into methane, described by the ADM1 variables X_{ac} and X_{H_2}, can be grouped into the variable X_2.

The conversion formulas can be therefore summarized as follows:

$$u_1 = X_{c,in} + X_{ch,in} + X_{pr,in} + X_{li,in} \tag{14.49}$$

$$u_2 = S_{su,in} + S_{aa,in} + S_{fa,in} \tag{14.50}$$

$$u_3 = 1000\,(S_{va,in}/208 + S_{bu,in}/160 + S_{pro,in}/112 + S_{ac,in}/64) \tag{14.51}$$

$$y_1 = 1000\,(S_{va}/208 + S_{bu}/160 + S_{pro}/112 + S_{ac}/64) \tag{14.52}$$

$$\begin{aligned} y_2 &= S_{su} + S_{aa} + S_{fa} + X_c + X_{ch} + X_{pr} + X_{li} \\ &\quad + (X_{su} + X_{aa} + X_{fa} + X_{C_4} + X_{pro}) + (X_{ac} + X_{H_2}) \end{aligned} \tag{14.53}$$

$$y_3 = (1 - Y_{ac})\rho_{11} + (1 - Y_{h_2})\rho_{12} \tag{14.54}$$

where Y_{ac} and Y_{h_2} are stoichiometric coefficients (gCOD/gCOD) quantifying the specific growth yields of acetoclastic methanogens (Y_{ac}) and of hydrogenotrophic methanogens (Y_{h_2}).

A sequence of steps of type '3-2-1-1'[6] in the ADM1 variable $X_{c,in}$ has been considered as test signal, resulting in a variation of the aggregate variable u_1 shown in Figure 14.2. An overall simulation interval of 40 days has been considered while, for the sake of realism, a white noise disturbance, with a signal-to-noise ratio of 30, has been added to the simulated outputs, sampled with a period of 1 day.

Figures 14.3–14.5 compare the simulated measurements y_1, y_2 and y_3 (denoted by circles), obtained by a simulation of the ADM1 model, with the simulated outputs of the identified modified AMOCO model $\hat{y}_1 = \hat{S}_2, \hat{y}_2 = \hat{S}_3$ and $\hat{y}_3 = \hat{r}_{CH_4}$ (solid line). As it is apparent, a good correspondence has been obtained.

[6] The numbers denoting the test signal indicate the duration of the 'high' and 'low' signal intervals, where 'high' means a concentration 10 times higher than the nominal value. Such a shape of signal is frequently used in the aeronautical field [18], where the period '2' corresponds to the semi-period of the expected mode to be identified, while in our case is equal to half t_{HR}.

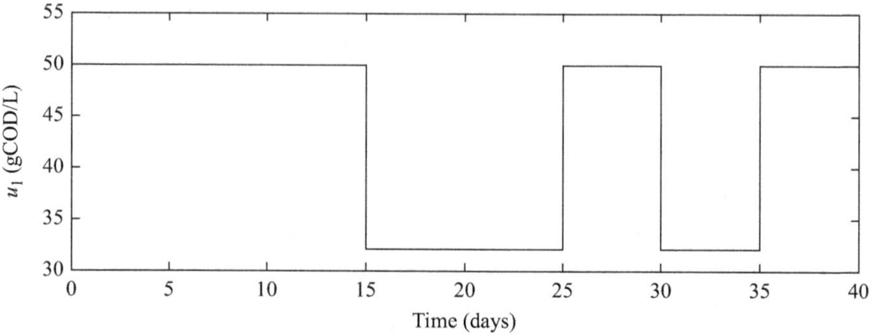

Figure 14.2 Input u_1 for identification

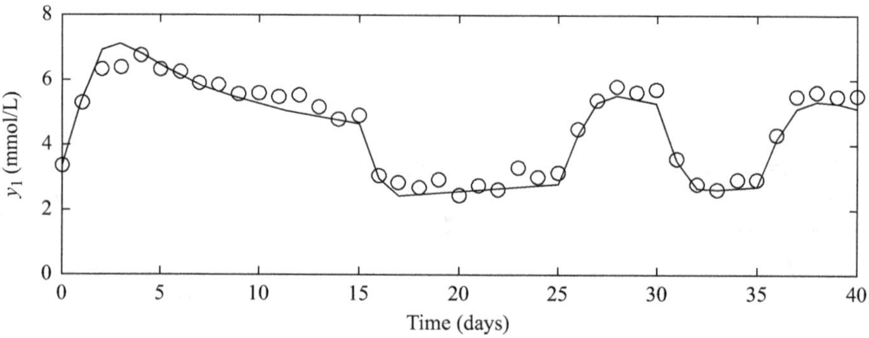

Figure 14.3 Simulated measurements y_1 (circles) and simulated model output $\hat{y}_1 = \hat{S}_2$ (solid line)

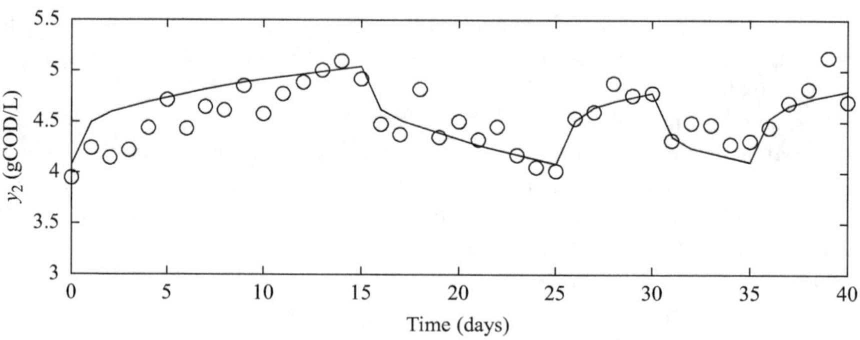

Figure 14.4 Simulated measurements y_2 (circles) and simulated model output $\hat{y}_2 = \hat{S}_3$ (solid line)

In terms of estimated parameters, the results of the identification are reported in Table 14.7, together with the initial parameter values. Note that, in order to test the performance of the identification algorithm, a much lower initial value for the

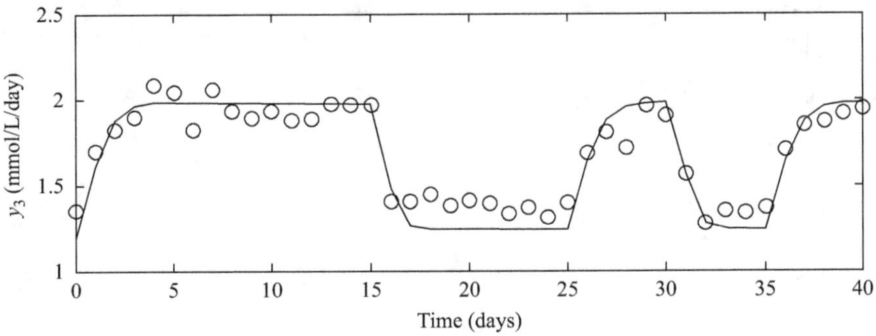

Figure 14.5 Simulated measurements y_3 (circles) and simulated model output $\hat{y}_3 = \hat{r}_{CH_4}$ (solid line)

Table 14.7 AMOCO initial and estimated parameters from identification based on ADM1 model simulations

	Initial value	**Final value**
k_0	1	0.7387
k_1 (gCOD gVS^{-1})	15	14.8410
k_2 (mmol gVS^{-1})	200	200.1184
k_3 (mmol gVS^{-1})	440	321.0709
k_6 (mmol gVS^{-1})	27	26.1949
μ_0 (d^{-1})	0.5	5.8874

parameter μ_0, with respect to the actual behaviour of the ADM1 model, has been considered. This test can be considered as representative of the case where a more easily hydrolysable load is introduced in the digester, marked just by a rapid decrease of the parameter μ_0.

A different variation of the ADM1 variable $X_{c,in}$ has been considered for the sake of validation of the identified model, obtained by increasing the said variable at a rate of 50%/day along 6 days (Figure 14.6). The obtained transients of the aggregate ADM1 model variables y_1, y_2 and y_3 (circles) and of the simulated outputs of the identified modified AMOCO model $\hat{y}_1$, $\hat{y}_2$ and $\hat{y}_3$ (solid lines) are shown in Figures 14.7–14.9, where a quite good matching between the ADM1 model and the identified modified AMOCO model can still be appreciated.

It must be finally pointed out that the choice of the number of identified parameters has been actually validated through a local identifiability analysis, based on the analysis of the Hessian condition number. Figure 14.10 shows the trend of the said condition number with respect to the identification steps, considering the assumed uncertain parameters (solid line), and the same trend obtained by identifying also

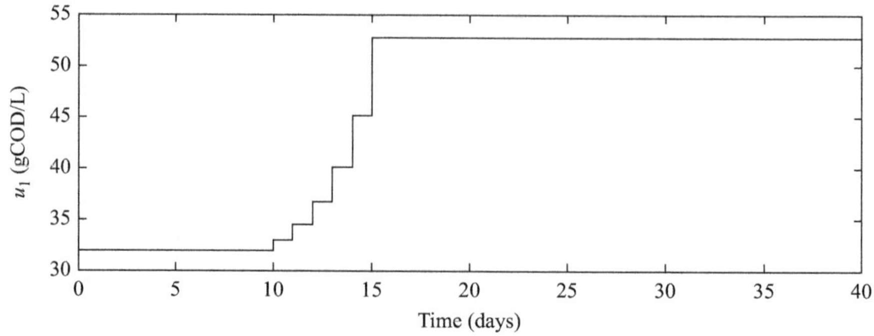

Figure 14.6 Input u_1 for validation

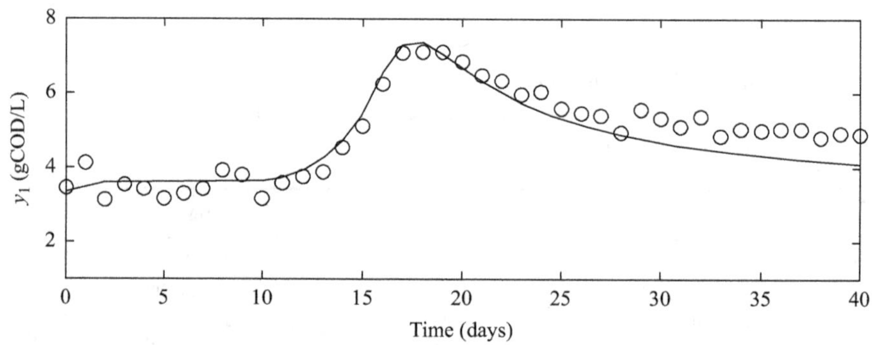

Figure 14.7 Simulated measurements y_1 (circles) and simulated model output $\hat{y}_1 = \hat{S}_2$ (solid line)

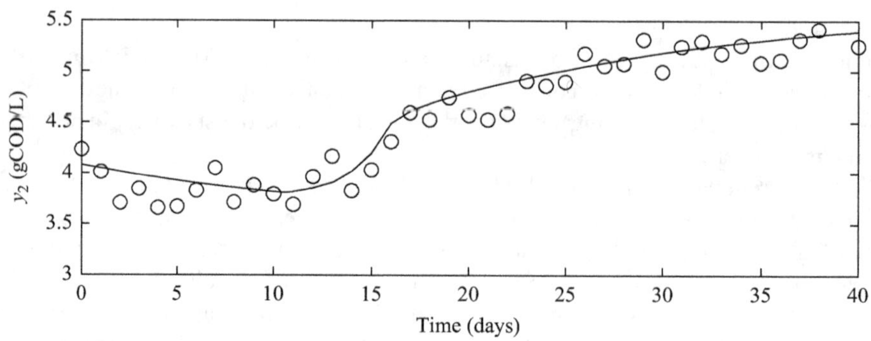

Figure 14.8 Simulated measurements y_2 (circles) and simulated model output $\hat{y}_2 = \hat{S}_3$ (solid line)

the parameters $\mu_{1,\max}$ and $\mu_{2,\max}$ (dashed line). As it is apparent, with a greater number of parameters the higher value of the condition number values indicates a worst identifiability condition.

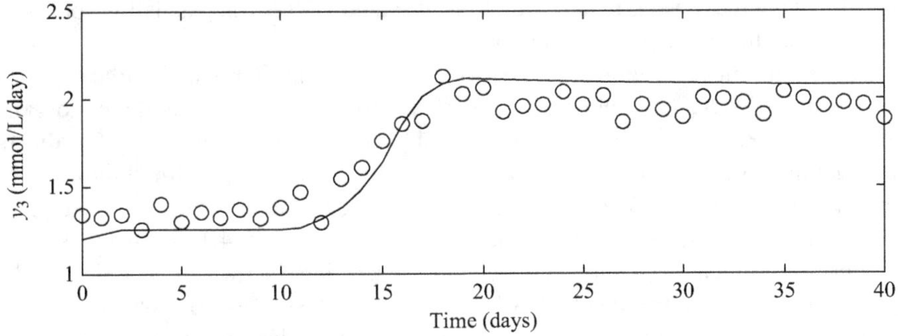

Figure 14.9 Simulated measurements y_3 (circles) and simulated model output $\hat{y}_3 = \hat{r}_{CH_4}$ (solid line)

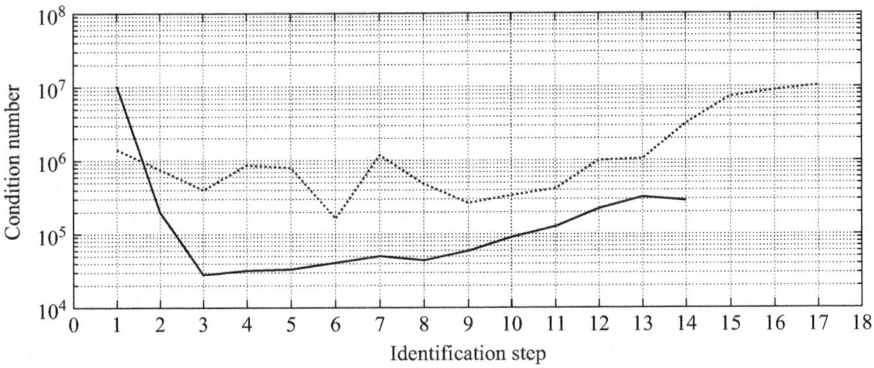

Figure 14.10 Condition number with 6 (solid line) and 8 (dashed line) parameters

14.6 Parameter identification based on experimental data

The case study consists in the anaerobic digestion of agricultural wastes, including mainly maize silage and cow manure. Sporadically, minor amounts of triticale, corn flour, soy flour and mashed potatoes are also included in the digester feed.

The plant consists in three completely mixed units, two of them serving as primary digesters, whose outflow is sent to a secondary digester. The overall working volume V is 6800 m^3, corresponding to an hydraulic retention time varying between 55 and 110 days, depending on the influent flow rate.

The monitoring campaign lasted 350 days and included the following influent and effluent data. The influent characteristics were assessed on an average twice a week and included the flow rate, the mass of each organic substrate and its total and volatile solids content. No COD data were available, and COD was therefore estimated by using COD/SV average literature data for each organic substrate. As for the effluent, twice a week data of volatile fatty acids, VFA, biogas daily production rate and methane content were available.

As for the model, minor modifications have been implemented with respect to the modified AMOCO model presented earlier. First, the dynamics and the inorganic carbon species and pH have been neglected, since no direct retrofit of the pH value on the dynamics of substrates and biomasses is included in this simplified model. This means that (14.19)–(14.23) are neglected. Furthermore, the flow rate q_{in} ($\mathrm{m^3\ d^{-1}}$) has been explicitly considered by substituting t_{HR} with V/q_{in} into (14.14)–(14.18). Finally, one more component, X_{ND}, has been included, describing the fraction of undegradable organic matter that is produced from X_0 during hydrolysis and that flows unchanged through the anaerobic reactor. Therefore the following dynamic equation has been added:

$$\frac{dX_{ND}}{dt} = -\frac{q_{in}}{V}X_{ND} + (1 - k_0)\mu_0 X_0 \tag{14.55}$$

where $(1 - k_0)$ is a stoichiometric coefficient describing the fraction of undegradable organic matter that is produced from X_0.

Since the influent flow rate is now considered as a further input, $q_{in} = u_4$, while the volatile fatty acid concentration $y_1 = x_3 = S_2$ ($\mathrm{mmol\ L^{-1}}$), the sum of the concentration of other substances $y_2 = x_1 + x_2 + x_6 + c(x_4 + x_5) = S_3$ ($\mathrm{gCOD\ L^{-1}}$), and the methane flow rate $y_3 = r_{CH_4}$ ($\mathrm{mmol\ L^{-1}\ d^{-1}}$) have been measured as outputs. The considered model can be summarized as follows:

$$\dot{x}_1 = \frac{u_4}{V}(u_1 - x_1) - \delta_2 x_1 \tag{14.56}$$

$$\dot{x}_2 = \frac{u_4}{V}(u_2 - x_2) + k_0\delta_2 x_1 - k_1\mu_{1,\max}\frac{x_2}{x_2 + K_{S1}}x_4 \tag{14.57}$$

$$\begin{aligned}\dot{x}_3 = {} & \frac{u_4}{V}(u_3 - x_3) + k_2\mu_{1,\max}\frac{x_2}{x_2 + K_{S1}}x_4 \\ & - k_3\mu_{2,\max}\frac{x_3}{x_3 + \delta_1 + x_3^2/K_{I2}}x_5\end{aligned} \tag{14.58}$$

$$\dot{x}_4 = -\frac{u_4}{V}x_4 + \mu_{1,\max}\left(\frac{x_2}{x_2 + K_{S1}} - k_{d1}\right)x_4 \tag{14.59}$$

$$\dot{x}_5 = -\frac{u_4}{V}x_5 + \mu_{2,\max}\left(\frac{x_3}{x_3 + \delta_1 + x_3^2/K_{I2}} - k_{d2}\right)x_5 \tag{14.60}$$

$$\dot{x}_6 = -\frac{u_4}{V}x_6 + (1 - k_0)\delta_2 x_1 \tag{14.61}$$

$$y_1 = x_3 \tag{14.62}$$

$$y_2 = x_1 + x_2 + x_6 + c(x_4 + x_5) \tag{14.63}$$

$$y_3 = r_{CH_4} = k_6 \mu_{2,\max} \frac{x_3}{x_3 + \delta_1 + x_3^2/K_{I2}} x_5 \tag{14.64}$$

where

$$x = [\,x_1 \quad x_2 \quad x_3 \quad x_4 \quad x_5 \quad x_6\,]^T = [\,X_0 \quad S_1 \quad S_2 \quad X_1 \quad X_2 \quad X_{ND}\,]^T \tag{14.65}$$

$$u = [\,u_1 \quad u_2 \quad u_3 \quad u_4\,]^T = [\,X_{0,in} \quad S_{1,in} \quad S_{2,in} \quad q_{in}\,]^T \tag{14.66}$$

$$\delta = [\,\delta_1 \quad \delta_2\,]^T = [\,K_{S2} \quad \mu_0\,]^T \tag{14.67}$$

Note that only two parameters were taken into account for identification, namely K_{S2} and μ_0, which were considered as more directly affecting the dynamic behaviour of the plant with reference to the considered experimental case test. The (fixed) numerical values of the remaining parameters are reported in Table 14.8 and the relevant LFT form is reported in Appendix B. In particular, it must be pointed out that the information content of the available data set was much lower with respect to the simulation test case, making the Hessian condition number to rapidly increase when considering more parameters to be identified.

Figures 14.11 and 14.12 show the input variations (u_1 and u_4), the other inputs are assumed negligible, while the results of the identification are shown in Figures 14.13–14.15.

As it is apparent, the identified model seems to be in a quite good agreement with the data, and the values of the identified parameters (Table 14.9) are reasonable from a physical point of view. On the other hand, there are some measured data that largely differ from the model's outputs, particularly with reference to S_2 and r_{CH_4}, and this could be ascribed, rather than to measurement noise, to an inaccurate characterization of the measured inputs, whose COD was inferred from VS data rather then directly measured.

Table 14.8 Fixed parameters for identification based on experimental data

$\mu_{1,\max}$ (d^{-1}) 0.206	$\mu_{2,\max}$ (d^{-1}) 0.2	K_{S1} (kgCOD m^{-3}) 1.096
K_{I2} (mmol L^{-1}) 434000	k_0 0.8	k_1 (gCOD gVS^{-1}) 31
k_2 (mmol gVS^{-1}) 460	k_3 (mmol gVS^{-1}) 423	k_6 (mmol gVS^{-1}) 400
k_{d1} 0.1	k_{d2} 0.1	

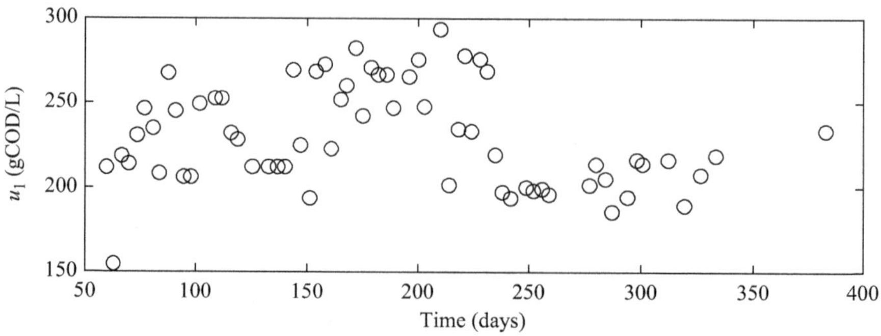

Figure 14.11 Measured u_1

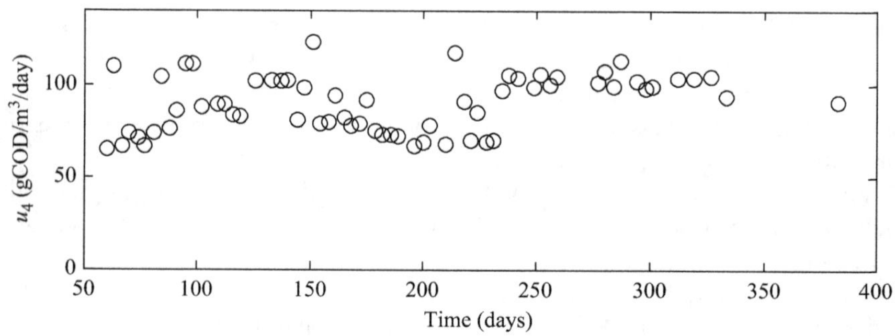

Figure 14.12 Measured u_4

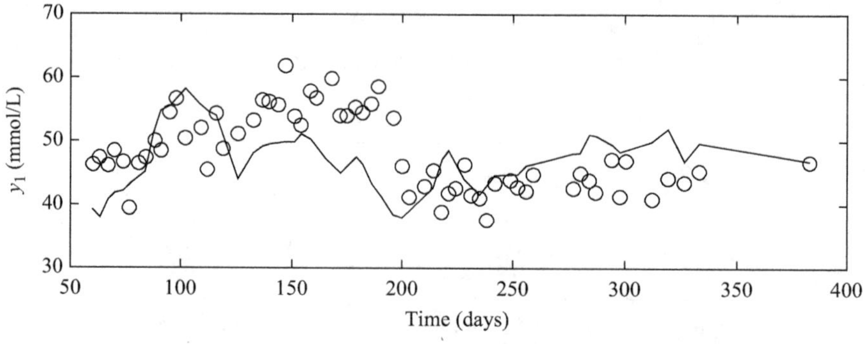

Figure 14.13 Measured y_1 (circles) and simulated model output $\hat{y}_1 = \hat{S}_2$ (solid line)

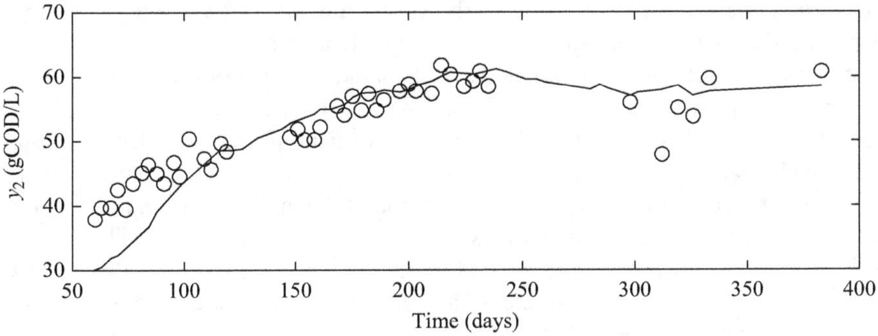

Figure 14.14 Measured y_2 (circles) and simulated model output $\hat{y}_2 = \hat{S}_3$ (solid line)

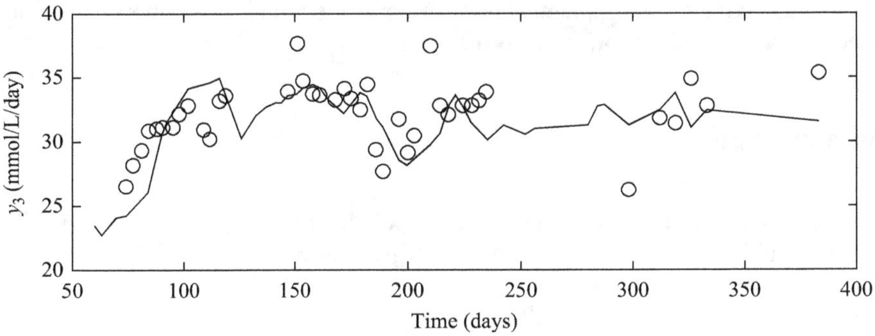

Figure 14.15 Measured y_3 (circles) and simulated model output $\hat{y}_3 = \hat{r}_{CH_4}$ (solid line)

Table 14.9 AMOCO initial and estimated parameters from experimental data

	Initial value	**Final value**
K_{S2} (mmol L^{-1})	250	230.05
μ_0 (d^{-1})	0.5	0.2516

14.7 Conclusion

Wastewater treatment and biogas production are largely based on a very complex anaerobic digestion process, involving several biochemical processes and species. Accurate and detailed models, such as the ADM1 model, cannot be used for online monitoring and control, so reduced order models have to be considered. To this aim, in this chapter, a modification of the well-known AMOCO model is first proposed in order to widen its field of applicability. Then, in order to perform parameter identification, an LFT formulation is derived, thanks to the use of a symbolic manipulation tool applied to an O-O model formulation. In fact, while being of reduced order, the modified AMOCO model is still non-linear so that the identification of (some of) its parameters can be suitably addressed through an LFT approach, which allows an efficient computation of gradients and Hessians of a maximum-likelihood cost function by simply simulating index-1, semi-explicit DAE systems. The approach has been applied to two case tests: in the first test, the data used for identification have been generated by a simulation of the ADM1 model, assuming waste-activated sludge as substrate, and in the second, the data have been collected on a real plant, used for anaerobic digestion of agricultural wastes. Quite good results have been obtained, suggesting for an adoption of the approach to online process monitoring and, above all, for the implementation of advanced control techniques. These results open interesting perspective to the implementation of control strategies to improve anaerobic processe stability and dynamic performances.

Acknowledgements

This work has been performed within the project La Fabbrica della Bioenergia. The authors would also like to thank Andrea Allegrini, Sonia Hassam and Alberto Leva for their valuable contribution.

Appendix A. LFT model for parameter identification based on ADM1 model simulation data

$$\begin{bmatrix} \dot{x}_1 \\ \dot{x}_2 \\ \dot{x}_3 \\ \dot{x}_4 \\ \dot{x}_5 \end{bmatrix} = \begin{bmatrix} -dx_1 + w_6 + du_1 \\ -dx_2 - w_1 - \mu_{1,\max} w_2 + du_2 \\ -dx_3 + \mu_{1,\max} w_3 + \mu_{2,\max} w_4 + du_3 \\ -(d + k_{d1}\mu_{1,\max})x_4 + \mu_{1,\max}\zeta_1 \\ -(d + k_{d2}\mu_{2,\max})x_5 + \mu_{2,\max}\zeta_2 \end{bmatrix} \quad \begin{bmatrix} z_1 \\ z_2 \\ z_3 \\ z_4 \\ z_5 \\ z_6 \end{bmatrix} = \begin{bmatrix} w_6 \\ \zeta_1 \\ \zeta_1 \\ -\zeta_2 \\ -\zeta_2 \\ -x_1 \end{bmatrix}$$

$$\begin{bmatrix} \omega_1 \\ \omega_2 \\ \omega_3 \\ \omega_4 \end{bmatrix} = \begin{bmatrix} x_5 \\ x_3 \\ x_4 \\ x_2 \end{bmatrix}, \quad \begin{bmatrix} y_1 \\ y_2 \\ y_3 \end{bmatrix} = \begin{bmatrix} x_3 \\ x_1 + x_2 + c(x_4 + x_5) \\ -\mu_{2,\max} w_5 \end{bmatrix}$$

$$\begin{bmatrix} w_1 \\ w_2 \\ w_3 \\ w_4 \\ w_5 \\ w_6 \end{bmatrix} = \begin{bmatrix} z_1\delta_1 \\ z_2\delta_2 \\ z_3\delta_3 \\ z_4\delta_4 \\ z_5\delta_5 \\ z_6\delta_6 \end{bmatrix}, \quad \begin{bmatrix} \zeta_1 \\ \zeta_2 \end{bmatrix} = \begin{bmatrix} \dfrac{\omega_3\omega_4}{K_{S1} + \omega_4} \\ \dfrac{\omega_1\omega_2}{K_{S2} + \omega_2 + \omega_2^2/K_{I2}} \end{bmatrix}$$

Appendix B. LFT model for parameter identification based on experimental data

$$\begin{bmatrix} \dot{x}_1 \\ \dot{x}_2 \\ \dot{x}_3 \\ \dot{x}_4 \\ \dot{x}_5 \\ \dot{x}_6 \end{bmatrix} = \begin{bmatrix} \zeta_3 \\ \zeta_4 \\ \zeta_5 \\ \zeta_6 \\ \zeta_7 \\ \zeta_8 \end{bmatrix}, \quad \begin{bmatrix} z_1 \\ z_2 \end{bmatrix} = \begin{bmatrix} x_3 \\ x_1 \end{bmatrix}, \quad \begin{bmatrix} w_1 \\ w_2 \end{bmatrix} = \begin{bmatrix} z_1\delta_1 \\ z_2\delta_2 \end{bmatrix}$$

$$\begin{bmatrix} \zeta_1 \\ \zeta_2 \\ \zeta_3 \\ \zeta_4 \\ \zeta_5 \\ \zeta_6 \\ \zeta_7 \\ \zeta_8 \end{bmatrix} = \begin{bmatrix} \mu_{1,\max}\dfrac{\omega_1}{\omega_1 + K_{S1}}\omega_2 \\ \mu_{2,\max}\dfrac{\omega_3^2}{\omega_3^2 + \omega_5 + \omega_3^3/K_{I2}}\omega_4 \\ \left(\dfrac{\omega_6}{V}\right)(\omega_7 - \omega_8) - \omega_9 \\ \left(\dfrac{\omega_6}{V}\right)(\omega_{10} - \omega_1) + k_0\omega_9 - k_1\omega_{11} \\ \left(\dfrac{\omega_6}{V}\right)(\omega_{12} - \omega_3) + k_2\omega_{11} - k_3\omega_{13} \\ -\left(\dfrac{\omega_6}{V}\right)\omega_2 + \omega_{11} - k_{d1}\mu_{1,\max}\omega_2 \\ -\left(\dfrac{\omega_6}{V}\right)\omega_4 + \omega_{13} - k_{d2}\mu_{2,\max}\omega_4 \\ -\left(\dfrac{\omega_6}{V}\right)\omega_{14} + (1 - k_0)\omega_9 \end{bmatrix}, \quad \begin{bmatrix} \omega_1 \\ \omega_2 \\ \omega_3 \\ \omega_4 \\ \omega_5 \\ \omega_6 \\ \omega_7 \\ \omega_8 \\ \omega_9 \\ \omega_{10} \\ \omega_{11} \\ \omega_{12} \\ \omega_{13} \\ \omega_{14} \end{bmatrix} = \begin{bmatrix} x_2 \\ x_4 \\ x_3 \\ x_5 \\ w_1 \\ u_4 \\ u_1 \\ x_1 \\ w_2 \\ u_2 \\ \zeta_1 \\ u_3 \\ \zeta_2 \\ x_6 \end{bmatrix}$$

$$\begin{bmatrix} y_1 \\ y_2 \\ y_3 \end{bmatrix} = \begin{bmatrix} x_3 \\ x_1 + x_2 + x_6 + c(x_4 + x_5) \\ k_6\zeta_2 \end{bmatrix}$$

References

[1] Allegrini, A.: *Anaerobic Digestion Modelling: A Comparison Between ADM1 and AMOCO*. Master's thesis, Politecnico di Milano (2010)

[2] Batstone, D.J., Keller, J., Angelidaki, I., Kalyuzhnyi, S.V., Pavlostathis, S.G., Rozzi, A., et al.: The IWA Anaerobic Digestion Model no. 1 (ADM1). *Water Science and Technology* **45**(10), 65–73 (2002)

[3] Bernard, O., Hadj-Sadok, Z., Dochain, D., Genovesi, A., Steyer, J.: Dynamical model development and parameter identification for an anaerobic wastewater treatment process. *Biotechnology and Bioengineering* **75**(4), 424–438 (2001)

[4] Blumensaat, F., Keller, J.: Modelling of two-stage anaerobic digestion using the IWA Anaerobic Digestion Model no. 1 (ADM1). *Water Research* **39**(1), 171–83 (2005)

[5] Casella, F., Lovera, M.: LPV/LFT modelling and identification: overview, synergies and a case study. In: *IEEE International Conference on Computer-Aided Control Systems, 2008. CACS 2008*. pp. 852–857 (2008)

[6] Demourant, F., Ferreres, G.: Closed loop identification of a LFT model. *Journal Européen des Systémes Automatisés* **36**(3), 449–464 (2002)

[7] Dennis, Jr., J.E., Schnabel, R.B.: Numerical methods for unconstrained optimization and nonlinear equations. *Classics in Applied Mathematics*. Soc for Industrial & Applied Math, 0898713641, **16** (1996)

[8] Doren, J.V., den Hof, P.V., Hansen, J.D., Bosgra, O.: Determining identifiable parameterisations for large-scale physical models in reservoir engineering. In: *2008 IFAC World Congress. Seoul*, South Korea (2008)

[9] Dötsch, H.G.M., Van den Hof, P.M.J.: Test for local structural identifiability of high-order non-linearly parametrized state space models. *Automatica* **32**(6), 875–883 (1996)

[10] Duff, I.S., Reid, J.K.: An implementation of Tarjan's algorithm for the block triangularization of a matrix. *ACM Transactions on Mathematical Software* **4**(2), 137–147 (1978)

[11] Ficara, E., Hassam, S., Allegrini, A., Leva, A., Malpei, F., Ferretti, G.: Anaerobic digestion models: a comparative study. In: *7th Vienna Conference on Mathematical Modelling, MATHMOD 2012* (2012)

[12] Fritzson, P.: *Principles of Object-Oriented Modeling and Simulation with Modelica 2.1*. Wiley-IEEE Computer Society, New York (2003)

[13] Galí, A., Benabdallah, T., Astals, S., Mata-Alvarez, J.: Modified version of ADM1 model for agro-waste application. *Bioresource Technology* **100**(11), 2783–2790 (2009)

[14] Graef, S.P., Andrews, J.F.: Stability and control of anaerobic digestion. *Journal of the Water Pollution Control Federation* **46**(4), 666–683 (1974)

[15] Hecker, S., Varga, A., Magni, J.F.: Enhanced LFR-toolbox for MATLAB. In: *Proceedings of 2004 IEEE International Symposium on Computer Aided Control Systems Design* (pp. 25–29). Taipei, Taiwan (2004)

[16] Hsu, K., Poolla, K., Vincent, T.: Identification of structured nonlinear systems. *IEEE Transactions on Automatic Control* **53**(11), 2497–2513 (2008)

[17] Hsu, K., Vincent, T., Wolodkin, G., Rangan, S., Poolla, K.: An LFT approach to parameter estimation. *Automatica* **44**(12), 3087–3092 (2008)

[18] Klein, V., Morelli, E.A.: *Aircraft System Identification: Theory and Practice*. AIAA Education Series, American Institute of Aeronautics and Astronautics, Reston, VA (2006)

[19] Lee, L.H., Poolla, K.: Identification of linear parameter-varying systems using nonlinear programming. *Journal of Dynamic Systems, Measurement, and Control* **121**, 71–78 (1999)

[20] Ljung, L. (ed.): System Identification: Theory for the User (2nd edn.). Prentice-Hall PTR, Upper Saddle River, NJ (1999)

[21] Mattsson, S.E., Elmqvist, H., Otter, M.: Physical system modeling with Modelica. *Control Engineering Practice* **6**(4), 501–510 (1998)

[22] Mattsson, S.E., Söderlind, G.: Index reduction in differential-algebraic equations using dummy derivatives. *SIAM Journal on Scientific Computing* **14**(3), 677–692 (1993)

[23] Rosen, C., Vrecko, D., Gernaey, K., Pons, M., Jeppsson, U.: Implementing ADM1 for plant-wide benchmark simulations in MATLAB/Simulink. *Water Science and Technology* **54**(4), 11–19 (2006)

[24] Stemann, S.W., Ristow, N.E., Wentzel, M.C., Ekama, G.A.: A steady state model for anaerobic digestion of sewage *sludges. Water SA* **31**(4), 511–528 (2005)

[25] Zhou, K., Doyle, J.C., Glover, K.: *Robust and Optimal Control*. Prentice-Hall, Upper Saddle River, NJ (1996)

Chapter 15

Modeling and parameter identification of a brake-by-wire actuator for racing motorcycles

Matteo Corno[1], *Fabio Todeschini*[2], *Giulio Panzani*[3] *and Sergio M. Savaresi*[4]

Abstract

This chapter presents a control-oriented model of an electro-hydraulic brake-by-wire actuator for racing motorcycles. Starting from a detailed description of the device, a first-principle model is derived. The model parameters are identified following a gray-box approach. Experimental results are used to validate the model and carry out a parameter sensitivity analysis.

15.1 Introduction

The need of automatically controlling wheel braking torque in road vehicles dates back to the introduction of the first Anti-lock Braking (ABS) Systems [1, 4, 28]. ABS systems are one of most important active safety systems installed on road vehicles. The designers of the first ABS systems had to face primitive braking actuators that admitted only discrete control actions (e.g., increase, hold, or release braking pressure [22]). With time, and with the introduction of higher level functionalities (e.g., electronic brake distribution [11] and electronic stability control [23, 20]), it became clear that a smooth and precise control of the braking torque would further improve safety and performance. Furthermore, with the introduction of (hybrid) electric vehicles, new requirements arise. To optimize regenerative braking, the friction brakes need to be continuously controlled (whereas other safety systems are active only during emergencies). At the same time, the more challenging domain of motor sports has

[1]Politecnico di Milano, Milano, Italy, e-mail: matteo.corno@polimi.it
[2]Politecnico di Milano, Milano, Italy, e-mail: fabio.todeschini@polimi.it
[3]Politecnico di Milano, Milano, Italy, e-mail: giulio.panzani@polimi.it
[4]Politecnico di Milano, Milano, Italy, e-mail: sergio.savaresi@polimi.it

started looking into the possibility of introducing some level of automatic control to improve racing performance.

All these factors make the case for brake-by-wire (BBW) a compelling one. BBW systems, by mechanically decoupling the braking system and the braking pedal/lever, allow for a genuinely automatic control of the braking pressure and with that higher performance and smoothness can be achieved. Over the years, the BBW philosophy has found several implementations. BBW systems can be generally clustered into two families: electro-hydraulic braking (EHB) [21, 6, 24] and electro-mechanical braking (EMB) [12, 13].

EHB systems maintain a significant portion of the components used in common hydraulic braking systems (such as calipers, hydraulic link, and accumulators) with the associated benefits in terms of cost and reliability. EMB systems are based on the direct electro-mechanical actuation of the caliper. EMBs yield a significant decrease in volume, weight, and component number, and can also offer faster response times [5]. However, they require a major redesign of the braking system. This, along with fail-safe operation concerns, still represents a significant obstacle for pure EMB designs.

Within this context, the chapter presents the study and modeling of an emerging BBW prototype specifically designed for racing motorcycles that hybridizes the EHB and EMB approaches. The main idea of this smart-actuator (Figure 15.1) is to retain the calipers and the braking lines of today's vehicles. However, instead of using electro-valves, accumulators, and pumps to regulate the braking pressure (as is done in EHB), an electro-mechanical actuator is directly connected to the piston of the master cylinder to generate the braking pressure. With this approach, the cost and the number of components can be kept relatively low: only the electro-mechanical actuator needs to be added to the braking system. This latter feature is also very beneficial for vehicles subject to packaging constraints (as motorbikes) [7]. Further, given the high bandwidth of the electro-mechanical actuator, the response times of the hybrid configuration are expected to be faster than those of a classical EHB design based on accumulators and valves.

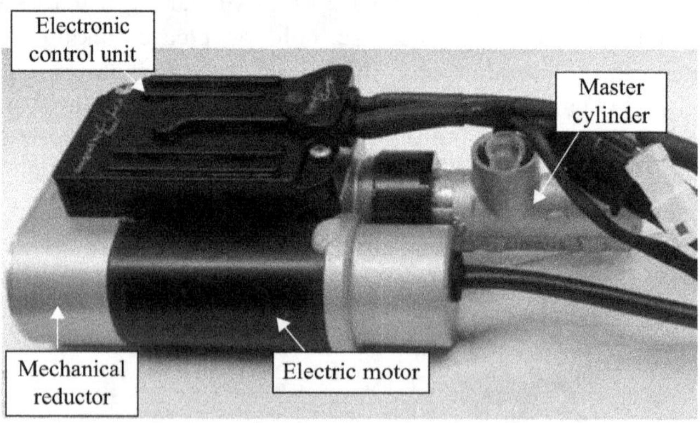

Figure 15.1 Brake-by-wire system prototype

The main goal of this chapter is to develop a control-oriented model useful to design advanced control systems and perform sensitivity analysis. The model has to be accurate, but at the same time easy to use for control system design objectives.

The remainder of this chapter is structured as follows: section 15.2 describes the actuator design and main features. Section 15.3 derives the first-principle model, whose parameters are then identified in section 15.4. Section 15.5 is finally devoted to the validation and analysis of some modeling choices.

15.2 System description

The BBW actuator was first introduced in [7]. Figure 15.2 portrays the entire system. Several different elements are observed:

- DC motor: The system is actuated by a 200 W DC brushed electric motor. The motor is powered by the vehicle electric power system at 12 V.
- Ball-screw transmission: The motor rotational motion is converted in linear displacement by a ball-screw mechanism, pushing the master cylinder piston and building up the necessary pressure.
- Master cylinder: The motor actuates the piston in the master cylinder. The piston displacement creates the pressure in the hydraulic circuit. The piston is also connected to a return spring that pushes the piston back in its resting position when no force is applied by the motor.
- Compensation chamber: In order to compensate for hydraulic fluid volume variations due to temperature, wear, and leakages, the system is equipped with a compensation chamber. The chamber is connected to the master cylinder; when the piston is completely retracted the chamber is connected to the main hydraulic circuit. When the piston moves, the chamber is shut off and isolated from the main hydraulic circuit. Such a mechanism is common in all braking systems.
- Pipeline: The pipeline hydraulically connects the master cylinder to the caliper and pads which are then pressed against the disk by the oil pressure.

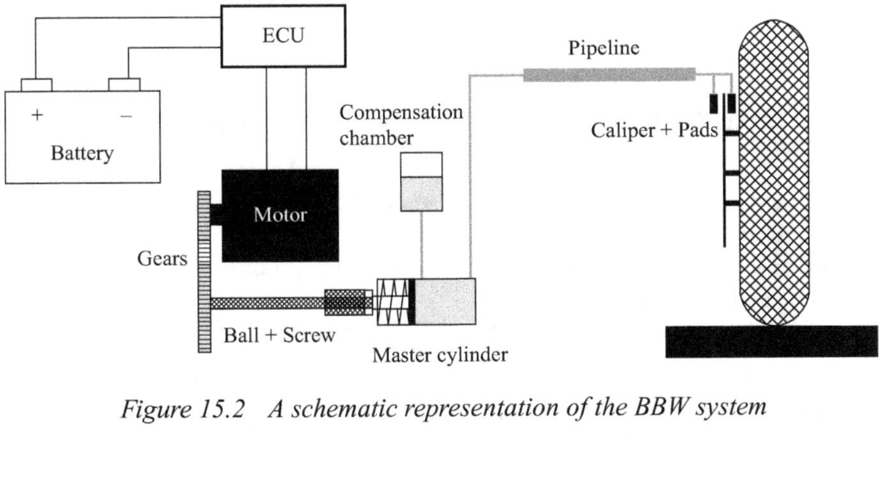

Figure 15.2 A schematic representation of the BBW system

Table 15.1 *Physical parameters of the BBW actuator*

Variable	Symbol	Value
inertia of motor + gear	J_m	1.1×10^{-5} kg m^2
mass of the master cylinder piston	m_c	1×10^{-2} kg
combined reduction ratio	G	3.294×10^3 rad/m
current/torque gain	k_m	16.8×10^{-3} Nm/A
area of the master cylinder	A_c	1.13×10^{-4} m^2
length of the master cylinder	L_c	29×10^{-3} m
bulk modulus of the brake fluid*	β	1.6×10^9 Pa
area of the pad's cylinder	A_p	3.22×10^{-3} m^2
stiffness of the pad*	k_p	1.28×10^8 N/m
pipeline diameter	–	8×10^{-3} m
pipeline length	–	0.8 m

*Nominal value.

- Caliper and pads: The braking torque is obtained by pushing the pads against the disk. The pads are mounted on the caliper, which is basically a hydraulic actuator.

Besides the above-mentioned hardware, the system is equipped with electronics and sensors. A pressure transducer (installed close to the master cylinder outlet) measures the pressure in the hydraulic system and a Hall effect encoder measures the rotational position of the motor which is easily translated into a linear displacement of the piston knowing the transmission kinematics. Finally, a Hall current sensor measures the current through the motor windings.

The smart-actuator is equipped with a programmable electronic unit that is employed to perform the identification experiments and to log the relevant variables.

From the system component datasheets and the mechanical design the values of some characteristics are known. These are summarized in Table 15.1.

15.3 Brake-by-wire modeling

From the description of the system, it is easy to separate the system into three different domains:

- electric domain: made up of the vehicle power system, ECU, current sensor, and DC motor;
- mechanical domain: made up of the DC motor, gears, ball-screw mechanism, piston, and master cylinder;
- hydraulic domain: hydraulic fluid, pipelines, and caliper.

The three domains are coupled and connected. In the following paragraphs, the modeling details of each components are discussed.

15.3.1 Electric domain modeling

The main actuator is a permanent magnet DC motor; a simple RL circuit augmented with a velocity-dependent counter-electromotive force suffices to describe its main dynamics (see Figure 15.3). The model equations are

$$\begin{cases} V(t) = R_{mot}i(t) + L_{mot}\dfrac{di(t)}{dt} + e(t) \\ e(t) = K_e\Omega(t) \\ T(t) - T_r(t) = J_{mot}\dfrac{d\Omega(t)}{dt} \\ T(t) = K_t i(t) \end{cases} \tag{15.1}$$

where $V(t)$ is the supplied voltage, $e(t)$ is the counter-electromotive force, and $i(t)$ is windings current. Further T, T_r, J_{mot} and Ω are respectively motor torque, load torque, and motor inertia, and motor angular velocity. The resistance R_{mot} and inductance L_{mot} are provided by the manufacturer; however, it is usually preferred to identify them from data. In this way any possible parasitic effect of the wiring and sensing devices is also taken into account by the model.

As shown by model (15.1) the motor torque is proportional to the current; it is therefore useful to have an inner-current closed-loop control. Such closed loop can be easily designed to have a bandwidth at least at order of magnitude faster than the pressure dynamics of interest. Once such current loop is implemented, motor current can be considered the actual input of the pressure dynamics.

15.3.2 Mechanical domain modeling

The main physical quantities that affect the mechanical dynamics are

- motor inertia J_{mot};
- motor torque T and motor angular velocity Ω;
- gear ratio K_{rid} and screw pitch ρ_{rid};
- piston mass M_{pist};

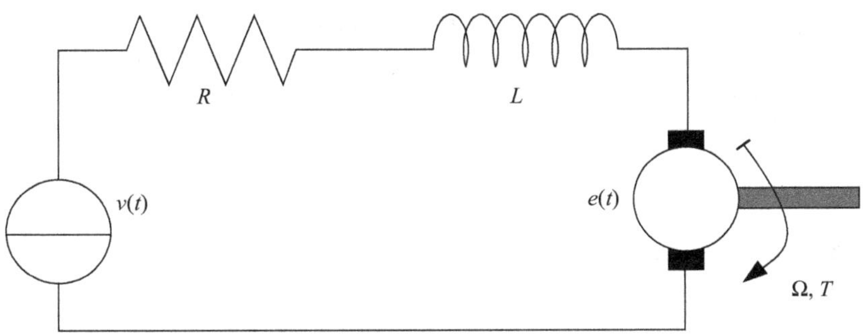

Figure 15.3 Equivalent electric circuit of the permanent magnet DC motor

- motor force at the piston F_{pist}, piston linear velocity v, and displacement x;
- load force acting on the piston F_r.

The mechanism is arranged in such a way that when the motor is fully retracted the ball-screw mechanism peg is not in contact with the piston. As a consequence the piston displacement x and velocity v are given by

$$\begin{cases} x = 0 & x_{mot} < x_{kiss} \\ x = x_{mot} - x_{kiss} & x \geq x_{kiss} \end{cases} \tag{15.2}$$

as well as the velocity discontinuity

$$\begin{cases} v = 0 & x_{mot} < x_{kiss} \\ v = \dot{x}_{mot} & x_{mot} \geq x_{kiss} \end{cases} \tag{15.3}$$

where x_{mot} is the position of the tip of the ball-screw mechanism and x_{kiss} the position where the tip of the ball-screw mechanism is in contact with the piston.

The motor angular velocity Ω is found to be proportionally related to the piston linear velocity v:

$$\Omega = K_{rid}\frac{2\pi}{\rho_{rid}}v \Rightarrow v = \frac{1}{K_{rid}}\frac{\rho_{rid}}{2\pi}\Omega = K\Omega \tag{15.4}$$

The instantaneous variation of the total kinetic energy of the system is

$$\sum_{i=1}^{m}\mathbf{F_i} \times \mathbf{v_i} + \sum_{j=1}^{n}\mathbf{T_j} \times \boldsymbol{\omega}_\mathbf{j} = \frac{dE_c}{dt} \tag{15.5}$$

from which

$$P_{mot} + P_{trasm} + P_{pist} = \frac{d}{dt}(E_{c,mot} + E_{c,pist}) \tag{15.6}$$

where P_{mot} is the motor power, P_{trasm} is the transmission power loss, and P_{pist} is the power due to the forces on the piston. The total kinetic energy is divided into $E_{c,mot}$ and $E_{c,pist}$, respectively the kinetic energy of the motor and of the piston.

The motor quantities are immediately derived as

$$P_{mot} = T\Omega \tag{15.7}$$

$$E_{c,mot} = \frac{1}{2}J_{mot}\Omega^2 \tag{15.8}$$

$$P_{pist} = F_{pist}v \tag{15.9}$$

$$E_{c,pist} = \frac{1}{2}M_{pist}v^2 \tag{15.10}$$

The load force F_{pist} in (15.9) is the summation of all forces acting on the piston; in total we account for three terms

$$F_{pist} = F_{frict} + F_{press} + F_{spring} \tag{15.11}$$

F_{frict} is due to friction in the master cylinder. F_{press} represents the force due to the hydraulic fluid, and F_{spring} is the force generated by the return spring. The term due to the pressure is further developed in

$$F_{press} = A_{pist}p_{MC} \tag{15.12}$$

The fluid pressure in the master cylinder p_{MC} is the interface of the mechanical domain and the hydraulic domain; its expression will be derived later. F_{spring} is due to two terms, a static preload and a position-dependent term

$$F_{spring} = F_{preload} + K_{spring}(x) \tag{15.13}$$

Modeling the term F_{frict} requires a more convoluted model. Several friction models have been proposed in the literature [18, 2, 15]; in the rest of the chapter, we will refer to the LuGre model [3]. The LuGre model represents a viable trade-off between the accuracy of the model (its ability to capture pre-sliding displacement, frictional lag, or the variable breakaway force), and the number of parameters on which it depends. It models the contact between the two surfaces as the interaction between a number of flexible appendixes, called bristles. When the two surfaces move with respect to one another, the bristles plastically deform. This deformation generates the friction force; if the relative displacement is increased the bounds between the bristles break down and the friction force decreases to only the viscous friction. The average bristle deformation is described by the differential equation [29]

$$\frac{dz}{dt} = v - \frac{|v|}{g(v)}z \tag{15.14}$$

where v is the relative velocity. The friction force dependency on the bristle deformation and the relative velocity is

$$F(z, v) = \sigma_0 z + \sigma_1 \frac{dz}{dt} + \sigma_2 v \tag{15.15}$$

where σ_0 and σ_1 are the bristle stiffness and damping and σ_2 the viscous friction coefficient. At steady state, (15.14) shows that the bristle deformation is

$$z = \frac{|v|}{v}g(v) = g(v)\,\mathrm{sign}(v) \tag{15.16}$$

From the above expression a couple of observations are due: (1) $g(v)$ is a positive function as it is multiplied by the relative velocity v. (2) $g(v)$ is the magnitude of

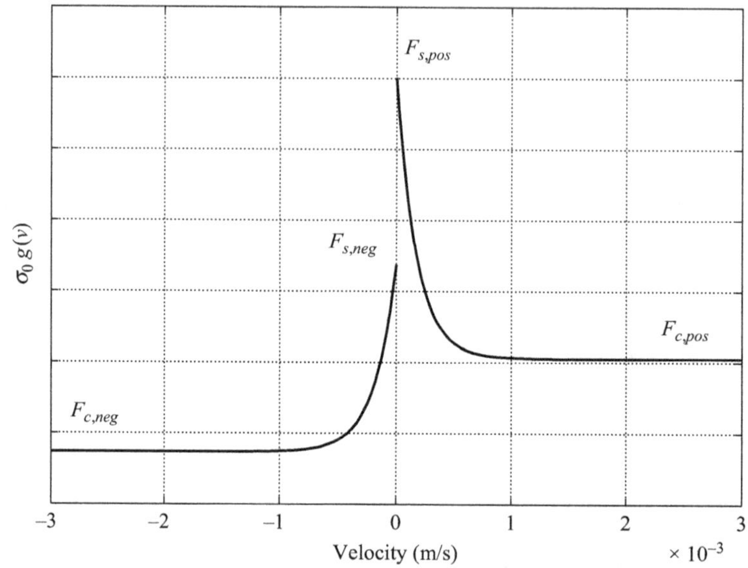

Figure 15.4 Example of $\sigma_0\, g(v)$ and the role of its parameters

the static friction force. In this context a modified asymmetric Stribeck curve [8] is adopted:

$$g(v) = \begin{cases} \dfrac{1}{\sigma_0}(F_{pr,neg} + f_{cfr,neg}p)(1 + (k_{brk,neg} - 1)e^{-c_{v,neg}v^2}), & v < 0 \\ \dfrac{1}{\sigma_0}(F_{pr,pos} + f_{cfr,pos}p)(1 + (k_{brk,pos} - 1)e^{-c_{v,pos}v^2}), & v \geq 0 \end{cases} \tag{15.17}$$

Equation (15.17) introduces new symbols: the term $F_c = F_{pr} + f_{cfr}p$ is what is usually related to the Coulomb friction. It is made pressure-dependent through the parameter f_{cfr}. The dependency on the pressure is needed because of the ball-screw mechanism. The higher the load on the mechanism, the more the balls will be under stress and the higher the friction will be. k_{brk} is the ratio between the breakaway force F_s and the Coulomb force F_c. c_v is the transition coefficient that describes the transition from micro-sliding regime to the sliding regime. Figure 15.4 plots $\sigma_0\, g(v)$ for a given value of the pressure p.

Besides the friction acting on the master cylinder piston, P_{trasm} accounts for an additional dissipative term.

$$P_{trasm} = F_{loss}\Omega = r_{loss}\Omega^2 \tag{15.18}$$

Note that for $x_{mot} < x_{kiss}$, P_{trasm} is the main dissipative term.

If $x_{mot} \geq x_{kiss}$ and manipulating (15.6) one obtains

$$T\Omega - r_{visc}\Omega^2 - F_r v = J_{mot}\Omega\dot{\Omega} + M_{pist}v\dot{v} \tag{15.19}$$

which can be expressed as a function of the angular velocity

$$T\Omega - r_{visc}\Omega^2 - F_r K\Omega = J_{mot}\Omega\dot{\Omega} + M_{pist}K^2\Omega\dot{\Omega} \tag{15.20}$$

Plugging in the expression of $\dot{\Omega}$ yields

$$\dot{\Omega} = \frac{T - F_r K - r_{visc}\Omega}{J_{mot} + M_{pist}K^2} \tag{15.21}$$

Note the following:

- The load force is scaled by the final gear ratio K. A small motor torque can therefore exert the required force.
- The mass of the piston M_{pist} is scaled by the square of the final gear ratio K; as a consequence it is negligible with respect to the motor inertia J_{mot}.

15.3.3 Hydraulic domain

The hydraulic domain is composed of the compensation chamber, the master cylinder, the pipeline, and the brake caliper. The model of each component is detailed out.

The hydraulic fluid is a DOT 4 fluid; its compressibility is described by the bulk modulus β. The bulk modulus relates the pressure variation with the volume variation as in (see also [26])

$$\beta = -V\frac{\partial p}{\partial V} \tag{15.22}$$

The bulk modulus is a property of the fluid mix. Despite all the efforts to perfectly bleed the hydraulic system, a certain amount of air is always mixed with the hydraulic fluid. This can be accounted by defining the total volume of the fluid as the sum of the fluid volume and air fluid:

$$V = V_{oil} + V_{air} \tag{15.23}$$

The ratio between the two volumes is called R:

$$R = \frac{V_{air}}{V_{oil}} \tag{15.24}$$

R is usually around a few percent (we will refer to $R = 0.01$). The effective bulk modulus β_{eff} of the oil–air mix is thus ([26])

$$\beta_{eff}(p, R) = \beta\left[\frac{R + \left(\frac{p}{p_{atm}} + 1\right)^{\frac{1}{\gamma}}}{\frac{R}{\gamma}\frac{\beta}{p_{atm} + p} + \left(\frac{p}{p_{atm}} + 1\right)^{\frac{1}{\gamma}}}\right] \tag{15.25}$$

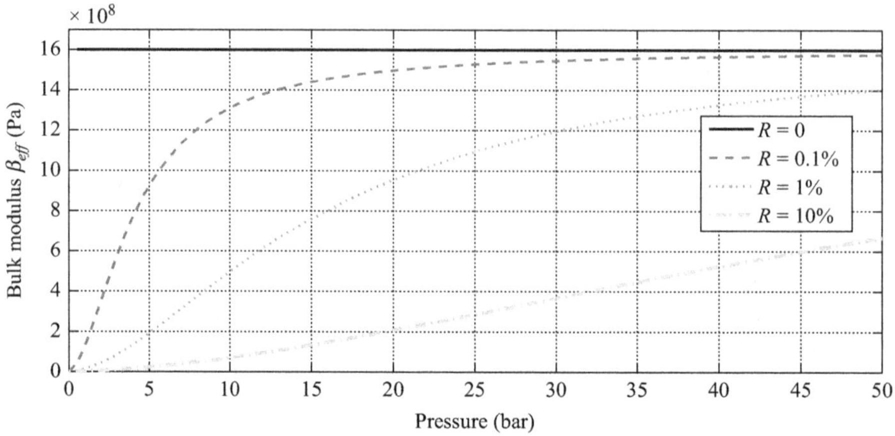

Figure 15.5 β_{eff} *as a function of fluid pressure for several values of R*

where p is the fluid pressure and p_{atm} is the atmospheric pressure. The coefficient γ is the adiabatic index. Figure 15.5 plots the resulting pressure dependence of β_{eff} on pressure for different air-to-oil ratios. Not surprisingly, as the air-to-oil ratio decreases the effective bulk modulus tends to that of oil; moreover, at high pressures the fluid becomes stiffer. In a normal motorcycle braking system, the maximum braking pressure is around 20÷30 bar, and a constant bulk modulus approximation would not be an accurate description of the fluid properties.

As the fluid is pushed by the master cylinder pressure, it is useful to analyze the type of flow in the system. The hydrodynamic flow is well described by the dimensionless Reynolds number

$$\mathrm{Re} = \frac{\rho u d}{\mu} \tag{15.26}$$

where d is the pipeline diameter, ρ, u, and μ are the fluid density, velocity, and dynamic viscosity. Depending on the value of Re, one may have the following:

- for $\mathrm{Re} \le 2300$ the flow is laminar;
- for $2300 < \mathrm{Re} < 3500$ the flow is in a transitional state;
- for $\mathrm{Re} \ge 3500$ the flow is turbulent.

Expression (15.26) is rewritten as a function of the more easily determined kinematic viscosity ν:

$$\nu = \frac{\mu}{\rho} \tag{15.27}$$

$$\mathrm{Re} = \frac{ud}{\nu} \tag{15.28}$$

The flow velocity is closely related to that of the piston in the master cylinder, which is at most 100 m/s. From this velocity a value of Re $\cong$ 0.64 is found, which sets the hydrodynamic regime well within the laminar flow. This result is not surprising as the braking system is basically a force actuator, rather than a displacement actuator. The caliper and pads are designed to have a very limited stroke, limiting the fluid flow.

15.3.3.1 Master cylinder

Figure 15.6 represents the main variables of interest for the model of the master cylinder. As already discussed in the mechanical domain section, when $x_{mot} < x_{kiss}$ hydrodynamics is not affected by the force exerted by the electric motor. The most relevant condition is therefore $x_{mot} \geq x_{kiss}$ or $x \geq 0$. In these conditions, the lumped mass conservation law applied to the master cylinder yields

$$\frac{dM}{dt} = \frac{d[\rho A_{pist}(L_{pist} - x)]}{dt} = -\rho(q_{cha} + q) \tag{15.29}$$

where q_{cha} and q are the volumetric flow respectively to the compensation chamber and the master cylinder outlet. Considering a rigid master cylinder and a compressible fluid

$$A_{pist}(L_{pist} - x)\frac{d\rho}{dt} + \rho A_{pist}\frac{d(L_{pist} - x)}{dt} = -\rho(q_{serb} + q) \tag{15.30}$$

Recalling that p and density ρ are related by

$$d\rho = \frac{\rho}{\beta_{eff}} dp \tag{15.31}$$

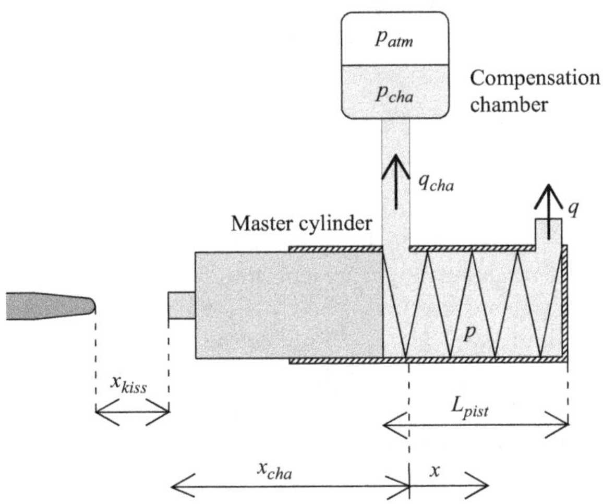

Figure 15.6 Schematics of the master cylinder model

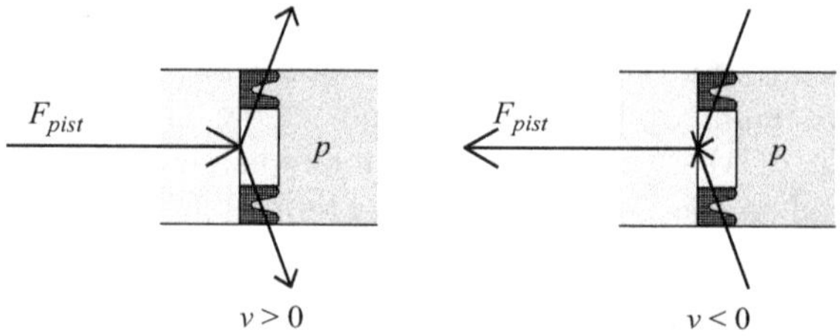

Figure 15.7 Lip seal behavior for positive and negative piston velocities

one obtains

$$A_{pist}(L_{pist} - x)\frac{\rho}{\beta_{eff}}\frac{dp}{dt} - \rho A_{pist}\frac{dx}{dt} = -\rho(q_{serb} + q) \tag{15.32}$$

Canceling ρ out and plugging the expression of p yield

$$\frac{dp}{dt} = \frac{\beta_e}{A_{pist}(L_{pist} - x)}(A_{pist}\dot{x} - q_{cha} - q - q_l) \tag{15.33}$$

Note that (as better explained in the following section) q_{cha} is null for $x > x_{cha}$. The term q_l accounts for the leakage through the piston seal. The leakage through the lip seal exhibits two different behaviors depending on whether the piston is moving in one direction or the other (see Figure 15.7). The leakage flow is modeled through two leakage C_l coefficients and the following expressions:

$$q_l = \begin{cases} C_{l,pos}(p - p_{cha}) & v \geq 0 \\ C_{l,neg}(p - p_{cha}) & v < 0 \end{cases} \tag{15.34}$$

15.3.3.2 Compensation chamber

As pointed out in section 15.2, the hydraulic fluid compensation chamber provides a fluid reservoir that compensates for changes in volume due to temperature and braking pad wear. Figure 15.8 schematically represents the compensation chamber model. The key component is a free surface tank where the surface pressure is the atmospheric pressure p_{atm}. The fluid level, h_{cha}, determines the pressure, p_{cha}, at the bottom of the tank, which is connected to the master cylinder by a very short duct. q_{cha} is the inlet volumetric flow to the tank. Being a free surface tank the fluid compressibility is neglected. Given these considerations and initially considering $x = 0$ the governing equation is

$$p_{cha} = p_{atm} + \rho g h_{cha} \tag{15.35}$$

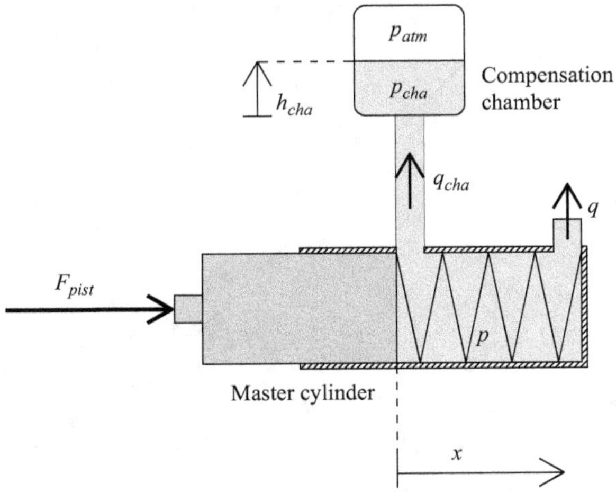

Figure 15.8 Schematics of the compensation chamber model

where g is the gravity acceleration. Using the lumped conservation of mass equation

$$\frac{dM}{dt} = w_{cha} \tag{15.36}$$

where M is the total fluid mass in the tank and w_{cha} is the inlet mass flow. Simple algebraic manipulation leads to

$$\frac{d(\rho A_{cha} h_{cha})}{dt} = \rho q_{cha} \Rightarrow A_{cha}\frac{dh_{cha}}{dt} = q_{cha} \tag{15.37}$$

As shown in Figure 15.8, the flow q_{cha} depends on the piston position: when $x = 0$, the duct to the tank is completely open; as x increases, the duct is partially closed till the point where the compensation chamber is shut off. The position, x_{cha}, corresponding to the full closure of the compensation chamber is easily measured at 2 mm. The piston position–dependend aperture is modeled as a load loss, i.e., a valve whose aperture is determined by x. Figure 15.9 plots the valve characteristics. The flow q_{cha} is then given by the valve equation

$$q_{serb} = C_e A_0 A_{\%}(x)\sqrt{\frac{2}{\rho}|p_{MC} - p_{cha}|}\,\mathrm{sign}(p_{MC} - p_{cha}) \tag{15.38}$$

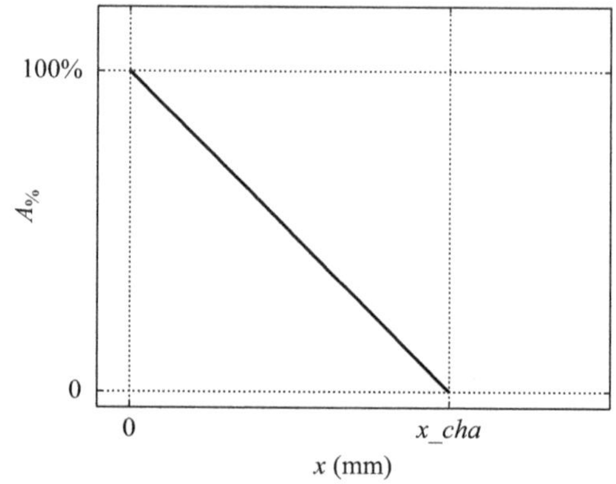

Figure 15.9 Characteristic of the valve connecting the master cylinder to the compensation chamber

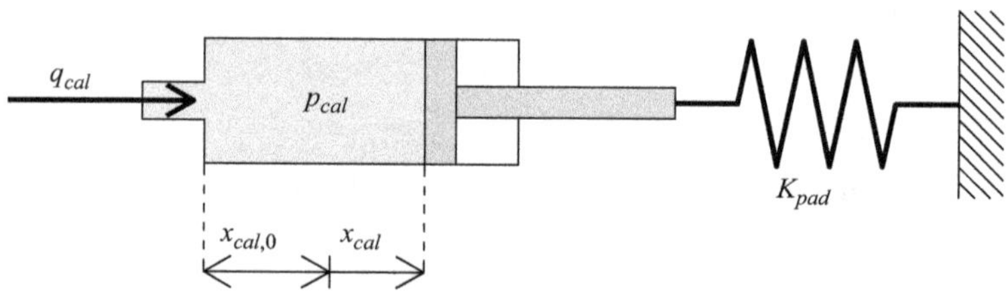

Figure 15.10 Caliper and braking pad schematics

The complete compensation tank model is obtained by the union of expressions (15.35), (15.37), and (15.38) into

$$\begin{cases} p_{cha} = p_{atm} + \rho g h_{cha} \\ \dfrac{dh_{cha}}{dt} = \dfrac{1}{A_{cha}} q_{cha} \\ q_{cha} = C_e A_0 A_{\%}(x) \sqrt{\dfrac{2}{\rho} |p_{MC} - p_{cha}|} \operatorname{sign}(p_{MC} - p_{cha}) \end{cases} \tag{15.39}$$

15.3.3.3 Caliper and braking pads

The model of the caliper is very similar to that of the master cylinder. Figure 15.10 represents the main elements of this submodel. The caliper can be modeled as a hydraulic piston with an inlet flow q_{cal} and internal pressure p_{cal}. Actually the braking system is composed of four calipers; for simplicity's sake they are lumped into one caliper with four times the area.

The braking pads can be modeled by their equivalent stiffness K_{past} [9], which depends on the area of the pads A_{pad} and their normalized stiffness K^{norm} (see [16]):

$$K_{pad} = K_{pad}^{norm}(2A_{pad}) \tag{15.40}$$

Plugging the caliper and pad parameters into the conservation of mass yields

$$\frac{dp_{cal}}{dt} = \frac{\beta_{eff}}{A_{cal}(x_{cal,0} + x_{cal})}(q_{cal} - A_{cal}\dot{x}_{cal}) \tag{15.41}$$

The caliper piston position x_{pad} is derived from the force balance neglecting the inertia of the piston (the mass of the piston is in the order of grams):

$$p_{cal}A_{cal} - K_{pad}x_{cal} = 0 \tag{15.42}$$

From which, the velocity of the caliper piston is

$$\dot{x}_{cal} = \frac{A_{cal}}{K_{pad}}\dot{p}_{cal} \tag{15.43}$$

Plugging the above expression into (15.41) yields

$$\dot{p}_{cal} = \frac{\beta_{eff}K_{pad}}{A_{cal}x_{cal,0}K_{pad} + A_{cal}^2 p_{cal} + \beta_e A_{cal}^2} q_{cal} \tag{15.44}$$

The above fractional term can be rewritten as

$$\frac{\beta_{eff}K_{pad}}{A_{cal}x_{cal,0}K_{pad} + A_{cal}^2 p_{cal} + \beta_e A_{cal}^2} = \frac{1}{\dfrac{V_{cal,0}}{\beta_{eff}} + \dfrac{A_{cal}^2 p_{cal}}{\beta_{eff}K_{pad}} + \dfrac{A_{cal}^2}{K_{pad}}} \tag{15.45}$$

Using the nominal parameters from the datasheet at 15 bar one obtains

$$\begin{cases} \dfrac{V_{cal,0}}{\beta_{eff}} \approx 6 \times 10^{-15} \\ \dfrac{A_{cal}^2 p_{cal}}{\beta_{eff}K_{pad}} \approx 8 \times 10^{-17} \\ \dfrac{A_{cal}^2}{K_{pad}} \approx 8 \times 10^{-14} \end{cases}$$

which indicates that the second term is negligible with respect to the others.

The model of the caliper and pad subsystem becomes

$$\dot{p}_{cal} = \frac{1}{\dfrac{V_{cal,0}}{\beta_{eff}} + \dfrac{A_{cal}^2}{K_{pad}}} q_{cal} \tag{15.46}$$

15.3.3.4 Pipeline

The pipeline connects the master cylinder to the caliper. The length of the pipeline is such that the transport phenomena cannot be neglected. The behavior of the fluid in the pipeline is described by the conservation of mass and momentum:

$$\begin{cases} \dfrac{\partial \rho A_{pipe}}{\partial t} + \dfrac{\partial w}{\partial x} = 0 \\ \dfrac{\partial \rho A_{pipe} u}{\partial t} + \dfrac{\partial \rho A_{pipe} u^2}{\partial x_p} + A_{pipe}\dfrac{\partial p}{\partial x_p} + \rho A_{pipe} g \dfrac{dz}{dx_p} + \dfrac{C_f}{2}\rho\omega|u|u = 0 \end{cases} \tag{15.47}$$

where A_{pipe} is the pipeline section, x_p is the position along the length L_{pipe} of the pipeline, u is the fluid velocity and z is the height of the pipeline, C_f is the so-called Fanning coefficient, and ω is the perimeter of the pipeline. The Fanning coefficient accounts for the friction along the pipeline surface and, in case of laminar flow, is

$$C_f = \frac{64}{\mathrm{Re}} = \frac{64\nu}{d_{pipe} u} \tag{15.48}$$

where d_{pipe} is the inner diameter of the pipeline. With simple substitutions, the friction term in the conservation of momentum can be rewritten as

$$\frac{C_f}{2}\rho\omega|u|u = \frac{32\nu\rho\pi L_{pipe}}{A_{pipe}^2} q = R_{pipe} q \tag{15.49}$$

The pipeline deformation is assumed negligible:

$$\begin{cases} \dfrac{\partial A_{pipe}}{\partial t} = 0 \\ \dfrac{\partial A_{pipe}}{\partial x_p} = 0 \end{cases} \tag{15.50}$$

The conservation of mass expression is further developed, accounting for β_{eff}:

$$A_{pipe}\frac{\rho}{\beta_{eff}}\frac{\partial p}{\partial t} + \frac{\partial w}{\partial x} = 0 \tag{15.51}$$

The conservation of momentum and mass yields a system of partial differential equations. This infinite-dimensional system can be transformed into a finite-dimensional one by means of the finite-difference method [25]. The finite-difference approximation divides the pipeline into N cells in which pressure and flow are assumed constant. If the dependence of p and q on t is omitted

$$\begin{aligned} \frac{\partial P}{\partial x} &\approx \frac{P(i+1) - P(i)}{\Delta x} \\ \frac{\partial q}{\partial x} &\approx \frac{q(i) - q(i-1)}{\Delta x} \end{aligned} \qquad \text{for } i = 2 \dots N \tag{15.52}$$

In (15.52) Δx represents the cell length. If the pressure is assumed constant within the single cell, the integral of the conservation of mass results in

$$\int_0^{L_{pipe}} A_{pipe} \frac{\rho}{\beta_{eff}} \frac{\partial p}{\partial t} dx_p + \int_0^{L_{pipe}} \frac{\partial w}{\partial x} dx_p = 0 \tag{15.53}$$

from which

$$\frac{dp}{dt} = \frac{\beta_{eff}}{\rho A_{pipe} L_{pipe}} (w_{in} - w_{out}) = \frac{\beta_{eff}}{A_{pipe} L_{pipe}} (q_{in} - q_{out}) \tag{15.54}$$

while the integral of the conservation of momentum yields

$$\int_0^{L_{pipe}} \left(\frac{\partial w}{\partial t} + w \frac{\partial u}{\partial x_p} + u \frac{\partial w}{\partial x_p} + A_{pipe} \frac{\partial p}{\partial x_p} \right.$$
$$\left. + \rho A_{pipe} g \frac{dz}{dx} + \frac{C_f}{2} \rho \omega |u| u \right) dx_p = 0 \tag{15.55}$$

Solving the integral and making the dependendency on the volumetric flow explicit (through (15.49)), one obtains

$$\frac{\rho L_{pipe}}{A_{pipe}} \frac{dq}{dt} + \frac{\rho q}{A_{pipe}^2} (q_{out} - q_{in}) + (p_{out} - p_{in}) + \rho g (z_{out} - z_{in}) + R_{pipe} q = 0 \tag{15.56}$$

Plugging (15.54) into (15.56) and writing $(z_{in} - z_{out}) = h_{pipe}$ one gets

$$\begin{cases} \dfrac{dp}{dt} = \dfrac{\beta_e}{A_{pipe} L_{pipe}} (q_{in} - q_{out}) \\ \dfrac{\rho L_{pipe}}{A_{pipe}} \dfrac{dq}{dt} + \dfrac{\rho q}{A_{pipe}^2} (q_{out} - q_{in}) + (p_{out} - p_{in}) + \rho g h_{pipe} + R_{pipe} q = 0 \end{cases} \tag{15.57}$$

The overall pipeline system, composed of N subsystems of the type described by (15.57), requires a series of boundary conditions in order to be interfaced with the rest of the system. The number of variables can be reduced and the integration scheme made more efficient by introducing a further approximation. The underlying idea is that of decoupling the conservation of mass model and the flow friction model. The pipeline is conceptually split into three parts [17]:

- Two parts describe the pressure dynamics considering the fluid as compressible. This model behavior is described by the first equation of the model (15.57):

$$\frac{dp}{dt} = \frac{\beta_{eff}}{A_{pipe} L_{pipe}} (q_{in} - q_{out}) \tag{15.58}$$

- The third part considers the fluid as incompressible and models the flow and friction it is subject to. This is described by the incompressible flow model

$$\frac{\rho L_{pipe}}{A_{pipe}} \frac{dq}{dt} + (p_{out} - p_{in}) + \rho g h_{pipe} + R_{pipe} q = 0 \tag{15.59}$$

The model can be better understood by considering the well-known electrical-hydrodynamic equivalence. In the electric-hydraulic equivalence, pressure is related to voltage, and flow to current; for example, in this equivalence, expression (15.58) represents a hydraulic capacitor with capacity C_{pipe}:

$$C_{pipe} = \frac{A_{pipe} L_{pipe}}{\beta_{eff}} \tag{15.60}$$

Analogously, the terms in (15.59) are

- inductor with inductance $I_{pipe} = \dfrac{\rho L_{pipe}}{A_{pipe}}$;
- potential difference $(p_{out} - p_{in})$;
- ideal voltage generator $U_{pipe} = \rho g h_{pipe}$;
- resistance $R_{pipe} = \dfrac{32 \nu \rho \pi L_{pipe}}{A_{pipe}^2}$.

The equivalent electrical circuit can be therefore arranged as shown in Figure 15.11. The interface variables flow q and pressure p refer to the master cylinder, while q_{pad} and p_{pad} to the caliper and pad. The symbol "$*$" indicates that the pipeline has been divided into its several components making sure that $V_{pipe} = A_{pipe} L_{pipe}$. This approach introduces a new parameter $L_{\%} = 0 \div 100\%$. It indicates the ratio between the fluid modeled as incompressible and the fluid modeled as compressible:

$$\begin{cases} L_p = L_{pipe} L_{\%} \\ L_q = \dfrac{L_{pipe}(1 - L_{\%})}{2} \end{cases} \tag{15.61}$$

where L_p represents the length of the pipeline considered compressible and L_q the incompressible one. As pointed out above, the fluid flow is very limited, this consideration leads to considering high values for the L parameter.

The several models described in the previous sections are assembled in the complete BBW model as shown in Figure 15.12.

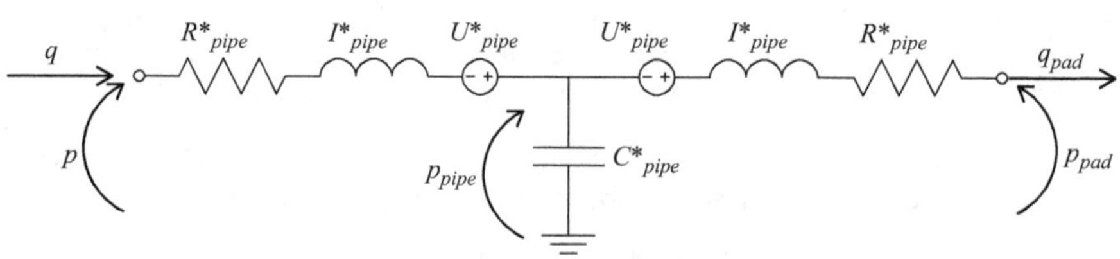

Figure 15.11 Electric equivalence of the half-pipeline

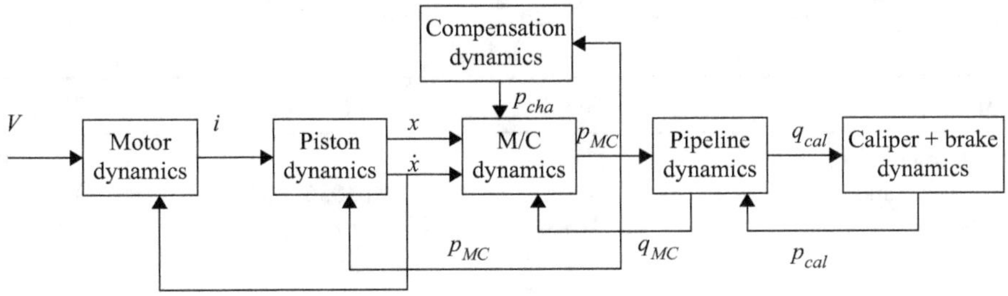

Figure 15.12 Overall model block diagram showing the main interface variables

15.4 Parameter identification

In the previous section the governing equations have been derived. The equations depend on a number of parameters. Some of them (lengths, masses, areas, gear ratios, and so on) are known or easily measured; others are either difficult to measure or represent a collection of more physical quantities that cannot be separated (e.g., friction parameters). The latter quantities are better identified from data in an approach usually referred to as *gray-box* identification. In this section the experiments and methods employed to identify such parameters are described in detail.

Table 15.2 summarizes the parameters that have been assumed known. The parameters to be identified are clustered into three different groups: electrical model parameters, mechanical parameters, and friction parameters.

15.4.1 Electric dynamics identification

The first subsystem to be identified is the electrical subsystem. It is important to start from this subsystem because it can be considered decoupled from all the others and the knowledge of this subsystem is necessary in order to design an inner current control loop. The current control is necessary to perform the other experiments. From the previous section, it can be concluded that the electrical equivalent model of the motor is a first-order linear system. The input of such model is the applied voltage and the output the measured current; its parameters are R_{mot} and L_{mot}. The identification is performed in the frequency domain executing a sine sweep test. The sine sweep test (see Figure 15.13) is performed with a stalled rotor in order to exclude the effect of the counter-electromotive force. The sine sweep experiment is the basis for the derivation of a nonparametric frequency response model of the motor $G_{mot}(j\omega)$. The experimental frequency response is computed as (see [27])

$$\hat{G}_{mot}(j\omega) = \frac{\hat{S}_{II}(j\omega)}{\hat{S}_{Iu}(j\omega)} \tag{15.62}$$

where $\hat{S}_{xy}$ denotes the cross-spectrum of (x, y) estimated using Welch's averaged periodogram. Once the experimental frequency response has been obtained, the

Table 15.2 Known parameters of the BBW system

Name	Symbol	Value	Unit
Mechanical domain			
Final gear ratio	K	0.3036×10^{-3}	m/rad
Piston mass	M_{pist}	10×10^{-3}	kg
Spring constant	K_{spring}	3000	N/m
Spring preload	$F_{preload}$	30	N
Screw-ball kiss point	x_{kiss}	1.4×10^{-3}	m
DOT 4 fluid			
Density ρ	1027	kg/m^3	
Bulk modulus	β	1.6×10^{9}	Pa
Kinematic viscosity	ν	5×10^{-4}	m^2/s
Compensation chamber			
Atmospheric pressure	p_{atm}	101325	Pa
Tank section	A_{cha}	1.26×10^{-3}	m^2
Valve coefficient	C_e	0.03	/
Valve nominal section	A_0	2.83×10^{-5}	m^2
Master cylinder			
Cylinder length	L_{pist}	29×10^{-3}	m
Piston area	A_{pist}	1.13×10^{-4}	m^2
Leakage coefficient $\nu \geq 0$	$C_{l,pos}$	1×10^{-18}	$m^3/s/Pa$
Leakage coefficient $\nu < 0$	$C_{l,neg}$	4×10^{-15}	$m^3/s/Pa$
Pipeline			
Inner section area	A_{pipe}	8.04×10^{-6}	m^2
Length	L_{pipe}	0.80	m
Inner diameter	d_{pipe}	3.20×10^{-3}	m
Pad and caliper			
Piston area	A_{cal}	3.22×10^{-3}	m^2
Volume at rest	$V_{cal,0}$	9.00×10^{-6}	m^3
Pad area	A_{pad}	4.40×10^{-3}	m^2
Pad stiffness	K_{pad}	1.23×10^{8}	N/m

parameters of a first-order system are fitted in order to minimize the difference between the analytical transfer function and the experimental one. Figure 15.14 plots the comparison between the experimental and identified frequency responses. The resulting identified model is

$$G_{i,id}(s) = \frac{0.0388}{1 + \frac{s}{2\pi \cdot f_p}} e^{-0.0005s} \quad \text{with} \quad f_p = 280.7 \text{ Hz} \tag{15.63}$$

The model is validated in the time domain through the multiple sinusoidal applied voltage excitation shown in Figure 15.15. The electric model very accurately captures the dynamics.

Once an accurate model of the motor electric dynamics is available, a current controller is designed with a bandwidth of 100 Hz sufficiently higher than the target

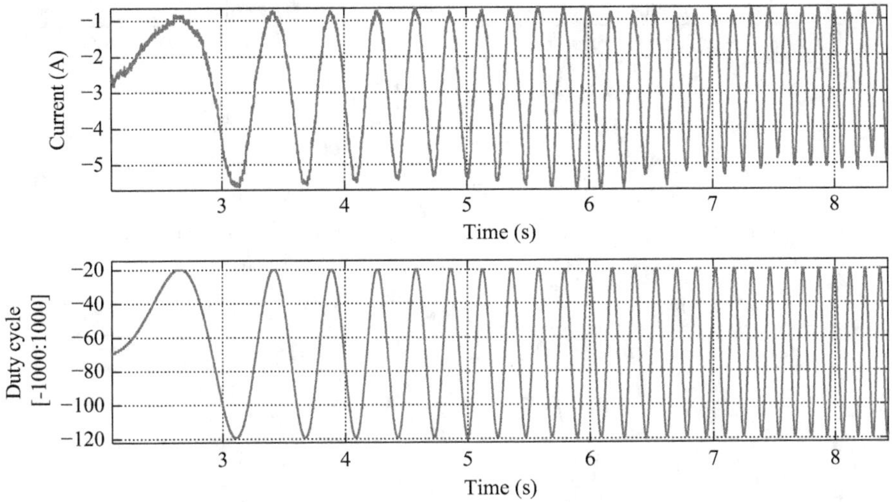

Figure 15.13 Sine sweep experiment performed to identify the electric parameters of the motor model: measured current and applied DC

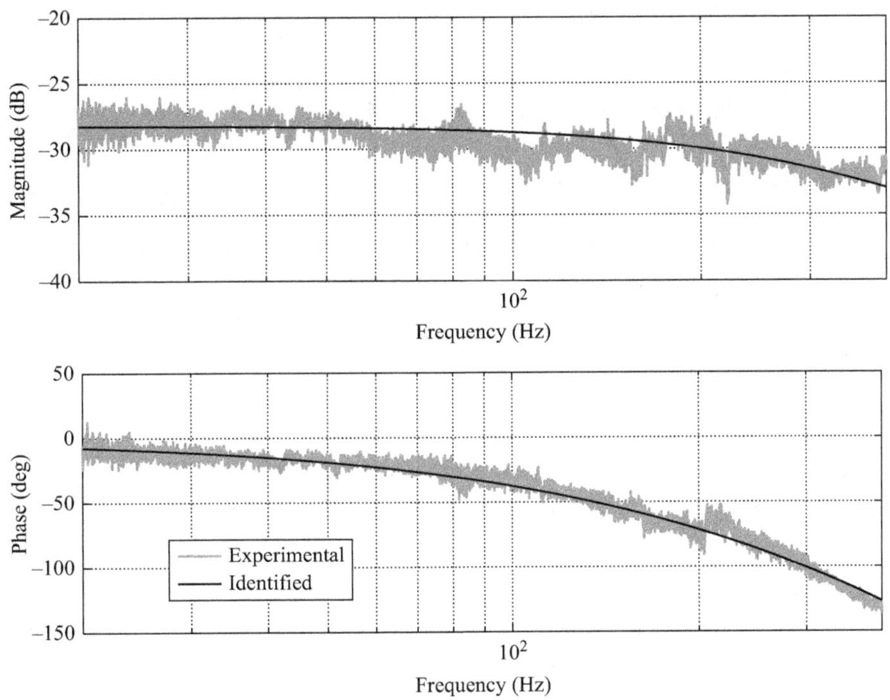

Figure 15.14 Sine sweep experiment performed to identify the electric parameters of the mode: frequency response

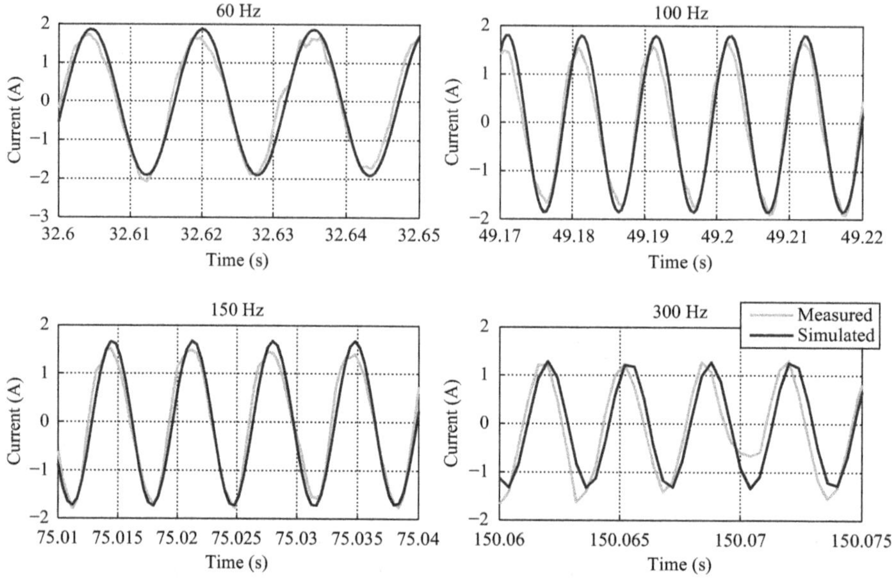

Figure 15.15 Comparison between measured and simulated current at four different frequencies

bandwidth which is around 15–20 Hz [5]. For this point on, it is assumed that such controller is in place and that the actual control variable is the motor reference current.

15.4.2 Motor mechanical dynamics – J_{mot} and r_{visc} – identification

The mechanical subsystem (15.21) depends on the following parameters:

- Final gear ratio K
- Piston mass M_{pist}
- Motor inertia J_{mot}
- Viscous friction r_{visc}

The first two parameters are available from direct measurement and not discussed further. The motor inertia is provided by the motor manufacturer; however, it is preferred to experimentally identify it because once the motor is installed additional elements are connected to the motor shaft whose inertia is hard to measure. Finally, r_{visc} needs to be identified from data as it is a lump factor that accounts for several phenomena: internal motor friction, gear friction, and ball and screw mechanism friction.

In order to identify the two parameters of interest the master cylinder has been detached from the master cylinder piston. The motor only actuates the transmission without the hydraulic load. As a consequence, the model becomes

$$\dot{\Omega} = \frac{T - r_{visc}\Omega}{J_{mot}} \tag{15.64}$$

which is transformed into the Laplace domain as

$$G_{I,\theta}(s) = \frac{\theta(s)}{I(s)} = \frac{K_t}{s(r_{visc} + sJ_{mot})} = \frac{\dfrac{K_t}{r_{visc}}}{s\left(1 + s\dfrac{J_{mot}}{r_{visc}}\right)} \tag{15.65}$$

As the above model has a pole at the origin, this calls for ad-hoc identification experiments. If the experiment is performed in open-loop, only a limited number of data points can be collected before the ball-screw mechanism reaches its mechanical limit. This problem can only be solved by performing closed-loop identification. A first-attempt PID position control is thus designed. The controller can be tuned by trial and error as it is used to keep the ball-screw mechanism within its operational limits and closed-loop performance is not a concern. Once the position control is in place, a number of closed loop experiments are performed: in particular a ramp-like reference position is fed to the controller while the reference current and position are logged. Four experiments are run with different rate of change of the reference position: 2.5, 5, 10, and 25 mm/s. Figure 15.16 plots the 2.5 mm/s test.

Three experiments (2.5, 10, and 25 mm/s) have been employed to identify the parameters, while the last one is left for validation. Equation (15.65) restricts the choice of the model family to

$$G_{I,\theta}(s;\theta) = \frac{K_p}{s(1 + sT)} \tag{15.66}$$

where

$$K_p = \frac{K_t}{r_{visc}} \qquad T = J_{mot} r_{visc} \tag{15.67}$$

The parameters are identified by minimizing the prediction error [14] between the simulated and measured motor position. The final values are $r_{visc} = 1.97 \times 10^{-4}$ N/m/s

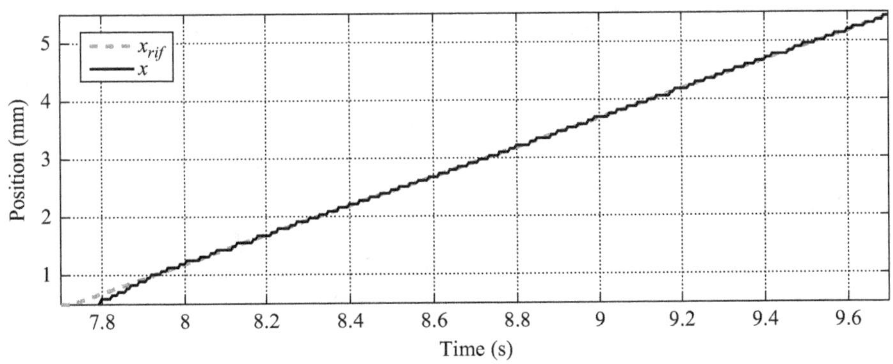

Figure 15.16 Inertia and viscous friction identification experiment at 2.5 mm/s

and $J_{mot} = 1.37 \times 10^{-5}$ kg/m^2. As expected J_{mot} is greater than the motor inertia provided by the manufacturer. Figure 15.17 shows the result of the validation experiment. The matching between the model and the data is satisfactory. The resulting time constant T is

$$T = \frac{J_{mot}}{r_{visc}} = \frac{1.37 \times 10^{-5}}{1.97 \times 10^{-4}} = 69.5\,\text{ms} \Rightarrow f_p = \frac{1}{2\pi}\frac{1}{T} = 2.3\,\text{Hz} \tag{15.68}$$

As in all identification experiment design, the input signal needs to be properly exciting. The ramp signal is not a sufficiently exciting signal; however, note that the input signal employed in the identification of (15.66) is the reference current and not the reference position. The spectrum of the reference current can be analyzed. Figure 15.18 plots the spectrum of the current signal in one of the identification experiments; it is clear that the current signal has components well beyond the frequency of the identified pole, showing that the signal is indeed exciting.

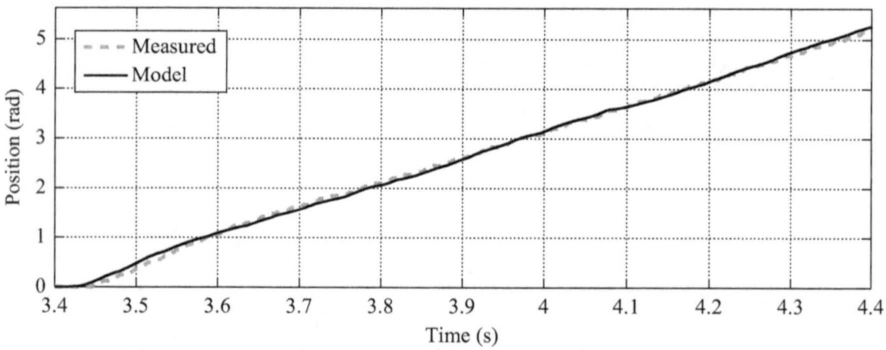

Figure 15.17 Inertia and viscous friction validation experiment at 5 mm/s

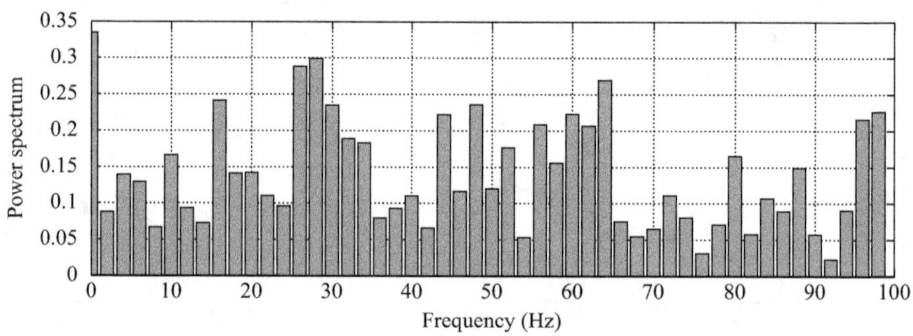

Figure 15.18 Power spectrum of the current signal in the 2.5 mm/s identification experiment

15.4.3 Friction model identification

Another important set of parameters to be identified is that related to the friction model; they are

- $F_{pr,neg}$ and $F_{pr,pos}$: Coulomb friction for zero pressure;
- $f_{cfr,neg}$ and $f_{cfr,pos}$: Coulomb pressure–dependency coefficient;
- $k_{brk,neg}$ and $k_{brk,pos}$: breakaway force to Coulomb force ratio;
- $c_{v,neg}$ and $c_{v,pos}$: transition coefficient;
- σ_0: bristle stiffness;
- σ_1: bristle damping;
- σ_2: viscous friction.

The asymmetry is considered only in the first four pairs of parameters. The remainder of the parameters have been considered symmetric.

A series of experiments have been executed to identify the friction parameters. A series of current ramps with increasing peak current are executed. The current, pressure, and speed measurements (obtained by differentiation of the encoder signal) are used to estimate the friction force. The collection of these data gives a static map relating the friction force and the piston velocity. This static map, along with the assumption of considering that the bristle dynamics have reached steady state, is the basis of the identification of the friction parameters. The identification is performed in two steps:

- The velocity–force maps obtained for different pressures are used to identify, through a least-squares approach, $f_{cfr,neg}$ and $f_{cfr,pos}$, i.e., the scaling factors. This is possible because the entire static map is scaled by the same factor.
- All the available experiments are scaled according to the pressure weighting factors; this effectively cancels out the pressure dependency. The resulting scaled data are then used to fit the remaining parameters (separating the positive velocity experiments from the negative velocity ones). A standard static optimization toolbox [10] is employed.

The proposed two-step identification method reduces the dimensionality of the optimization problem. Figure 15.19 plots some of the experiment employed and the obtained curves, showing a satisfactory agreement between the measures and the friction model. The friction model has been identified statically, meaning that the bristles are assumed to have reached steady state; as a consequence the parameters σ_1 and σ_0 are not identified with this method.

15.4.4 Final parameter identification

Previously some experiments have been designed to separately identify as many parameters as possible; a few parameters, namely σ_0, σ_1, R, and L, still need to be identified. Given the small number of remaining parameters, these are identified by using a time-domain *brute force* method. The current ramp experiments are fed

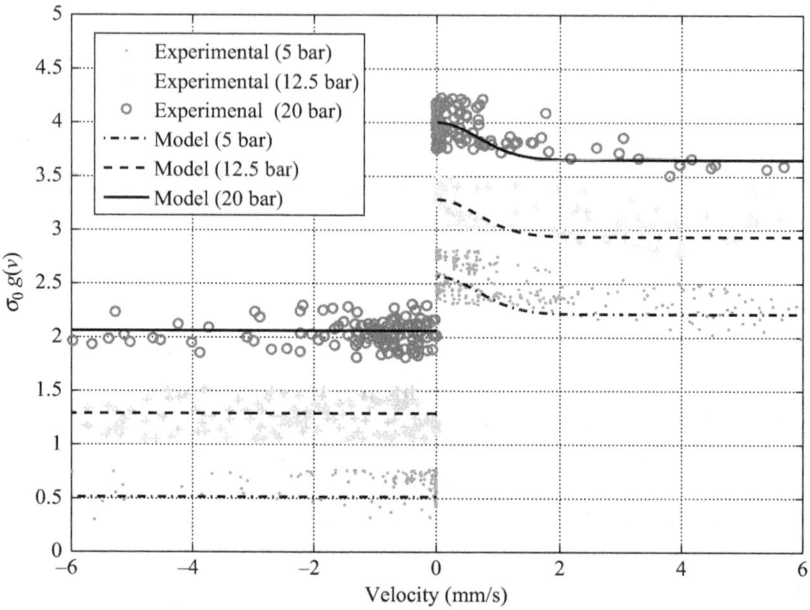

Figure 15.19 Friction characteristic map (experimental data and model)

into an optimization routine that minimizes the simulation error. In the following section, the total model is validated and the importance of these parameters is discussed underlying their effect and sensitivity.

15.5 Validation and analysis

In this section several experiments are employed to validate the proposed model. The section is divided into two parts; in the first part the model is validated against quasi-static and dynamic data measured from the BBW system. In the second part some of the modeling choices will be justified by discussing their effect in simulation.

15.5.1 Validation

Two different sets of experiments will be employed to validate the overall model of the brake-by-wire system: quasi-static and dynamic tests. In the quasi static tests a slow increasing-then-decreasing reference current ramp is applied to the system and the position and pressure are measured and compared. Figure 15.20 plots the results of such an experiment. As clearly seen, the system is heavily affected by stick-slip events that the model is capable of capturing. Also note that the stick-slip events are by nature not very repeatable so the fitting between the measured data and the

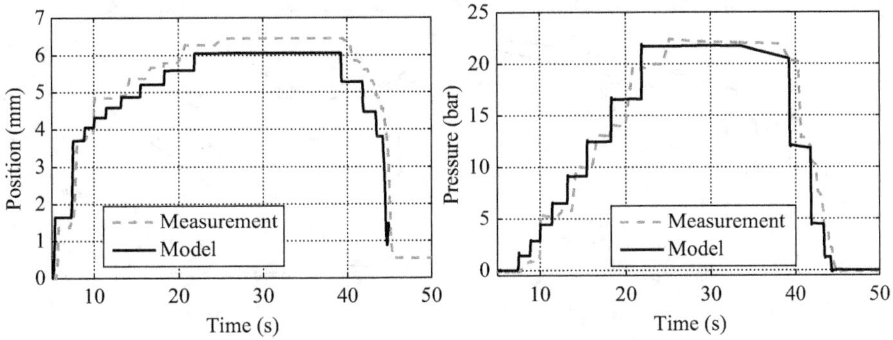

Figure 15.20 Comparison between measured and simulated position and pressure during a quasi-static current increase and decrease

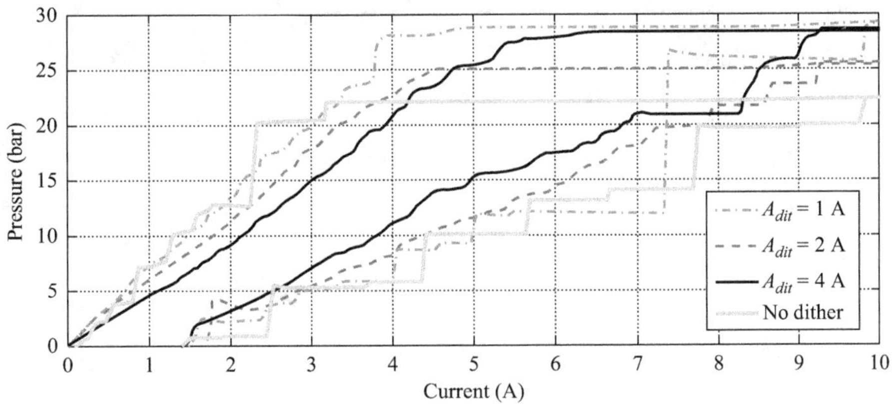

Figure 15.21 Effect of different current dither amplitudes on the pressure response to a quasi-static current increase and decrease

simulated one is not perfect. However, the frequency and amplitude of the stick-slip are correctly captured.

The system is clearly friction dominated, the predominance of the stick-slip phenomenon makes it difficult to further validate the model. One solution is that of overimposing a dither on the current ramp. The dither, by keeping the piston in motion, minimizes the stick-slip phenomenon and allows for a more detailed analysis. Figure 15.21 shows the pressure–current map obtained from the quasi-static experiments for different levels of dither. As the dither amplitude increases the stick-slip phenomenon almost completely disappears. The dither is capable of reducing the stick-slip phenomenon but it does not remove the hysteresis.

Introducing the dither in the real system and in simulation, the validation experiments become more readable. Figure 15.22 shows the comparison between the

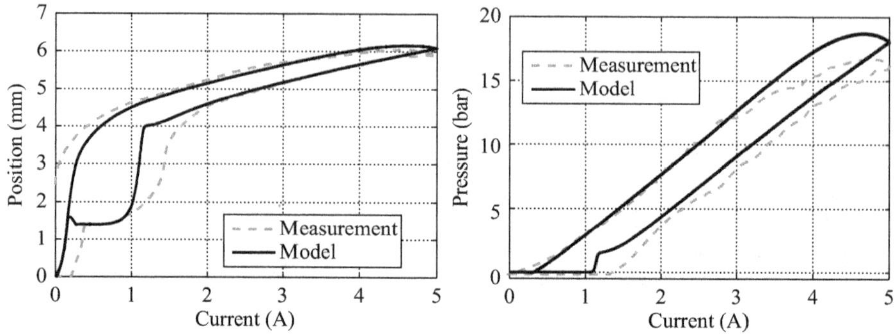

Figure 15.22 Comparison between measured and simulated position and pressure during a quasi-static current increase and decrease, with dither

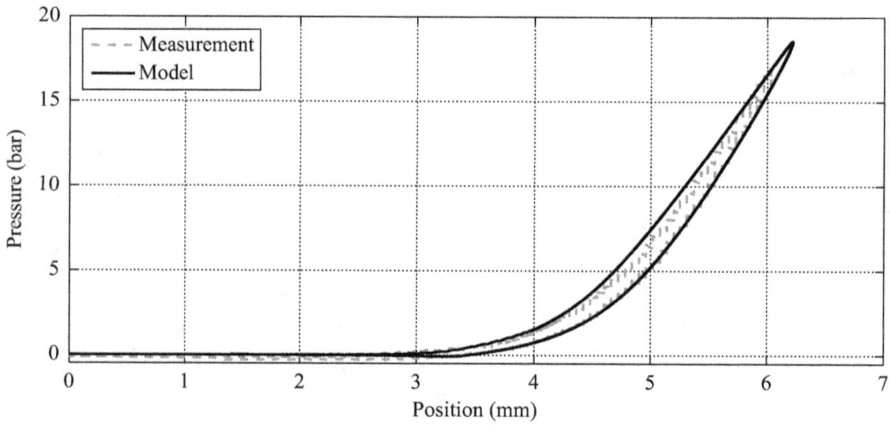

Figure 15.23 Comparison between the measured pressure–position map and the simulated one for a ramp experiment

simulated pressure and position and the actual ones during the quasi-static test performed with dither. The figure shows a satisfactory agreement both in pressure and position. The same data can also be looked at from a different perspective, namely the pressure–position map. This map is particularly useful when implementing feed-forward compensation schemes for pressure control [7]. Figure 15.23 plots this quasi-static characteristic. From the figure some remarks are due:

- the characteristic is highly nonlinear;
- despite the nonlinearity the model accurately captures the characteristic;
- the characteristic exhibits a hysteresis that is also present in the model.

All in all the model is capable of capturing the quasi-static behavior of the system.

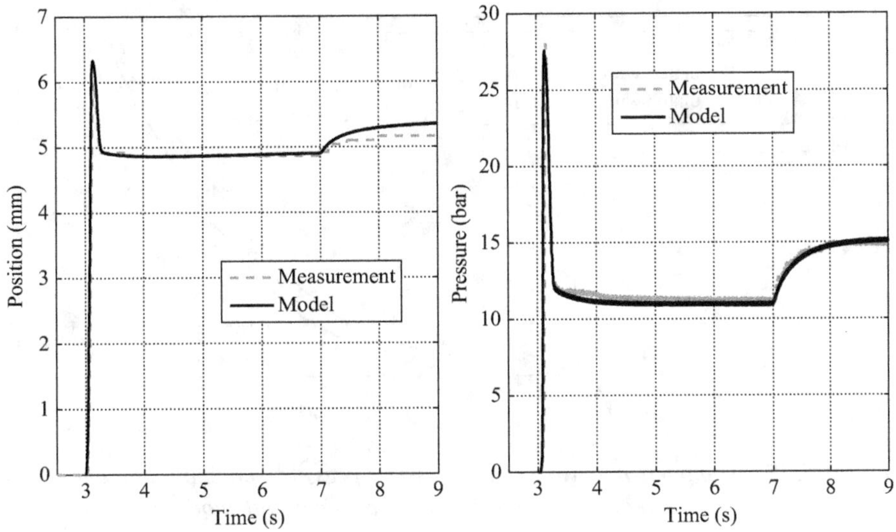

Figure 15.24 Comparison between the measured pressure–position map and the simulated one for a double step experiment

To validate the dynamic behavior of the system a current step experiment is performed. A double step is applied, from 0 to 3 A followed by a step from 3 A to 4 A. This is a very challenging experiment; it dynamically excites the system and, starting from the resting position, it tests the piston dynamics in its full range through the compensation chamber. From the results reported in Figure 15.24, it is clear that

- the system exhibits a considerable initial overshoot. This is due to the presence of the compensation chamber. The compensation chamber introduces a discontinuity in the system. When current is initially applied, the piston is free to gain momentum until it shuts the compensation chamber off and pressure is built. At this point the piston *hits* the fluid causing the overshoot;
- the model captures the initial overshoot very accurately both in terms of pressure and position. This initial overshoot is very critical from the control point of view [7] because it affects the first part of the braking maneuver. Having a model that captures that dynamics is paramount to the design of pressure control schemes;
- the model describes the pressure dynamics during the second step very accurately; the position dynamics is described less accurately in terms of steady state gain, but the transient is correctly described.

In conclusion, the proposed model is capable of capturing the most important features from the control point of view: hysteresis, nonlinearity, initial overshoot, and transients.

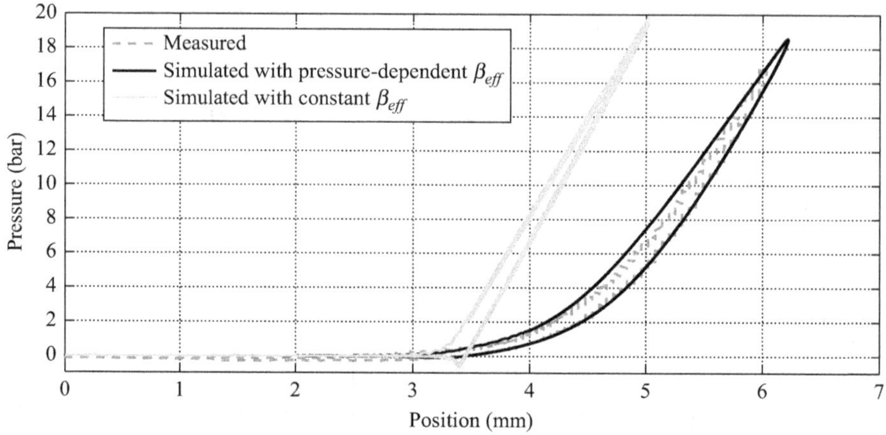

Figure 15.25 Comparison between the measured pressure–position map and the simulated ones with a pressure-dependent bulk modulus and a constant bulk modulus

15.5.2 Discussion on modeling choices

In designing the model, a series of choices have been made. Several of these choices involve the complexity of a model, for example, opting for a dynamic friction model over a static one. In the following paragraphs some of these choices will be justified by comparing their effect in terms of dynamic and quasi-static responses.

In modeling the fluid dynamics, a pressure–dependent bulk modulus has been adopted in (15.25). This choice is not very common in the modeling of hydraulic systems; in this case, given the relatively low operating pressure, a pressure–dependent bulk modulus is essential. Figure 15.25 plots the quasi-static pressure–position characteristic. The use of a constant bulk modulus leads to the impossibility of having an accurate matching. The above sensitivity can also be employed to fine-tune the air-to-fluid ratio R; as this parameter mainly affects the shape of the pressure–position nonlinearity.

Another important addition to the model is the leakage through the lip seal. Its effect can be once again evaluated using the quasi-static map. Figure 15.26 shows its role. The leakage has a direct effect on the amplitude of the hysteresis: without leakage, the model does not have any hysteresis; when the leakage is modeled, the hysteresis amplitude grows with the leakage parameter.

Modeling friction is probably one of the more daunting tasks. A dynamical friction model has been preferred over a more traditional static one [19]. Figure 15.27 effectively shows the reason for this choice. The figure plots the two quasi-static pressure–current and position–current characteristics. The response simulated with the Stribeck model does not effectively capture the hysteresis, especially in the pressure response. This also suggests that these kinds of quasi-static experiments are helpful to fine-tune σ_0 and σ_1.

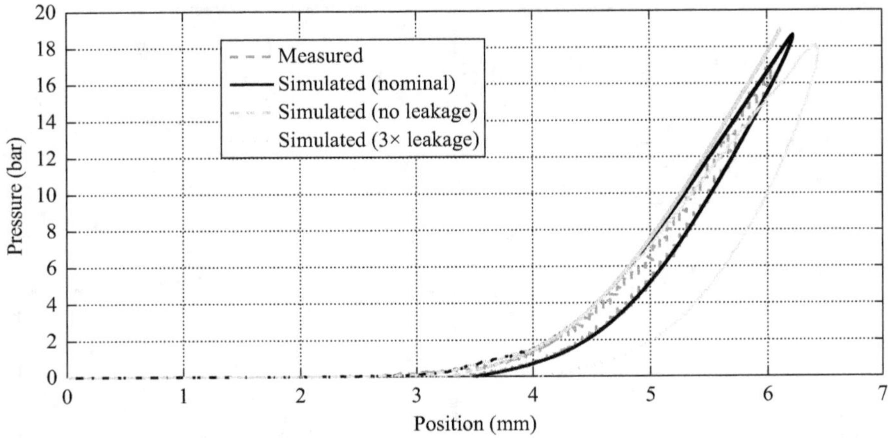

Figure 15.26 Comparison between the measured pressure–position map and the simulated ones for different levels of leakage

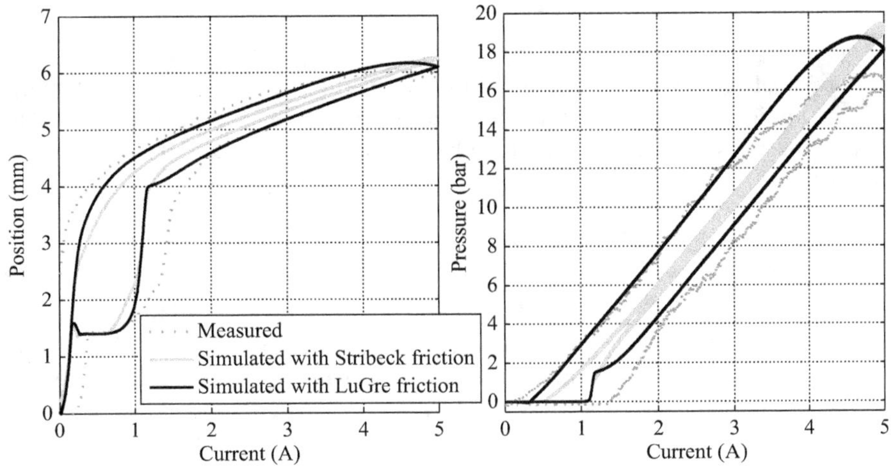

Figure 15.27 Comparison between the current and pressure behavior simulated with different friction models

The final choice to be discussed is the pipeline model. The pipeline model describes the propagation dynamics, so its effects are better described by looking at the step responses. Figure 15.28 shows the effect of several choices of L. As L decreases, the response becomes slower and slower. As expected it does not have an influence on the final steady state value. This analysis shows that the step responses are the best experiments to identify L.

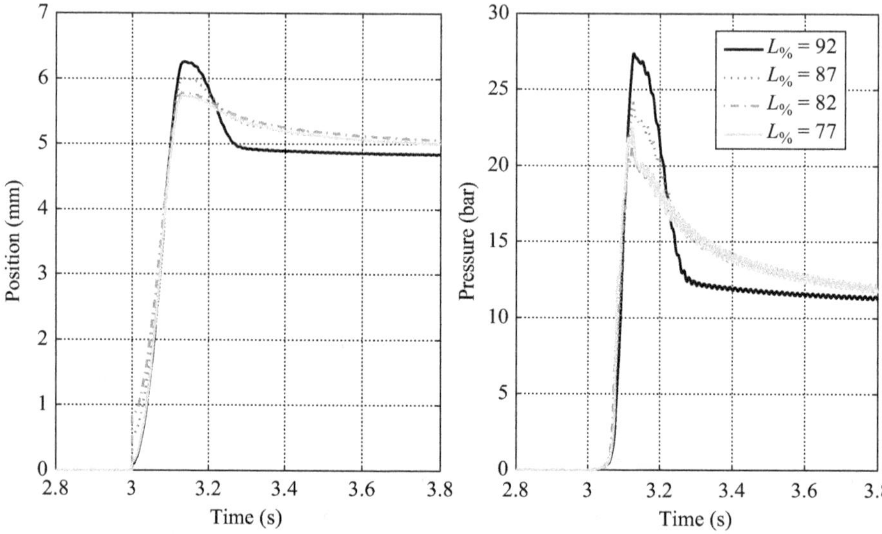

Figure 15.28 Comparison between the current and pressure behavior simulated for different values of the ratio L

One final remark refers to the intrinsic variability in the system. The braking pads are designed to wear and as they wear their geometric properties vary. The wear in the pads can be modeled by adapting the volume of the braking caliper according to

$$V_{wear} = A_{pz} \rho_{wear} \tag{15.69}$$

As the pads wear the caliper piston volume increases and the additional fluid is provided by the compensation chamber; however, this change of volume has an effect on the position–pressure map as shown in Figure 15.29. The figure shows that the model is easily adapted to account for wear and changing operating conditions.

15.6 Conclusions

This chapter presented the modeling, identification and analysis of a novel Brake-by-Wire system for racing motorcycles. The model is developed following a modular approach: the three main domains have been isolated and modeled. This allows for a high level of reusability and adaptability of the code.

For each domain the main governing equations have been derived. Subsequently a gray-box identification approach has been followed to identify the unknown physical parameters. The use of a gray-box (as opposed to a black-box) technique is to be

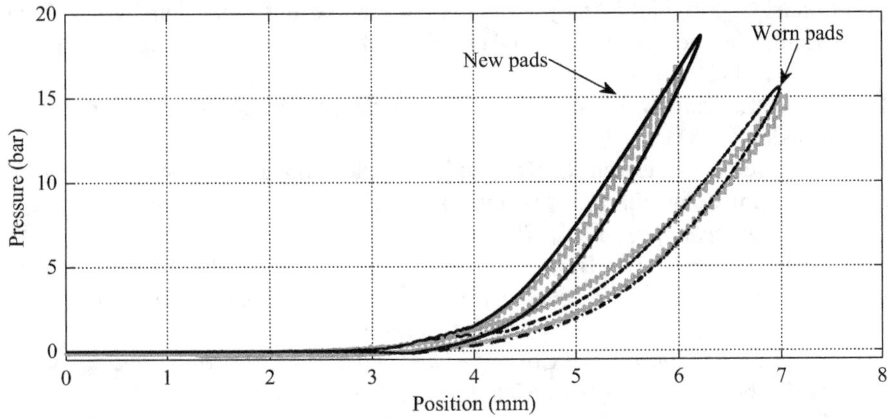

Figure 15.29 Comparison between the current and pressure behavior simulated and measured with different pads

preferred when the system under analysis is subject to wear and/or a physical parameter sensitivity analysis is required. The versatility of the model is shown by several sensitivity analysis.

References

[1] *Antilock Brake Review*. Warrendale, PA (1992). SAE Standard: J2246

[2] Armstrong, B., Dupont, P., Canudas de Wit, C.: A survey of models, analysis tools and compensation methods for the control of machines with friction. *Automatica* **30**(7), 1083–1138 (1994)

[3] Astrom, J., Canudas-de Wit, C.: Revisiting the LuGre friction model. *IEEE Control Systems Magazine* **28**(6), 101–114 (2008)

[4] Corno, M., Gerard, M., Verhaegen, M., Holweg, E.: Hybrid ABS control using force measurement. *IEEE Transactions on Control Systems Technology* **20**(5), 1223–1235 (2012). doi 10.1109/TCST.2011.2163717

[5] Corno, M., Savares, S.M., Balas, G.: On linear parameter varying (LPV) slip-controller design for two-wheeled vehicles. *International Journal of Robust and Nonlinear Control* **19**(12), 1313–1336 (2009)

[6] D'Alfio, N., Morgando, A., Sorniotti, A.: Electro-hydraulic brake systems: design and test through hardware-in-the-loop simulation. *Vehicle System Dynamics* **44**(1), 378–392 (2006)

[7] Dardanelli, A., Alli, G., Savaresi, S.: Modeling and control of an electro-mechanical brake-by-wire actuator for a sport motorbike. In: *5th IFAC Symposium on Mechatronic Systems*, pp. 524–531 (2010)

[8] de Castro, R., AraÃžjo, R., Freitas, D.: Adaptive compensation of the Stribeck friction in brake-by-wire actuators. In: *The 13th Mechatronics Forum International Conference* (2012)

[9] Duan, C., Singh, R.: Analysis of the vehicle brake judder problem by employing a simplified source-path-receiver model. *Automobile Engineering* **225, Part D**, 141–149 (2010)

[10] Gill, P.E., Murray, W., Wright, M.H.: *Practical Optimization*. Academic Press, London (1981)

[11] Johansen, T., Petersen, I., Kalkkuhl, J., Lüdemann, J.: Gain-scheduled wheel slip control in automotive brake systems. *IEEE Transactions on Control Systems Technology* **11**(6), 799–811 (2003)

[12] Kwak, J., Yao, B., Bajaj, A.: Analytical model development and model reduction for electromechanical brake system. In: *International Mechanical Engineering Congress and Exposition* (2004)

[13] Line, C.: *Modelling and Control of an Automotive Electromechanical Brake*. Ph.D. thesis, University of Melbourne (2007)

[14] Ljung, L.: *System Identification: Theory for the User*. Prentice-Hall, Englewood Cliffs, NJ (1987)

[15] Mare, J.C.: Friction modelling and simulation at system level: a practical view for the designer. *Proceedings of the Institution of Mechanical Engineers, Part I: Journal of Systems and Control Engineering* **226**(6), 728–741 (2012)

[16] Nonaka, H., Nishizawa, Y., Kurita, Y., Oura, Y.: *Considering the Dynamic Pad Stiffness in FEM Analysis of Disk Brake Squeal*. SAE Technical Paper 2010-01-1716 (2010)

[17] Novak, N.N., Filipović, V.Z., Dubonjić, L.M.: Design of controllers with fixed order for hydraulic control system with a long transmission line. *FME Transactions* **38**, 79–86 (2010)

[18] Olsson, H., Åström, K.J., Canudas de Wit, C., Gäfvert, M., Lischinsky, P.: Friction models and friction compensation. *European Journal of Control* **4**(3), 176–195 (1998)

[19] Panzani, G., Corno, M., Savaresi, S.M.: On adaptive electronic throttle control for sport motorcycles. *Control Engineering Practice* **21**(1), 42–53 (January 2013)

[20] Rajamani, R.: *Vehicle Dynamics and Control*. Mechanical Engineering Series, Springer, New York, NY (2006)

[21] Reuter, D.F., Lloyd, E.W., Zehnder, J.W., Elliott, J.A.: Hydraulic design considerations for EHB systems. In: *SAE World Congress* (2003)

[22] Savaresi, S., Tanelli, M.: *Active Braking Control Systems Design for Vehicles*. Springer-Verlag, London, UK (2010)

[23] Sawase, K., Sano, Y.: Application of active yaw control to vehicle dynamics by utilizing driving/braking force. *JSAE Review* **20**(2), 289–295 (1999)

[24] Sorniotti, A., Repici, G.: Hardware in the loop with electro-hydraulic brake systems. In: *9th WSEAS International Conference on Systems* (2005)

[25] Strikwerda, J.: *Finite Difference Schemes and PDE*. Wadsworth & Brooks/Cole, Pacific Grove, CA (1989)

[26] Thomas, P.: *Simulation of Industrial Processes (For Control Engineers)*. Butterworth Heinemann, Jordan Hill, Oxford (1999)

[27] Wellstead, P.: Non-parametric methods of system identification. *Automatica* **17**, 55–69 (1981)

[28] Wellstead, P., Petit, N.: Analysis and redesign of an antilock brake system controller. In: *Proceedings of the Institute of Electrical and Electronics Engineers*, pp. 413–426 (1997)

[29] de Wit, C.C., Olsson, H., Åström, K.J., Lischinsky, P.: A new model for control of systems with friction. *IEEE Transactions on Automatic Control* **40**(3) (March 1995)

Chapter 16

LPV modeling and identification of a 2-DOF flexible robotic arm from local experiments using an H_∞-based glocal approach

Daniel Vizer[1], *Guillaume Mercère*[2], *Edouard Laroche*[3] *and Olivier Prot*[4]

Abstract

This chapter presents a series of methodological contributions for the identification of flexible manipulators from experimental data. The goal of this identification procedure consists more precisely in obtaining reliable LPV models written as linear-fractional representations (LFR) for a 2-DOF robotic manipulator having structural flexibilities. The case of a two-segment arm initially designed for cardiac robotized surgery is more specifically considered in simulation. In order to reach this goal, the H_∞-based optimization technique described in Chapter 9 is applied. This methodology is indeed very flexible and allows us to derive models with or without structure. Thus, as far as the model structure is concerned, two different cases are considered in this chapter. First, a specific attention is paid to a fully parameterized LPV-LFR, the parameters of which are estimated on the gathered I/O data sequences exclusively. Second, the prior information derived from the study of the nonlinear equations governing the behavior of the robotic manipulator is used to build the structure of the LPV-LFR and an LPV physically structured state-space form is identified from the same I/O data sequences as those used for the fully parameterized state-space form. This study proves that using the synergy between an analytic and an experimental approach can be really helpful for the identification of an LPV flexible robotic manipulator model.

16.1 Introduction

During the last decade, the demand for accurate controllers for systems having highly nonlinear behavior has raised drastically [15]. In order to design a controller that is

[1]University of Technology and Economics of Budapest, Budapest, Hungary and University of Poitiers, Poitiers Cedex, France, e-mail: vizer@iit.bme.hu
[2]University of Poitiers, Poitiers Cedex, France, e-mail: guillaume.mercere@univ-poitiers.fr
[3]University of Strasbourg, Illkirch Cedex, France, e-mail: laroche@unistra.fr
[4]University of Limoges, Limoges Cedex, France, e-mail: olivier.prot@unilim.fr

able to satisfy all the requirements, it is compulsory to have an adequate model of the system under study. That is the case with the robotic systems for instance where the manipulators may have strong nonlinearities in certain working conditions. The application of robotic systems has been spread out in many practical fields during the last decade [6]. Nowadays, such systems are used not only in the manufacturing industry but also, among others, in the healthcare and space industries (see, e.g., [1, 20]). These latter fields require the use of lightweight robots causing the appearance of structural flexibilities contrary to heavier industrial robots for which the rigid body assumption is satisfied in general. Robotic systems used in the aforementioned fields also have more stringent precision requirements than their industrial counterparts. Furthermore, such systems are manufactured in a significantly smaller quantity resulting sometimes in really unique robotic structures. Therefore, the identification as well as the control of such robotic systems have to be dealt with an emphasized attention.

The models of robotic systems are usually white-box models based on first principles and laws of physics governing the behavior of the system. These dynamic models are often derived by employing the Euler–Lagrange or Appell equations and the virtual work principle [7]. However, the application of these techniques requires high-level skills in robotics because the involved kinematics are unique for each robot structure. This is all the more true when the user wants to have access to physical parameters of the system which are imperfectly known. Furthermore, such white-box models may be too complex in the end. In order to circumvent these problems, robotic system identification efforts are now performed in the manufacturing industry [12]. However, a direct identification of a nonlinear black-box model is often complicated because

- strong nonlinearities (e.g., inherent flexibilities) can be effective in particular working conditions;
- the development of a global nonlinear model structure can rely on strong assumptions such as a uniform density of the manipulator segments or the nature of the deformations if any.

Because a linear time-invariant model may not be sufficient when the system is used in a large robot workspace (because, e.g., the nonlinearities effects may vary with the operating conditions), LPV models are more and more introduced in robotics (see, e.g., [24, 13, 3]). The development of LPV model identification for the experimental modeling of robots is advocated for two main reasons. First, from an identification point of view, the introduction of such a structure allows the use of standard tools dedicated to LTI models for the estimation of models with a structural flexibility able to picture time-varying as well as nonlinear dynamics. Second, from a control viewpoint, the construction of a reliable LPV model can be seen as a standard but essential initial step for many new control law determination techniques developed in robotics [4]. Note that specific LPV model-based controller design techniques (see, e.g., [21, 2, 17, 11] in aeronautics) can be found in the literature. However, their direct application or adaptation to robotic systems is not an easy task because of the stronger nonlinear nature of the robotic systems in comparison with standard aeronautical processes. The determination of such reliable LPV models is still a challenging problem as it is shown in [22, 16].

In this chapter, the goal is to obtain reliable LPV models written as linear-fractional representations (LFR) for a 2-DOF robotic manipulator having structural flexibilities. In order to reach this goal, the H_∞-based optimization technique described in Chapter 9 is applied. As a glocal method (see Chapter 9 for a definition), this H_∞-based optimization technique will yield a consistent linear fractional LPV model of the robotic system described in section 16.2 from local experiments (acquired for fixed values of the scheduling variables) without any interpolation phase. As far as the model structure is concerned, two different cases are considered hereafter. First, a specific attention is paid to a fully parameterized LPV-LFR, the parameters of which are estimated on the gathered I/O data sequences exclusively. Second, the prior information derived from the study of the nonlinear equations governing the behavior of the robotic manipulator is used to build the structure of the LPV-LFR and an LPV physically structured state-space form is identified from the same I/O data sequences as those used for the fully parameterized state-space form. Notice right now that, for these LPV models, the assumption of flexible segments is considered to deal with the flexibilities encountered with the specific 2-DOF robotic manipulator used hereafter to validate the technique developed in Chapter 9. The identification as well as the validation of these estimated fully parameterized and structured LPV-LFR are performed by using simulated data sets acquired on a simulator available in MATLAB®-Simulink® built from the nonlinear equations governing the behavior of the flexible arm described hereafter.

The chapter is organized as follows. In section 16.2 the modeling and structural aspects of the flexible robotic manipulator under study are presented by considering flexible segments exclusively. In the aforementioned section, linear fractional LPV models are more precisely derived from the standard first principles. Then, the identification of a fully parameterized as well as a physically structured LPV model is performed in section 16.3. Finally, section 16.4 concludes this chapter.

16.2 Modeling of a flexible robotic manipulator

16.2.1 Description of the 2-DOF robotic manipulator

In the current study, the considered system is a horizontal arm composed of two flexible segments as depicted[1] in Figure 16.1(a). Such a structure can be found, e.g., when considering the two first rotoid joints of a SCARA manipulator. Under specific working conditions, for instance when fast displacements are requested, this type of manipulator may have significant flexibilities. Indeed, even if these deformations only yield short displacements of the end-effector, this is sufficient to restrict the bandwidth of the control loop, as shown in [5]. Therefore, a model of these flexible modes is

[1]Notations are explained in section 16.2.3.

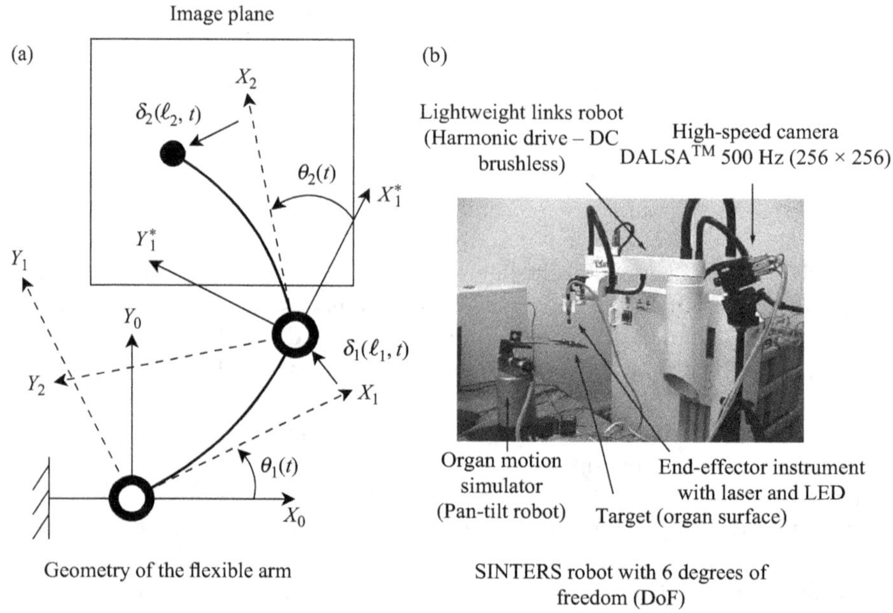

Figure 16.1 Flexible robotic manipulator under study

necessary in order to design an image-based control loop where the position of the end-effector is considered as a measured output. For instance, these flexible characteristics are satisfied by a prototype designed by SINTERS and used in [9] (see Figure 16.1(b)). This robot is lightweight because it is designed to attain fast dynamics in order to compensate the heart tissue motion for cardiac surgery. As a result, it is observed that the bandwidth is restricted by flexible modes that can be attributed to flexible segments. In this chapter, the modeling aspects of flexible segments are introduced. As explained in section 16.1, such a physical study is performed to obtain a reliable LPV model structure used, afterwards, as an input of the H_∞-based optimization technique developed in Chapter 9.

For the system depicted in Figure 16.1(b), both joints are torque-controlled and the joint positions ϕ_1 and ϕ_2 are measured by encoders. The aim is to control the position of the end-effector which is measured by a video camera. Deformations of the structure are considered but are not measured. In the following, both segments are assumed to have the same width $\ell_1 = \ell_2 = 0.5\,\mathrm{m}$ and respective masses of 7.5 and 5 kg. Their sections are squares with a side length of 5 cm. The material is also assumed to have a Young modulus equal to 5 MPa.

A general scheme of the test-bed used hereafter is presented in Figure 16.2(a). On the real system (available at the ICube Lab of the University of Strasbourg), each joint is actuated by a DC motor. The motor speed of each joint is locally controlled by a commercial drive. The joint velocity controllers are tuned in order to satisfy a good trade-off between high bandwidth and low noise on the current references. A dedicated

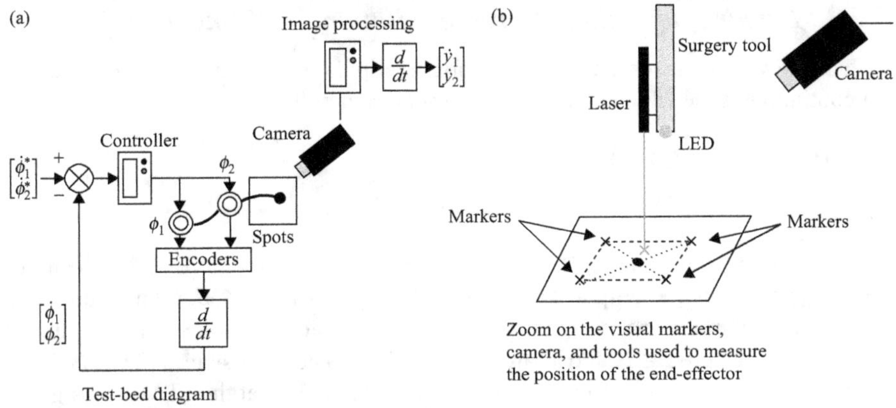

Figure 16.2 Diagrams of the system under study

computer is also used to communicate with the system and to control the motors via an I/O board. The end-effector could hold any small tool, e.g., a scalpel when this robotic manipulator is used in a surgical theater (see [8] for such an applicative example). As shown in Figure 16.2(b), the position of the end-effector is measured thanks to a LED, a laser, and optical markers. A CCD camera is used to measure the relative position between the organ and a tool held by the end-effector. A second computer is dedicated to the image acquisition and the image processing. Details and characteristics for the high-speed cameras and the real-time implementation of the data acquisition and image processing are available in [8, 5]. The measurement of the relative translation of the instrument with respect to the heart is performed by measuring

- the vector between the target center of mass (black ellipsoid in Figure 16.2(b)) and the laser spot (light gray cross in Figure 16.2(b));
- the distance in the image between the LED spot (light gray circle in Figure 16.2(b)) and the target center of mass (black ellipsoid in Figure 16.2(b)).

The first degree of freedom (#1) of the arm is used for the vertical motion. The second and third ones (#2 and #3) are linked to the relative position of the target center of mass (black ellipsoid in Figure 16.2(b)), and the laser spot (light gray cross in Figure 16.2(b)). For this validation step, the studied motion is restricted to these two last degrees of freedom (#2 and #3).

As explained previously, the spot position measurements are made with the help of a CCD camera. In practice, this camera must be located so that it does not disturb the surgeon and his staff. Furthermore, it is difficult to move during the operation mainly for safety reasons. These practical conditions highly reduce the working field of the robot. Because of these constraints, the use of a global technique [22] requiring a persistent excitation of the inputs as well as the scheduling variables is not conceivable. On the contrary, a local approach [22] seems to be well-suited for the LPV model identification of such a system.

16.2.2 Linear fractional LPV representation: a reminder

In what follows, the afore-described robotic manipulator is assumed to be modeled by a continuous-time LPV state-space representation defined as

$$\dot{x}(t) = A(p(t), \Theta)x(t) + B(p(t), \Theta)u(t) \tag{16.1a}$$

$$y(t) = C(p(t), \Theta)x(t) \tag{16.1b}$$

where, as explained in Chapter 9, $x(t) \in \mathbb{R}^{n_x}$ is the state vector, $u(t) \in \mathbb{R}^{n_u}$ is the input vector, $y(t) \in \mathbb{R}^{n_y}$ is the output vector and the state matrices (A, B, C) are functions of a measurable time-varying signal $p(t) \in \mathbb{K} \subseteq \mathbb{R}^{n_p}$ used hereafter as the scheduling variable vector where $t \in \mathbb{R}$ and $\mathbb{K}$ are the so-called scheduling spaces [23]. As far as the structure of (A, B, C) matrices is concerned, we assume that the LPV model given by (16.1) depends on a parameter vector $\Theta \in \mathbb{R}^N$ and can be reformulated as a linear fractional transformation (LFT) description resulting in the following state-space matrices [26]:

$$A(p(t), \Theta) = A_0(\Theta) + B_w(\Theta)\Delta(p(t))(I - D_{zw}(\Theta)\Delta(p(t)))^{-1}C_z(\Theta) \tag{16.2a}$$

$$B(p(t), \Theta) = B_0(\Theta) + B_w(\Theta)\Delta(p(t))(I - D_{zw}(\Theta)\Delta(p(t)))^{-1}D_{zu}(\Theta) \tag{16.2b}$$

$$C(p(t), \Theta) = C_0(\Theta) + D_{yw}(\Theta)\Delta(p(t))(I - D_{zw}(\Theta)\Delta(p(t)))^{-1}C_z(\Theta) \tag{16.2c}$$

The matrix $\Delta(p(t))$ is a $p(t)$-dependent matrix while the matrices $(A_0, B_w, \dots, D_{yu})$ are time-invariant matrices with appropriate dimensions (see Chapter 9 for more details and definitions of the involved matrices and signals). According to the prior knowledge available on the physical laws governing the behavior of such a system, these matrices can be fully parameterized or physically structured. Getting accurate physically structured matrices is the main goal of the following developments.

16.2.3 Nonlinear and linearized dynamic models

As explained previously, the structure of the LPV representation given in (16.2) is highly linked to the nonlinear equations governing the behavior of the system to identify. The structure of this LPV-LFR can be indeed directly extracted from the (Jacobian) linearization of these nonlinear equations. These nonlinear and linearized dynamic models are introduced in this section.

Under the assumption of Euler–Bernoulli beam, the dynamic equations of a flexible arm can be derived by using the assumed-mode method where the deformation field is decomposed into a finite sum of elementary deformations [19]. In the current case, small deformations are considered and only one mode is chosen for the transverse deformation field. For segment #k, $k = 1, \dots, n_\phi$, the deformation field writes $\delta_k(x, t) = x^2 v_k(t)$, where x represents the abscissa along the segment and $v_k(t)$ is the deformation. Therefore, the resulting deformation at the end of the segment of length ℓ_k is $\delta_k(\ell_k, t) = \ell_k^2 v_k(t)$. The dynamic model has been derived from the Virtual Work Principle using the DynaFlex toolbox developed on Maple (see [19]). By denoting

$\phi = \begin{bmatrix}\phi_1 & \phi_2\end{bmatrix}^\top \in \mathbb{R}^{n_\phi}$ and $v = \begin{bmatrix}v_1 & v_2\end{bmatrix}^\top \in \mathbb{R}^{n_v}$, the resulting model relies on a generalized position vector $q = \begin{bmatrix}\phi^\top & v^\top\end{bmatrix}^\top \in \mathbb{R}^{n_q}$, with $n_q = n_\phi + n_v$, and writes (see Figure 16.1(a) for the notations)

$$\mathcal{M}(q(t))\ddot{q}(t) = \mathcal{F}(q(t), \dot{q}(t)) + \mathcal{G}u(t) \tag{16.3}$$

where[2] $\mathcal{M}(q)$ is the inertia matrix, $\mathcal{F}(q, \dot{q})$ is a generalized force vector that includes the Coriolis and centrifugal effects (see [14] for details about the mathematical expressions of the matrix $\mathcal{M}$ and the vector $\mathcal{F}$). The torque vector $u = \begin{bmatrix}u_1 & u_2\end{bmatrix}^\top$ has only effects on the dynamics of the rigid positions ϕ_1 and ϕ_2, corresponding to

$$\mathcal{G} = \begin{bmatrix} I_{n_\phi \times n_\phi} \\ 0_{n_v \times n_\phi} \end{bmatrix} \tag{16.4}$$

The x and y positions of the end-effector can be written from the geometric model, resulting in the nonlinear measurement equation $z = g(q)$, i.e.,

$$\begin{aligned} z_1 = {} & \left(\ell_1 - \frac{2}{3}\ell_1^3 v_1^2\right)\cos(\phi_1) - \ell_1^2 v_1 \sin(\phi_1) \\ & + \left(\ell_2 - \frac{2}{3}\ell_2^3 v_2^2\right)\cos(\phi_{12}) - \ell_2^2 v_2 \sin(\phi_{12}) \end{aligned} \tag{16.5a}$$

$$\begin{aligned} z_2 = {} & \left(\ell_1 - \frac{2}{3}\ell_1^3 v_1^2\right)\sin(\phi_1) + \ell_1^2 v_1 \cos(\phi_1) \\ & + \ell_2^2 v_2 \cos(\phi_{12}) + \left(\ell_2 - \frac{2}{3}\ell_2^3 v_2^2\right)\sin(\phi_{12}) \end{aligned} \tag{16.5b}$$

where $\phi_{12} = \phi_1 + \phi_2 + 2\ell_1 v_1$.

In order to fix the structure of the global LPV model required by the technique introduced in this chapter, a standard Jacobian linearization can be applied to the generalized second-order model given in (16.3). More precisely, for a set of working points $(q_0, \dot{q}_0)$, we get[3]

$$M(p)\ddot{q}(t) = D(p)\dot{q}(t) + K(p)q(t) + Gu(t) \tag{16.6}$$

[2]For the sake of conciseness, the time index t is dropped when this index is not necessary for the understanding of the equations. Notice, however, that the signals and the deformations are still time-varying.
[3]In (16.6) as well as in the following equations, the equilibrium values are omitted in order to shorten the notations.

where $p = p(q_0, \dot{q}_0)$ is the scheduling variable vector and where

$$M(p(q_0,\dot{q}_0)) = \mathcal{M}(q_0) \quad D(p(q_0,\dot{q}_0)) = \frac{\partial \mathcal{F}(q_0,\dot{q}_0)}{\partial \dot{q}} \tag{16.7a}$$

$$K(p(q_0,\dot{q}_0)) = \frac{\partial \mathcal{F}(q_0,\dot{q}_0)}{\partial q} - \mathcal{M}^{-1}(q_0)\frac{\mathcal{M}(q_0)}{\partial q}\mathcal{M}^{-1}(q_0)\mathcal{F}(q_0,\dot{q}_0) \quad G = \mathcal{G} \tag{16.7b}$$

are the inertia, the damping, the stiffness, and the control matrices of the linearized model, respectively. Notice that G does not depend on the scheduling variable p. In this work, we focus on the identification of a model that includes the variability of the behavior with respect to the positions ϕ_k, $k = 1, 2$. Then, the other phenomena can be neglected and, consequently,[4] $q_0 = \begin{bmatrix}\phi_0^\top & 0_{1\times n_v}\end{bmatrix}^\top$ and $\dot{q}_0 = 0_{n_q\times 1}$. Like for most of the methods considered in the literature, the inertia matrix $M(p)$ is inverted. This property is generally used in the literature (see, e.g., [10]) leading to the following local equation

$$\ddot{q} = M^{-1}(p)D(p)\dot{q} + M^{-1}(p)K(p)q + M^{-1}(p)Gu \tag{16.8}$$

Notice that this matrix inversion leads to matrices $M^{-1}(p)D(p)$, $M^{-1}(p)K(p)$, and $M^{-1}(p)G$ which satisfy a fractional dependence on the scheduling variable p. This hint justifies the use of a linear fractional LPV description of the system instead of a more standard affine LPV model. Now, by considering

$$\chi = \begin{bmatrix}q^\top & \dot{q}^\top\end{bmatrix}^\top \in \mathbb{R}^8 \tag{16.9}$$

as state vector, the following local linearized state equation can be deduced[5]

$$\dot{\chi} = A(p)\chi + B(p)u \tag{16.10}$$

with

$$A(p) = \begin{bmatrix} 0_{n_q\times n_q} & I_{n_q\times n_q} \\ M^{-1}(p)K(p) & M^{-1}(p)D(p)\end{bmatrix} \tag{16.11a}$$

$$B(p) = \begin{bmatrix} 0_{n_q\times n_u} \\ M^{-1}(p)G\end{bmatrix} \tag{16.11b}$$

For this state-space form, the state equation only depends on ϕ_2[6] whereas the output equation depends on both ϕ_1 and ϕ_2.

[4]This choice means that the Coriolis effects are neglected. The validity of this assumption will be evaluated simultaneously with the identified model.

[5]In the sequel, $z_\bullet$, $\phi_\bullet$, $v_\bullet$, $q_\bullet$, …, are considered as variations around a nominal value.

[6]The system behavior is invariant by a rotation of angle ϕ_1. Therefore, the matrices A and B do not depend on ϕ_1.

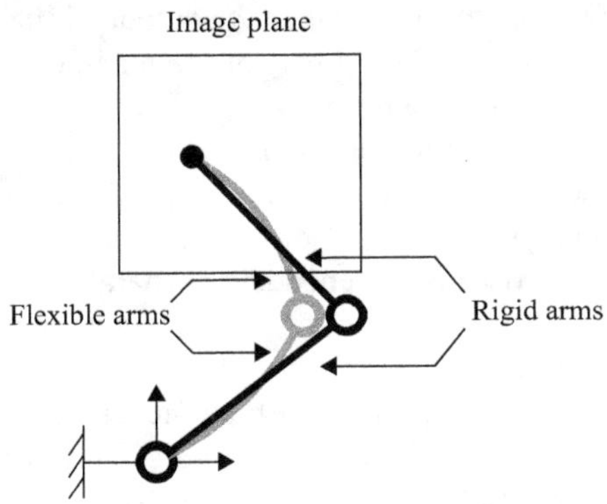

Figure 16.3 Flexible manipulator (gray) and virtual rigid manipulator (black) with the same image position of the end-effector

As shown in (16.5), the output equations are highly nonlinear with respect to ϕ_1 and ϕ_2. This complexity can be reduced by using efficiently the Jacobian of the rigid geometric model (see (16.5) for a definition of g), i.e.,

$$J(\phi_0) = \frac{\partial g\left(\begin{bmatrix}\phi_0^\top & 0_{1\times n_v}\end{bmatrix}^\top\right)}{\partial \phi} \tag{16.12}$$

Except for the singular positions, this Jacobian is invertible and it is possible to define a new measurement vector

$$\alpha = \alpha_0 + J^{-1}(\phi_0)(z - z_0) \tag{16.13}$$

where, in the following, α_0 and z_0 are assumed to be equal to 0 without any limitation for the applicability of the approach. The entries of this new measurement vector α are the angular positions of a fictitious rigid arm that would have the same geometry and the same measurement z as the flexible one (see Figure 16.3). The use of α instead of z allows the simplification of the measurement equation. Let us now consider $y = \alpha$, i.e.,

$$y = C\chi \tag{16.14}$$

where

$$C = \begin{bmatrix} C_1 & 0_{n_\phi \times n_q} \end{bmatrix} \tag{16.15}$$

and

$$C_1 = \begin{bmatrix} 1 & 0 & \ell_1 & 0 \\ 0 & 1 & \ell_1 & \ell_2 \end{bmatrix} \tag{16.16}$$

Thus, the structure of C_1 is quite simple when the output y is handled instead of z. By assuming that the geometry of the rigid part of the manipulator is available, C can moreover be considered as known a priori.

By looking closer at the equations available in [14], it is clear that the matrices M, D, and K are affine functions of $\cos(\phi_2)$ and $\sin(\phi_2)$. Then, the scheduling variable is chosen as $\cos(\phi_2)$, i.e., $p = \cos(\phi_2)$. Notice that this angular position is easy to measure on a flexible robot because an encoder is generally located at the motor side of the joints. This availability is paramount when the experimental modeling of the LPV model is considered.

Industrial manipulators are equipped with low-level joint-velocity control loops in order to reduce the effects of the friction that occurs in the gear-boxes and therefore obtain a simpler behavior (see Controller in Figure 16.2(a)). In this particular case, the inner loop is assumed to be as a standard static output feedback. More precisely

$$u = \Lambda \left(\dot{\phi}^* - \dot{\phi}\right) \tag{16.17}$$

where $\dot{\phi}^* = \begin{bmatrix}\dot{\phi}_1^* & \dot{\phi}_2^*\end{bmatrix}^\top$ is the vector of the speed references. The state-space representation given in (16.10) becomes

$$\begin{aligned}\dot{\chi} &= \begin{bmatrix} 0_{n_q \times n_\phi} \\ M^{-1}(p) G \Lambda \end{bmatrix} \dot{\phi}^* \\ &+ \begin{bmatrix} 0_{n_q \times n_q} & I_{n_q \times n_q} \\ M^{-1}(p) K(p) & M^{-1}(p)\left(D(p) - G\Lambda \begin{bmatrix} I_{n_\phi \times n_\phi} & 0_{n_\phi \times n_v}\end{bmatrix}\right)\end{bmatrix} \chi \end{aligned} \tag{16.18}$$

where $\Lambda = \operatorname{diag}(\lambda_1, \lambda_2)$.

Considering the position y as the output measurement has one drawback, the model is unstable due to the pure integration between speed and position. Therefore, it is generally recommended to use the velocity $\overline{y} = \dot{y}$ instead.[7] By using the state matrices introduced beforehand, the output equation rewrites

$$\overline{y} = C\dot{\chi} = \overline{C}\chi \tag{16.19}$$

where

$$\overline{C} = \begin{bmatrix} 0_{n_\phi \times n_q} & C_1 \end{bmatrix} \tag{16.20}$$

By analyzing the relations given in (16.18), it appears that this state-space form is not minimal because the two first states ϕ_1 and ϕ_2 have no effects on the output. A minimal realization of order 6 can be extracted by considering

$$\overline{\chi} = \begin{bmatrix} v^\top & \dot{\phi}^\top & \dot{v}^\top \end{bmatrix}^\top \tag{16.21}$$

[7] In practice, the velocity is estimated from the discrete-time measurement of the position.

More precisely

$$\dot{\overline{\chi}} = A_0(p)\overline{\chi} + B_0(p)\dot{\phi}^* \tag{16.22a}$$

$$\overline{y} = C_0\overline{\chi} \tag{16.22b}$$

where[8]

$$A_0(p) = \begin{bmatrix} 0_{n_v \times n_\phi} & 0_{n_v \times n_v} & I_{n_v \times n_v} \\ \left(M^{-1}(p)K(p)\right)(\,:, n_\phi + 1 : end) & -M^{-1}(p)G\Lambda & 0_{n_q \times n_v} \end{bmatrix} \tag{16.23a}$$

$$B_0(p) = \begin{bmatrix} 0_{n_v \times n_\phi} \\ M^{-1}(p)G\Lambda \end{bmatrix} \quad C_0(p) = \begin{bmatrix} 0_{n_\phi \times n_v} & C_1 \end{bmatrix} \tag{16.23b}$$

This linearized state-space representation can now be related to the LPV-LFR given in (16.1) and (16.2). Indeed, with straightforward calculations, it can be seen that such an LPV description can be transformed into a linear fractional transformation description such as the one given in (16.2) by considering the structured matrices of (16.23) as the A_0, B_0, and C_0 entries involved in (16.2) with, as additional terms, fully parameterized matrices B_w, C_z, D_{zu}, and D_{yw}. Such a gray-box LPV-LFR form is of great interest as far as linear fractional LPV model identification is concerned as shown in the next section.

16.3 Identification results

In this section, the identification procedure introduced in Chapter 9 is applied to estimate reliable linear fractional LPV models of the afore-introduced 2-DOF robotic flexible manipulator. The work performed in the previous section gives access to a reliable LPV model structure. Now, the steps composing the glocal identification procedure introduced in Chapter 9 can be tackled.

As explained in Chapter 9, like any glocal approach based on local experiments, the first step of the procedure consists in selecting a set of N_{op} local I/O data to carry out the local LTI model identification. Herein, seven constant values of the scheduling variable ϕ_2 in the set $\{k\pi/8 : k = 1, \ldots, 7\}$ are selected, then seven noise-free I/O data sets are generated. This noise-free I/O data generation is made possible because, herein, a Simulink model of the robotic manipulator having flexible segments is available to carry out the experiments. The inputs of the simulator, as shown in Figure 16.2(a), are the angular position references $\boldsymbol{\phi}^* = \begin{bmatrix}\phi_1^* & \phi_2^*\end{bmatrix}^\top$ and are chosen as two uncorrelated pseudo-random binary sequences built so that all the dynamics of the system is well-excited locally, i.e., with the constraint that the value of the scheduling variable is kept constant (as much as possible). The outputs of the model are $\overline{y}$, i.e., the angular velocities of the fictitious rigid robot ($\dot{\alpha}$) having the same geometry of the flexible robot and the joint velocities ($\dot{\phi}$) derived from the joint

[8]For our system, $D_0(p) = 0_{n_q \times n_q}$.

positions measured by encoders. This noise-free I/O data is generated once for each value of $\phi_2 \in \{k\pi/8 : k = 1, \ldots, 7\}$. Then, a Monte Carlo simulation of size 100 is used for the estimation of the local models. This Monte Carlo simulation is carried out by adding to the noise-free outputs 100 realizations of two uncorrelated zero-mean white Gaussian noises satisfying a signal-noise ratio[9] equal to 30 dB. The length of each local data set is equal to 30,000 samples with a sampling period of 0.1 ms. For the identification, a down-sampling is performed with a rate equal to 10. A sample of noisy I/O data is given in Figure 16.4.

The second step of the local identification procedure developed in Chapter 9 consists in estimating reliable (fully parameterized) local models. In order to reach this goal, the PI-MOESP algorithm [25] is applied to perform the local estimation from the local I/O data sequences followed by a discrete-to-continuous-time transformation. This discrete-to-continuous-time domain conversion is performed by using the bilinear Tustin approximation to the derivative (e.g., the function `d2cm` available in MATLAB).

Once 100 local models are estimated for each operating point, 7 local models can be computed for each value of N_{op} by averaging these 100 local models. These seven average local models are then locally validated by considering two complementary tools:

- two I/O fit measurements (see (16.24)) evaluated on N_{op} new noise-free data sets generated for this task (cross-validation);
- a comparison of the frequency responses (magnitude Bode plots) of the estimated local model and an analytic local model calculated from a linearization of the nonlinear equations governing the behavior of the system.

For $i \in [1, n_y]$, the following fit measurements[10] are introduced in order to quantify the model quality on validation data (i.e., a data set different from the one used for the estimation)

$$\mathrm{BFT}_i = 100 \times \max\left(1 - \frac{\|\bar{y}_i - \hat{\bar{y}}_i\|_2}{\|\bar{y}_i - \mathrm{mean}(\bar{y}_i)\|_2}, 0\right) \tag{16.24a}$$

$$\mathrm{VAF}_i = 100 \times \max\left(1 - \frac{\mathrm{var}\left(\bar{y}_i - \hat{\bar{y}}_i\right)}{\mathrm{var}(\bar{y}_i)}, 0\right) \tag{16.24b}$$

Table 16.1 (see also the time responses in Figure 16.5 for a qualitative validation for $\phi_2 = \frac{5\pi}{8}$) gathers these fit measurements for $\phi_2 \in \{k\pi/8 : k = 1, \ldots, 7\}$. From these values, it can be concluded that, for each operating point, the estimated local LTI model describes the actual system quite well. The reader must keep in mind that the system behavior is highly nonlinear. Having fit measurements in the range [67%, 98%]

[9]The signal-to-noise ratio is defined as follows: $SNR = 10 \log\left(\frac{\mathrm{cov}\{\bar{y}_i\}}{\mathrm{cov}\{v\}}\right)$, $i \in [1, n_y]$ where v stands for the noise acting on the noise-free output $\bar{y}_i$.

[10]$\bar{y}_i$ stands for the i-th system output and $\hat{\bar{y}}_i$ for its estimate. var($\bullet$) is the variance of $\bullet$.

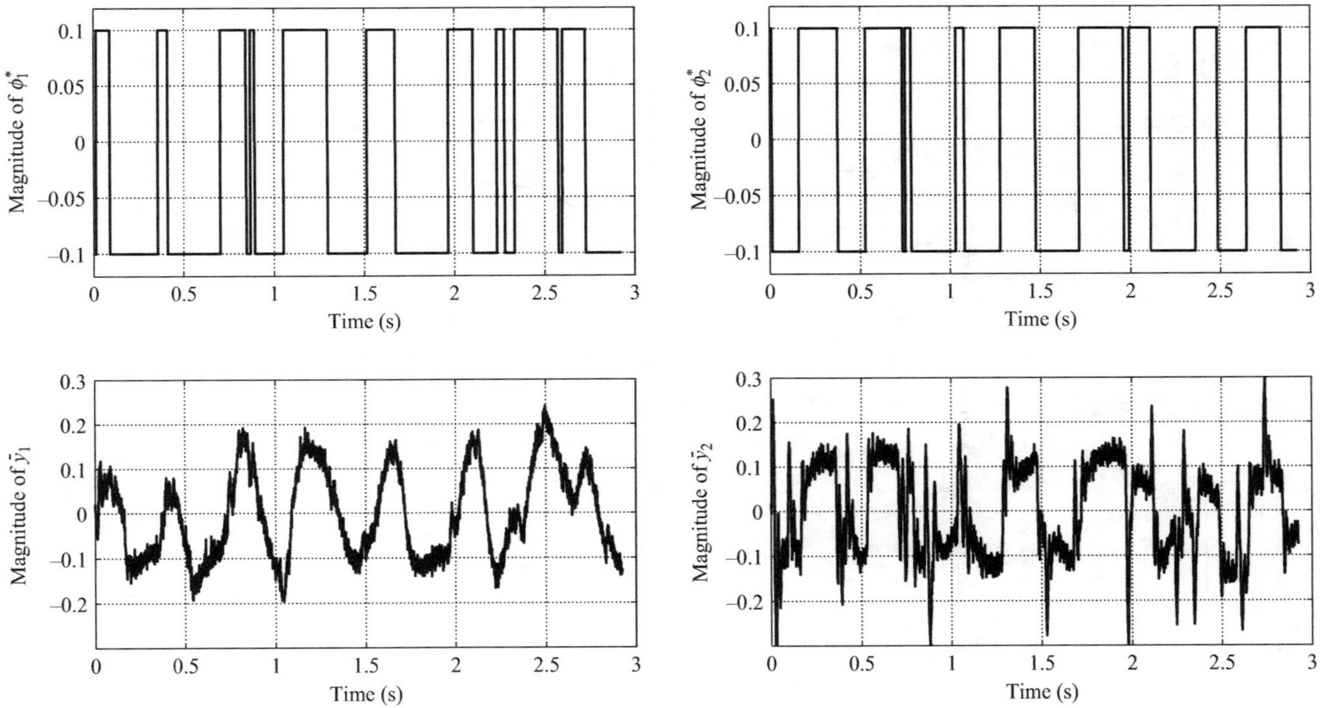

Figure 16.4 I/O data sample

Table 16.1 Performance metrics for the estimated LTI models on validation data

ϕ_2	$\frac{\pi}{8}$	$\frac{2\pi}{8}$	$\frac{3\pi}{8}$	$\frac{4\pi}{8}$	$\frac{5\pi}{8}$	$\frac{6\pi}{8}$	$\frac{7\pi}{8}$
BFT_1 (%)	85.83	91.23	90.23	92.09	89.65	88.42	91.01
BFT_2 (%)	67.42	98.18	85.07	97.83	88.97	89.82	97.23
VAF_1 (%)	97.98	99.23	99.04	99.37	98.93	98.66	99.2
VAF_2 (%)	89.39	99.96	97.77	99.9	98.78	98.96	99.9

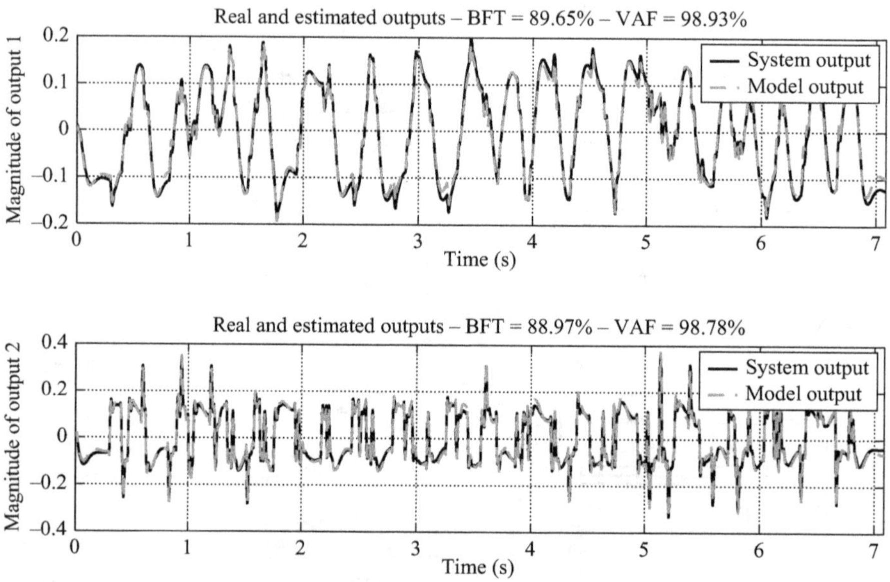

Figure 16.5 Comparison of the time responses of the system (-) and the estimated local model (- -) for $\phi_2 = \frac{5\pi}{8}$

is, according to the authors' experience, more than suitable. Notice also that the weakest values are obtained for $\phi_2 = \frac{\pi}{8}$, i.e., for a position close to singularity. This initial conclusion is enhanced by comparing the frequency responses of these estimated LTI models with the frequency responses of local analytic models. Figure 16.6 shows the magnitude Bode plots in the least favorable case, i.e., for $\phi_2 = \frac{\pi}{8}$ of

- the estimated local model (- -);
- a local analytic model of the flexible system (-) obtained by linearizing the nonlinear equations given in section 16.2.3;
- a local analytic model of an equivalent rigid system (-o).

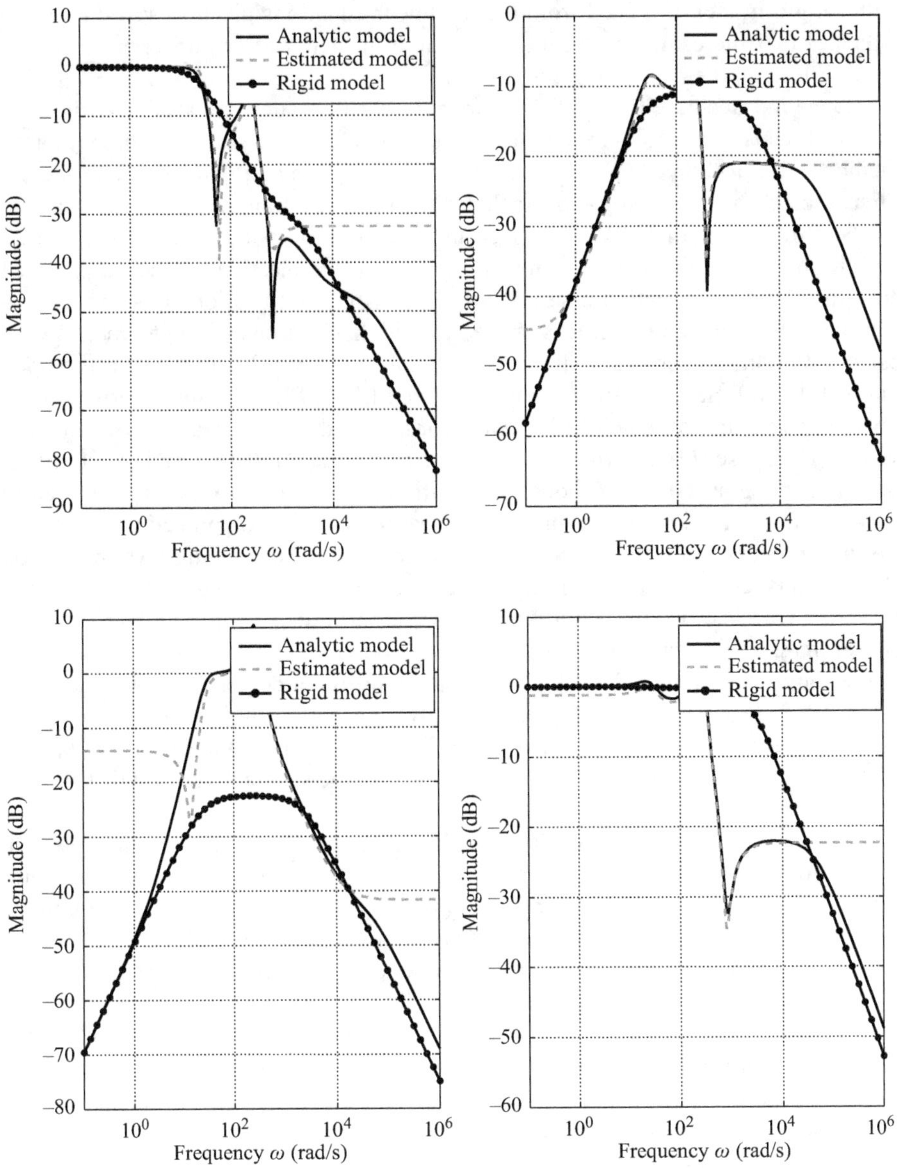

Figure 16.6 Comparison of the frequency responses of the analytic model (-), a rigid model (-o), and the locally estimated model (- -) for $\phi_2 = \frac{\pi}{8}$

These frequency plots show that the estimated local model is able to approximate the system behavior in a relatively wide range of frequencies (up to 1.10^{-4} rad/s). Notice, however, that the estimated local model has difficulties to picture the gain of the system when the coupling transfers $\overline{y}_1 \leftrightarrow u_2$ and $\overline{y}_2 \leftrightarrow u_1$ are considered. The comparison with a rigid model (obtained from the nonlinear equations given in section 16.2.3) indicates that the estimated local models are a lot better at capturing precisely the flexibilities and the behavior of the system.

Once N_{op} accurate local fully parameterized LTI models are available, the final LPV-LFR can be extracted from the available local information by using the H_∞-based method introduced in Chapter 9. Two specific cases[11] are more precisely considered. First, a black-box framework is tackled. By black-box framework it is meant that (*i*) the structure of the LPV-LFR is assumed to be fully parameterized and (*ii*) no prior information about the parameter values is available. Under such a practical constraint, the initialization phase of the underlying optimization algorithm can be very complicated. Hereafter, it is chosen to initialize the unknown parameters with pseudo-random values drawn from the standard uniform distribution on the open interval ± 1000. Second, a physically structured LPV-LFR state-space form is identified, the structure of which is derived according to the considerations described in section 16.2.3 (see (16.23) for the description of the $A_0(\Theta)$, $B_0(\Theta)$, and $C_0(\Theta)$ matrices). For this gray-box LPV model, a specific initialization process is suggested to avoid local minima inherent to such an H_∞-based optimization procedure. First, one working point (herein $\phi_2 = \pi/8$) is selected and the corresponding local black-box LTI model is restructured (according to the local analytic parameterization described in section 16.2.3) by using the null-space-based algorithm developed in [18]. Then, the entries of this restructured local LTI model are used as initial values for the parameters of the sought structured LPV/LFR. It is important to notice that only the matrices[12] A_0 and B_0 in (16.2) can be initialized by applying such a technique because they are the only ones that are physically structured. The initial values of the entries of B_w, C_z, D_{zu}, D_{yw}, D_{zw}, and D_{yu} are randomly generated from the standard uniform distribution on the open interval ± 1000.

Because of the complexity of the sought fully parameterized and structured LFT descriptions, only 20 Monte Carlo runs[13] are carried out for each LPV model structure. The values of the fit measurement $\|\breve{G}(s, \overline{p}, \Theta)\|_\infty$ (see Chapter 9 for a definition) are used to quantify the performance of the H_∞-based method introduced in Chapter 9 to build the final LPV-LFR from the available local information. The figures gathered in Tables 16.2 and 16.3 show that good fully parameterized and structured LFR can be obtained by applying the procedure described in Chapter 9, but finding the global minimum of the cost function involved in the H_∞-based optimization problem described in Chapter 9 is far from being easy. Although we cannot ensure that the global minimum is reached, it is essential to verify if the best local minima give

[11]As far as the input and output weighting functions $W_O(s)$ and $W_I(s)$ are concerned (see Chapter 9 for a definition), they have been chosen equal to the identity matrix for both cases in what follows.

[12]The components of the matrix $C_0(\Theta)$ being known a priori, its entries are not estimated but fixed in the following.

[13]For the fully parameterized LPV-LFR, we need about 970 iterations to reach the final local optimum (about 8.10^3 s on a recent laptop), while 1150 iterations are required for the structured LPV-LFR (which corresponds to 1270 s on the same computer). These computing time differences can be explained by the fact that, in the structured LPV-LFR case, fewer unknown parameters are sought which reduce the computing time of each iteration drastically.

Table 16.2 Numerical results for the estimated fully parameterized LPV-LFR

	Best	**Worst**	**Mean**
$\|\breve{G}(s,\bar{p},\Theta)\|_\infty$	0.2173	1.7141	0.7816

Table 16.3 Numerical results for the estimated structured LPV-LFR

	Best	**Worst**	**Mean**
$\|\breve{G}(s,\bar{p},\Theta)\|_\infty$	0.1379	0.7178	0.2446

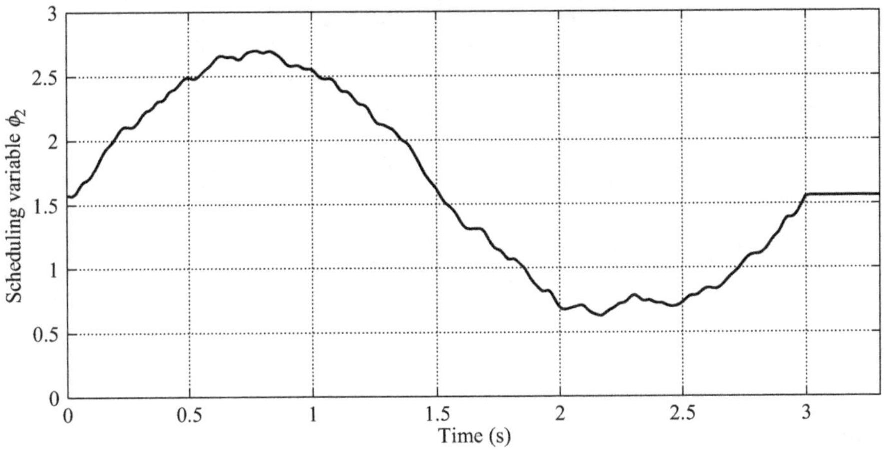

Figure 16.7 Evolution of ϕ_2 during the LPV model's validation

access to reliable fully parameterized and structured LPV-LFR able to reproduce the behavior of the nonlinear system under study.

In order to prove that the final estimated LPV-LFR models are relevant, the last step of the identification procedure developed in this chapter aims at validating these estimated black-box and gray-box LPV models on the whole functioning range of the system to identify. More precisely, the validation procedure consists in (*i*) exciting the system (as well as the models) so that the whole range of ϕ_2 is visited[14] (see Figure 16.7) and (*ii*) comparing the outputs of the nonlinear simulator

[14]Note that this evolution is relatively fast because the reachable range of values for ϕ_2 is visited in 3 s.

Table 16.4 Performance metrics (BFT and VAF) for the estimated (fully parameterized (fullyp.) and structured (struc.)) LPV/LFR on validation data

	BFT (%)	**VAF (%)**
$\breve{y}_1$ (fullyp.)	83.5	97.3
$\breve{y}_2$ (fullyp.)	89.9	98.9
$\breve{y}_1$ (struc.)	96.1	99.8
$\breve{y}_2$ (struc.)	85.4	97.8

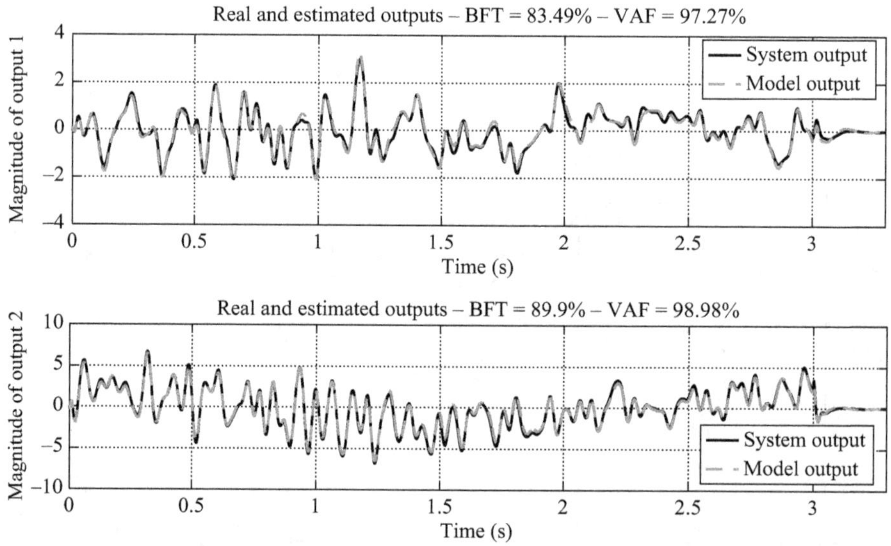

Figure 16.8 Comparison of the time responses of the system (-) and the fully parameterized LPV/LFR (- -)

with the model's outputs. The LPV models having the best fit measurement $\|\breve{G}(s, \overline{p}, \Theta)\|_\infty$ have been selected for this validation phase. The VAF and BFT indexes (see (16.24)) available in Table 16.4 as well as in Figures 16.8 and 16.9 show that the designed LPV models are able to capture the behavior of the nonlinear and flexible robotic manipulator studied in this chapter.

16.4 Conclusions

This chapter is devoted to the evaluation of a novel methodology for the identification of flexible manipulators. The case of a two-segment arm initially designed for

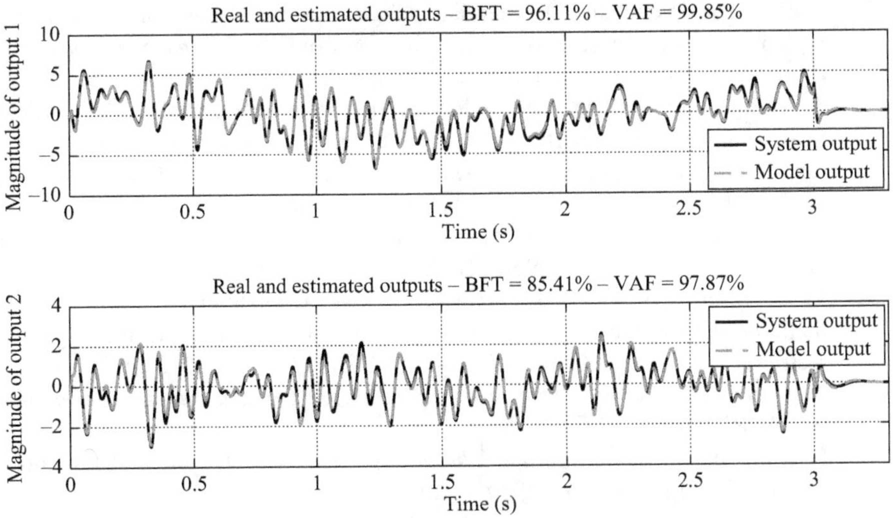

Figure 16.9 Comparison of the time responses of the system (-) and the structured LPV/LFR (- -)

cardiac robotized surgery has been considered in simulation. The H_∞-based glocal methodology introduced in Chapter 9 has been used in order to derive LPV models from local experiments (i.e., experiments with small displacements around given positions). This methodology is very flexible and allows us to derive models with or without structure. In this chapter, two linear fractional LPV models have been estimated, then validated from data generated from a simulator built from the nonlinear equations governing the behavior of the flexible arm to identify. A black-box as well as a gray-box identification procedure have been developed, then compared to reach this goal. By black-box framework, it is meant that (*i*) the structure of the LPV-LFR is assumed to be fully parameterized and (*ii*) no prior information about the parameter values is available. On the contrary, the gray-box framework assumes that a physically based LPV model is estimated, the structure of which is obtained by studying and linearizing the mechanical nonlinear equations satisfied by the system to identify. Both models would suit for controller synthesis. Reducing the number of parameters to estimate when accounting for the structure has allowed us to reduce significantly the computation time while having little effect on accuracy of the identified model. Overall, the results emphasize the potential of the methodology for the identification of robotic systems.

References

[1] M.A. Ardestani and M. Asgari. Modeling and analysis of a novel 3-DOF spatial parallel robot. In *Proceedings of the International Conference of Mechatronics and Machine Vision in Practice*, Auckland, New Zealand, November 2012.

[2] P. Blue, L. Güvenc, and D. Odenthal. Large envelope flight control satisfying H_∞ robustness and performance specifications. In *Proceedings of the American Control Conference*, San Francisco, CA, June 2001.

[3] S. Boonto and H. Werner. Closed-loop identification of LPV models using cubic splines with application to an arm-driven inverted pendulum. In *Proceedings of the American Control Conference*, Baltimore, MD, June 2010.

[4] J. Craig. *Introduction to Robotics: Mechanics and Control*. Pearson/Prentice Hall, Upper Saddle River, New Jersey 2005.

[5] L. Cuvillon, E. Laroche, J. Gangloff, and M. de Mathelin. A multivariable methodology for fast visual servoing of flexible manipulators moving in a restricted workspace. *Advanced Robotics*, 26:1771–1797, 2012.

[6] A. Dutta, ed. *Robotic Systems – Applications, Control and Programming*. Intech, Rijeka, Croatia 2012.

[7] C. Dym and I. Shames. *Solid Mechanics: A Variational Approach*. Augmented Edition. Springer, New York, 2013.

[8] J. Gangloff, R. Ginhoux, M. de Mathelin, L. Soler, and J. Marescaux. Model predictive control for compensation of cyclic organ motions in teleoperated laparoscopic surgery. *IEEE Transactions on Control Systems Technology*, 14:235–246, 2006.

[9] R. Ginhoux, J. Gangloff, M. de Mathelin, L. Soler, M. Arenas Sanchez, and J. Marescaux. Active filtering of physiological motion in robotized surgery using predictive control. *IEEE Transactions on Robotics*, 21:67–79, 2005.

[10] S. Hashemi, H. Abbas, and H. Werner. LPV modelling and control of a 2-DOF robotic manipulator using PCA-based parameter set mapping. In *Proceedings of the IEEE Conference on Decision and Control*, Shangai, P.R. China, December 2009.

[11] X. He, J. Zhao, and G. Dimirovski. A blending method control of switched LPV systems with slow-varying parameters and its application to an F-16 aircraft model. In *Proceedings of the Chinese Control Conference*, Mianyang, China, May 2011.

[12] K. Kozlowski. *Modelling and Identification in Robotics*. Springer, London 1998.

[13] N. Lachhab, H. Abbas, and H. Werner. A neural network based technique for modelling and LPV control of an arm-driven inverted pendulum. In *Proceedings of the IEEE Conference on Decision and Control*, Cancun, Mexico, December 2008.

[14] E. Laroche and L. Cuvillon. *Dynamical Model of a Planar 2-DOF Arm*. Technical report, Strasbourg University, Laboratoire des Sciences de l'Image de l'Informatique et de la Télédétection, 2007.

[15] W. Levine. *The Control Handbook: Control System Applications*. CRC Press, Boca Raton 2010.

[16] P. Lopes dos Santos, T. Azevedo Perdicoúlis, C. Novara, J. Ramos, and D. Rivera. *Linear Parameter-Varying System Identification: New*

Developments and Trends. Advanced Series in Electrical and Computer Engineering. World Scientific Publishing, Singapore 2011.

[17] B. Lu, F. Wu, and K. SungWan. Switching LPV control of an F-16 aircraft via controller state reset. *IEEE Transactions on Control Systems Technology*, 14:267–277, 2006.

[18] O. Prot, G. Mercère, and J. Ramos. A null-space-based technique for the estimation of linear-time invariant structured state-space representations. In *Proceedings of the IFAC Symposium on System Identification*, Brussels, Belgium, July 2012.

[19] P. Shi and J. McPhee. *DynaFlex Users' Guide*. Systems Design Engineering, University of Waterloo, 2002.

[20] W. Shin and D. Kwon. Surgical robot system for single-port surgery with novel joint mechanism. *IEEE Transactions on Bio-medical Engineering*, 60:937–944, 2013.

[21] M. Spillman, P. Blue, L. Lee, and S. Banda. Robust gain scheduling example using linear parameter-varying feedback. In *Proceedings of the IFAC World Congress*, San Francisco, CA, June 1996.

[22] R. Tóth. *Identification and Modeling of Linear Parameter-Varying Systems*. Springer-Verlag. Lecture Notes in Control and Information Sciences 403, Springer-Verlag, Berlin, Heidenberg 2010.

[23] R. Tóth, J. Willems, P. Heuberger, and P. Van den Hof. The behavioral approach to linear parameter-varying systems. *IEEE Transactions on Automatic Control*, 56:2499–2514, 2011.

[24] J. van Helvoort, M. Steinbuch, P. Lambrechts, and R. van de Molengraft. Analytical and experimental modelling for gain scheduling of a double scara robot. In *Proceedings of the IFAC Symposium on Mechatronic Systems*, Sydney, Australia, September 2004.

[25] M. Verhaegen and V. Verdult. *Filtering and System Identification: A Least Squares Approach*. Cambridge University Press, New York 2007.

[26] K. Zhou, J. Doyle, and K. Glover. *Robust and Optimal Control*. Prentice Hall, Englewood Cliffs, New Jersey 1996.

Index

www.ingramcontent.com/pod-product-compliance
Lightning Source LLC
LaVergne TN
LVHW011905200926
848484LV00002B/532
* 9 7 8 1 8 4 9 1 9 6 1 4 7 *